Quantity	Symbol	Dimensions	Units	
Power	P	FL/t	watts	W
Pressure	p	F/L^2	pascal ($=N/m^2$)	$Pa = kg/m \cdot s^2$
Radiant energy	E	Q/t	watt	W
Radiant energy flux	W	$Q/L^2 t$	watt/meter^2	W/m^2
Reflectivity	ρ	$[-]$	dimensionless	$[-]$
Solar constant	I_0	$Q/L^2 t$	watt/meter^2	W/m^2
Solar insolation	I	$Q/L^2 t$	watt/meter^2	W/m^2
Specific heat (constant pressure)	c_p	Q/MT	joule/kilogram·degree Celsius	$J/kg \cdot \Delta_1 °C$
Specific heat (constant volume)	c_c	Q/MT	joule/kilogram·degree Celsius	$J/kg \cdot \Delta_1 °C$
Surface coefficient of convection	h_c	$Q/L^2 T$	watt/meter^2·degree Celsius	$W/m^2 \cdot \Delta_1 °C$
Temperature—difference	ΔT	T	Kelvin or degree Celsius	ΔK or $\Delta °C$
Temperature—level	T	T	Kelvin or degree Celsius	K or °C
Temperature—"per degree"		$1/T$	Kelvin or degree Celsius	$1/\Delta_1 K$ or $1/\Delta_1 °C$
Thermal conductance	$\overline{\mathcal{K}}$	Q/tT	watt/degree Celsius	$W/\Delta_1 °C$
Thermal conductance	$\mathcal{K}$	$Q/tL^2 T$	watt/meter^2·degree Celsius	$W/m^2 \cdot \Delta_1 °C$
Thermal conductivity	k	Q/tLT	watt/meter·degree Celsius	$W/m \cdot \Delta_1 °C$
Thermal diffusivity	α	L^2/T	meter/second^2	m/s^2
Thermal resistivity	r	tLT/Q	meter·degree Celsius/watt	$m \cdot \Delta °C/W$
Thermal resistance	$\overline{\mathcal{R}}$	tT/Q	degree Celsius/watt	$\Delta °C/W$
Thermal resistance	$\mathcal{R}$	$tL^2 T/Q$	meter^2·degree Celsius/watt	$m^2 \cdot \Delta °C/W$
Time	t	t	seconds	s
Transmittance (overall coefficient)	U	$Q/tL^2 T$	watt/meter^2·degree Celsius	$W/m^2 \cdot \Delta_1 °C$
Velocity	V	L/t	meter/second	m/s
Viscosity, absolute	μ	M/L	pascal·second	Pa·s
Viscosity, kinematic	ν	L^2/t	meter^2/second	m^2/s
Volume	∇	L^3	meter^3	m^3
Wavelength	λ	L	micrometer	μm

Applied Heat Transfer

Applied Heat Transfer

James P. Todd
VERMONT TECHNICAL COLLEGE

Herbert B. Ellis
CALIFORNIA STATE POLYTECHNIC UNIVERSITY
POMONA, CALIFORNIA

HARPER & ROW, PUBLISHERS, New York
Cambridge, Philadelphia, San Francisco,
London, Mexico City, São Paulo, Sydney

1817

Sponsoring Editor: Cliff Robichaud
Project Editor: Pamela Landau
Production Manager: Marion Palen
Compositor: Science Typographers, Inc.
Printer and Binder: Murray Printing Company, Inc.
Art Studio: Vantage Art, Inc.
Cover Design: Lorene Bodenstadt

Applied Heat Transfer

Library of Congress Cataloging in Publication Data

Todd, James P.
    Applied heat transfer

    Includes index.
    1. Heat—Transmission.   I. Ellis, Herbert B.
II. Title.
TJ260.T59   621.402'2   82-946
ISBN 0-06-046635-9   AACR2

Contents

v

Preface

This textbook makes an applied approach to the introduction of basic principles and equations related to the transfer of thermal energy. Four important objectives in writing this book for students and practicing engineers were defined at the outset of this effort. The first objective is to provide the engineering technology student with the vocabulary, concepts, and applications of steady state heat transfer. The second objective is to expose the student to a methodology for the solution of applied heat transfer problems which will be advantageous in a career of engineering, science, or technology. The third objective is to give the reader an adequate background for more advanced studies. The fourth objective is to provide practicing engineers and technologists with a convenient and useful reference book. To accomplish these objectives, the authors have combined an extensive instructional experience at the college level with over 40 years of practical engineering experience in industry.

The book is organized in a conventional manner. The presentation assumes that the reader has a basic knowledge of mathematics, physics, and fluid mechanics. Calculus is not required as a prerequisite; however, it is suggested that it be required as a corequisite. This basic knowledge is usually acquired during the first two years of an engineering technology curriculum. The International System of Units (SI) is applied throughout the book in accordance with the current shift to SI (metric) units in this country.

An introduction to heat transfer concepts is presented first, followed by steady state heat transfer applications. Example problems in each chapter illustrate the use of the equations and suggested methodology in the solution of actual problems. At the end of each chapter there are practice problems underscoring the key points in the chapter.

Five appendixes contain all the reference material necessary to solve both the problems in the text and a variety of problems encountered in actual engineering practice. Appendix A provides physical and thermodynamic data. Appendix B presents tables of heat transfer data for metals, building materials, insulation, other solids, common liquids, and common gases. Appendix C presents conversion factors. Appendix D presents an approach to problem solving. Appendix E contains mathematical tables and dimensional data.

The development of applied heat transfer course material accomplished over a period of six years at the California State Polytechnic University, Pomona, is herein acknowledged. The unstinting support of students, faculty, and staff is gratefully appreciated.

JAMES P. TODD
HERBERT B. ELLIS

Chapter 1
Introduction

1-1 NATURE OF HEAT

According to Francis Bacon (1561–1626) in *Novum Organum*, "The very essence of heat ... is motion and nothing else." Heat is energy and the equivalence of heat and work was established in a series of famous experiments by James Prescott Joule (1818–1889). Temperature is a measurement of the motion energy level within a molecule. The greater the motion energy level within the molecule, the higher the temperature of the molecule. When all molecular motion ceases and there is no motion energy within the molecule, the temperature of the molecule is absolute zero.

Whenever a temperature gradient exists, thermal energy (heat) is transferred. The free flow of heat or thermal energy is always along a temperature gradient from a higher temperature to a lower temperature, in accordance with the second law of thermodynamics. As stated by Clausius, "Heat cannot, of itself, pass from a lower to a higher temperature reservoir." Temperature gradients exist throughout the world around us. Temperature gradients are inherent in material aspects of our everyday life, such as power generation, propulsion and transportation, chemical and metallurgical processing, heating and refrigeration.

1

An understanding of the laws of the flow of thermal energy (heat) is important to all fields of engineering because these laws are of controlling importance in the design, construction, and operation of the many diverse forms of heat exchange apparatus required in our scientific and industrial technology. Every engineer can expect to be confronted, from time to time, with problems relating to the efficient transmission or the control and regulation of thermal energy.

To facilitate the solution of heat transfer problems, much effort has been expended by investigators to obtain test data and to develop basic theories and equations that will provide relationships for the solution of whatever heat transfer problem is encountered. Test data have been correlated in terms of dimensionless parameters so as to provide useful empirical equations for general application. Theoretical considerations and models have been used to develop theoretical equations. Such theoretical equations are usually limited in the scope of their application because real material properties and fluid flow deviate from the theoretical models used in the development of the equations. To extend the scope of application of theoretical equations, empirical factors and coefficients are often applied to the basic equation.

1-2 BASIC MODES OF HEAT TRANSFER

Heat is transferred by three basic modes. In two of these modes the heat is transferred through material, whereas in the third mode it is transmitted through empty space and vacuum as well as through certain materials transparent to thermal radiation. The physical mechanisms and laws for the heat transfer are different for each of these basic modes, which are designated as:

- conduction
- convection
- radiation

1-2.1 Conduction

In conduction heat flow, the thermal energy is transmitted by direct molecular communication without appreciable displacement of the molecules. Conduction occurs through bodies of solids, liquids, and gases and from one body to another when they are in physical contact with each other. A common example of heat transfer by conduction is when one picks up a hot plate and burns his fingers where they are in contact with the hot plate. The heat energy is conducted from the hot plate to the colder fingers where there is physical contact between the two.

1-2.2 Convection

In convection heat flow, the thermal energy is primarily transmitted through appreciable displacement (physical circulation of the molecules within the convective body). Convection occurs only in fluids (liquids or gases), and convective heat transfer is usually between a solid surface and a fluid. The fluid molecules in contact with the solid surface are heated or cooled by conduction, and are swept away from the solid surface by convective flow patterns or flow turbulence, and subsequent thermal energy transfer by conduction with adjacent fluid molecules occurs. Or conversely, molecules at the temperature of the fluid body are swept toward a solid surface, and upon physical contact thermal energy transfer is effected. An example of heat transferred by convection is the chilling effect of a cold wind on a warm body. Without the flow of cold air past a warm body, the primary mode of energy transfer would be conduction, which would be much less.

When there is no phase change in the fluid (a liquid remains a liquid or a gas remains a gas), there are two categories of convection. When the convective flow patterns in the fluid are a result of the effect of the thermal gradient upon the density of the fluid, the resulting convective heat transfer process is called *free convection*. If, on the other hand, the convective flow patterns in the fluid are a result of mechanical pumping, the resulting heat transfer process is called *forced convection*.

When there is a phase change in the fluid (a liquid becomes a gas, or a gas becomes a liquid), there are two categories of convection, namely, *boiling* (evaporation) and *condensation*. Boiling may occur under conditions of either free or forced convection. A commonly observed example of free convection boiling heat transfer comes from an open pan of water on a stove. As energy is progressively added to the water from the surface of the hot pan, first a vapor comes from the surface, followed by the formation of bubbles on the bottom of the pan which rise up through the liquid and discharge as steam at the surface when these bubbles break. On the other hand, condensation is a forced convective process only. The phase change of the gas to a liquid on the condensation surface always causes a forced gas flow toward the condensation surface. An example of condensing heat transfer occurs whenever a cold water pipe "sweats." Moisture from warm moist air is condensed and wets the outer surface of the colder pipe.

1-2.3 Radiation

In radiation heat transfer, the thermal energy is transmitted by electromagnetic waves, differing from light (visible spectrum) only by their respective wavelengths. These electromagnetic waves (radiant energy) pass readily through vacuum and empty space, and can pass through gases, liquids, and some solids. The emitted radiant energy is not a continuous variable, but is formed by finite batches (or "quanta") of energy associated with step changes in the energy level of the emitting molecule of the body. The rate of this radiant energy emission, and hence the rate of radiant heat transfer, is proportional to the fourth power of the absolute temperature. This is in contrast to the direct proportionality of conductive and convective heat transfer to the temperature difference or temperature gradient. The transmitted radiant energy is absorbed to varying extents on or near the surface of bodies upon which it impinges, depending upon the absorptivity of the body. Sunshine is a common example of radiant heat transfer. The radiant energy of the sun is transmitted through space, and the earth's atmosphere. This transmitted energy is absorbed according to the characteristics of the surface upon which it is incident. Dark clothing usually absorbs a major portion of this radiant energy.

1-3 DIMENSIONS AND UNITS

Before attempting to study any engineering subject, dimensions and the units for these dimensions must be understood. Dimensions describe complex physical parameters in terms of simple or basic definitions. Units are the basis for measurements that evaluate or "quantize" the dimensions of the physical parameter.

1-3.1 Dimensions

A dimension can be defined as a term that describes certain qualities or characteristics of an entity (such as mass, M; length, L; area, A; etc.). In this context, a large number of dimensions are possible. To reduce such a large number of dimensions, many can be expressed in terms of basic dimensions. For example, length, area, and volume are dimensions describing certain characteristics of an object. But area A can be defined as a length squared $[L^2]$ and volume $[\mathcal{V}]$ as a length cubed $[L^3]$. These dimensions can be stated in terms of the basic dimension length $[L]$ as follows.

$$[A] = [L^2]$$
$$[\mathcal{V}] = [L^3]$$

Note the use of the bracket which is often used to identify a dimension as in the above equations. These should be read as: the dimension of area is equivalent to the dimension of length squared; or the dimension of volume is equivalent to the dimension of length cubed.

By using this technique, a large number of dimensions can be reduced to a smaller number of basic dimensions. All other dimensions expressed in terms of these basic dimensions are known as secondary or derived dimensions. Thus in the previous examples, area and volume are derived (dimensions) in terms of the basic dimension of length.

1-3.2 Units

While a dimension provides a description, a unit is a definite standard or measure of a dimension. A unit may be defined as a particular amount (or quantity) to be measured. For example, kilometer (km), meter (m), and millimeter (mm) are all different units used to specify definite lengths. The common dimension of all these units is the dimension of length $[L]$.

1-3.3 SI (Metric) Units

This book is written using SI (metric) units because of the current adoption of SI (metric) units for academic and industrial use in the United States. Usage has shown the SI units to be very convenient, simplifying many calculations.

The SI (metric) system is designated SI for Systèm International d'unites (see note below Table 1-1) and (metric) is usually added to the SI designation. A set of seven basic units, listed in Table 1-1, was officially adopted in a resolution of the Eleventh General Conference on Weights and Measures in Paris, 1960. The secondary SI (metric) units commonly used are listed in Table 1-2, below.

These basic units measure quantities that could vary considerably in magnitude. To avoid awkwardly large or small figures common prefixes representing multiples and submultiples are given in (Table 1-3).

Note that the multiples and submultiples are in increments of three orders of magnitude. Other orders of magnitude that do not follow this pattern are not part of the SI system.

Table 1-1   BASIC SI (METRIC) UNITS

DIMENSION	UNITS	
PHYSICAL PARAMETER	NAME	ABBREVIATION
Length	meter	m
Mass	kilogram	kg
Time	second	s
Electric current	ampere	A
Temperature	kelvin	K
Amount of substance	mole	mol
Luminous intensity	candela	cd

[a] "The International System of Units (SI)," NBS Special Publications 330, 1974 Edition, U.S. Department of Commerce/National Bureau of Standards, Washington, D.C. 20234.

Table 1-2 SECONDARY SI (METRIC) UNITS

DIMENSION	UNITS	
PHYSICAL PARAMETER	NAME	SYMBOL
Force	Newton $=[\mathrm{kg}\cdot\mathrm{m}/\mathrm{s}^2]$	N
Pressure	pascal $=[\mathrm{N}/\mathrm{m}^2]=[\mathrm{kg}/\mathrm{m}\cdot\mathrm{s}^2]$	Pa
Energy, heat, work, enthalpy	joule $=[\mathrm{N}\cdot\mathrm{m}]=[\mathrm{kg}\cdot\mathrm{m}^2/\mathrm{s}^2]$	J
Power, heat transfer rate	watt $=[\mathrm{J}/\mathrm{s}]=[\mathrm{N}\cdot\mathrm{m}/\mathrm{s}]=[\mathrm{kg}\cdot\mathrm{m}^2/\mathrm{s}^3]$	W

Heat transfer is involved with parameters that have various dimensions, such as heat, mass, force, area, length, time, and temperature, as well as various combinations of these dimensions. Each of these parameters is characterized by its dimensions. Each dimension is evaluated by a system of units. The symbols, dimensions, and SI (metric) units for the common parameters encountered in heat transfer problems are listed in Table 1-4.

The SI unit of temperature measurement is the Kelvin [K]. The Kelvin temperature scale is an "absolute" temperature scale as its starts from absolute zero. The Kelvin [K] is the preferred temperature scale to describe temperature levels, and the Kelvin temperature unit [ΔK] is preferred to describe differences or changes in temperature levels. However, in engineering and nonscientific areas wide use is still being made of the Celsius [°C] scale, which is based on 0°C as the temperature level at which ice and liquid water are in equilibrium at atmospheric pressure.

The student will encounter four different parameters involving temperature, namely:

- temperature level, [°C]
- temperature difference, [Δ°C]
- temperature gradient, [Δ°C/m]
- "per degree," [Δ_1°C]

The temperature level is the normal temperature reading of a thermometer. It is a measure of how hot or cold a substance is.

Table 1-3 PREFIXES TO SI (METRIC) UNITS

NAME	SYMBOL	MULTIPLY BY
	Multiples	
kilo	k	10^3
mega	M	10^6
giga	G	10^9
tera	T	10^{12}
	Submultiples	
milli	m	10^{-3}
micro	μ	10^{-6}
nano	n	10^{-9}
pico	p	10^{-12}

Table 1-4 COMMON HEAT TRANSFER PARAMETERS, DIMENSIONS, UNITS AND SYMBOLS

PARAMETER			SI (METRIC) SYSTEM	
NAME	TEXT SYMBOL	DIMENSIONS	UNIT NAME	UNIT SYMBOL
Absorptivity	α	$[-]$	dimensionless	$[-]$
Acceleration	a, g	L/t^2	meter/second^2	m/s^2
Air mass ratio	IM	$[-]$	dimensionless	$[-]$
Area	A	L^2	meter^2	m^2
Coefficient of thermal expansion	β	$1/T$		$1/\Delta_1$°C
Density	ρ	M/L^3	kilogram/meter^3	kg/m^3
Electrical:				
Current	I	A	ampere	A
Resistance	R	Q/tA^2	ohm	$\Omega = kg\cdot m^2/s^3\cdot A^2$
Voltage	V	Q/tA	volts	$V = kg\cdot m^2/s^3\cdot A$
Elevation (head)	Z	L	meter	m
Emissivity	ε	$[-]$	dimensionless	$[-]$
Force	F	F	newton	$N = kg\cdot m/s^2$
Free convection modulus	α	$1/L^3T$		$1/m^3\Delta_1$°C
Heat	Q	Q	joule	J
Heat flow rate (heat transfer rate)	$\dot{Q}$	Q/t	watt (=joule/second)	W
Heat flux	$\dot{Q}/A$	Q/t^2	watt/meter^2	W/m^2
Isentropic exponent (ratio of specific heats)	γ	$[-]$	dimensionless	$[-]$
Latent heat of vaporization	r	Q/M	joule/kilogram	J/kg
Length	x, L	L	meter	m
Mass	M	M	kilogram	kg
Mass flow rate	$\dot{M}$	M/t	kilogram/second	kg/s
Mass flow rate per unit area	G	M/L^2t	kilogram/second·meter^2	kg/s·m^2

Table 1-4 (*Continued*)

| PARAMETER | | | SI (METRIC) SYSTEM | |
NAME	TEXT SYMBOL	DIMENSIONS	UNIT NAME	UNIT SYMBOL
Power	P	FL/t	watts	W
Pressure	p	F/L^2	pascal ($=N/m^2$)	$Pa = kg/m \cdot s^2$
Radiant energy	E	Q/t	watt	W
Radiant energy flux	W	$Q/L^2 t$	watt/meter^2	W/m^2
Reflectivity	ρ	$[-]$	dimensionless	$[-]$
Solar constant	I_0	$Q/L^2 t$	watt/meter^2	W/m^2
Solar insolation	I	$Q/L^2 t$	watt/meter^2	W/m^2
Specific heat (constant pressure)	c_p	Q/MT	joule/kilogram · degree Celsius	$J/kg \cdot \Delta_1 °C$
Specific heat (constant volume)	c_v	Q/MT	joule/kilogram · degree Celsius	$J/kg \cdot \Delta_1 °C$
Surface coefficient of convection	h_c	$Q/L^2 T$	watt/meter^2 · degree Celsius	$W/m^2 \cdot \Delta_1 °C$
Temperature—difference	ΔT	T	Kelvin or degree Celsius	ΔK or $\Delta °C$
Temperature—level	T	T	Kelvin or degree Celsius	K or $°C$
Temperature—"per degree"		$1/T$	Kelvin or degree Celsius	$1/\Delta_1 K$ or $1/\Delta_1 °C$
Thermal conductance	$\overline{\mathscr{K}}$	Q/tT	watt/degree Celsius	$W/\Delta_1 °C$
Thermal conductance	$\mathscr{K}$	$Q/tL^2 T$	watt/meter^2 · degree Celsius	$W/m^2 \cdot \Delta_1 °C$
Thermal conductivity	k	Q/tLT	watt/meter · degree Celsius	$W/m \cdot \Delta_1 °C$
Thermal diffusivity	α	L^2/T	meter/second^2	m/s^2
Thermal resistivity	$\overline{r}$	tLT/Q	meter · degree Celsius/watt	$m \cdot \Delta °C/W$
Thermal resistance	$\overline{\mathscr{R}}$	tT/Q	degree Celsius/watt	$\Delta °C/W$
Thermal resistance	$\mathscr{R}$	tL^2T/Q	meter^2 · degree Celsius/watt	$m^2 \cdot \Delta °C/W$
Time	t	t	seconds	s
Transmittance (overall coefficient)	U	$Q/tL^2 T$	watt/meter^2 · degree Celsius	$W/m^2 \cdot \Delta_1 °C$
Velocity	V	L/t	meter/second	m/s
Viscosity, absolute	μ	M/L	pascal · second	Pa·s
Viscosity, kinematic	ν	L^2/t	meter^2/second	m^2/s
Volume	$\forall$	L^3	meter^3	m^3
Wavelength	λ	L	micrometer	μm

The temperature difference is simply the difference between two temperature levels. For example, if a substance at a temperature (level) of 25°C were to be heated to a temperature (level) of 35°C, the temperature difference (in this case an increase) is 10 Δ°C. A temperature level [°C] can be increased (or decreased) by a temperature difference. In units [°C]+[Δ°C]=[°C].

The temperature gradient is the temperature difference per unit length of a heat transfer path. For example, if two heat transfer paths had the same initial and final temperatures, they would have the same temperature difference across the path. If one were shorter in length than the other, it would have a larger temperature gradient along the path.

The "per degree" temperature parameter is associated with values for various thermodynamic properties for a change in temperature (level) of one (1) degree. An example is specific heat where the value used is the amount of heat required to change the temperature of a unit mass of substance one (1) degree Kelvin. Common arrangement of [Δ°C] and [Δ₁°C] units after equation substitutions are shown below with their reduction:

Ratios:

$$\frac{[\Delta°C]}{[\Delta_1°C]} = [-]$$

$$\frac{[-]}{[1/\Delta_1°C]} = [\Delta°C] = [\Delta K]$$

$$\frac{[°C]}{[°C]} \text{ or } \frac{[°C]}{[K]} \text{ or } \frac{[K]}{[°C]} \text{ is } not \text{ proper, and therefore does not cancel}$$

$$\frac{[\Delta K]}{[K]} \text{ or } \frac{[\Delta K]}{[\Delta_1 K]} \text{ or } \frac{[K]}{[\Delta_1 K]} \text{ or } \frac{[K]}{[K]} = [-]$$

Addition and subtraction:

$$°C \pm \Delta°C = °C$$
$$K \pm \Delta K = K$$

°C+°C is *not* proper, except for averaging where $(°C+°C)/2 = °C$

In the solution of any heat transfer equation, it is mandatory that a consistent set of dimensional units be used throughout. For example, the units of time for all the parameters in the equation must be consistently seconds, minutes, or hours; units of length must be consistently centimeters or meters, and so forth.

1-4 CONVERSION OF UNITS

This was written in a transition period wherein a large quantity of published and manufacturers' data may be in units other than SI (metric). Consequently, the student can expect to be faced with the need to make numerous conversions of units. Conversions from one unit system to another are simple if the following procedure is used. The general conversion equation is

$$\psi = \psi' F_c \ [\text{converted units}] \tag{1-1}$$

where

ψ = value of any parameter in converted units

ψ' = value of the parameter in initial units

F_c = parameter conversion factor between units

The first step is to write Eq. (1-1) indicating the desired units for the parameter ψ. The second step is to list the given value of the parameter ψ' and its units. The third step is to list the equation of equality between the given and desired units of the parameter. Equations of equality for various conversions are given in Appendix C-4. The final step is to derive from the equation of equality the conversion factor (F_c) with the required ratio of units. The values of the parameter ψ' and the conversion factor F_c are then substituted into Eq. (1-1); the numerical values in one substitution, and the units in a separated substitution.

The use of the general conversion equation is shown by the following illustrative problems.

Example Problems

A. Convert a heat flow rate ($\dot{Q}$) of 10 B/hr to watts.

SOLUTION

EQUATION FOR CONVERTED PARAMETER

$$\dot{Q} = \dot{Q}' F_c \ [\text{W}] \quad [\text{Eq. (1-1)}]$$

PARAMETERS

$$\dot{Q}' = 10 \ [\text{Btu/hr}] \quad \text{(given)}$$

Equation of Equality

$$1 \ [\text{Btu/hr}] = 0.293 \ [\text{W}] \quad \text{Appendix C-4}$$

$$F_c = 0.293 \ \frac{[\text{W}]}{[\text{Btu/hr}]}$$

SUBSTITUTION

$$\dot{Q} = (10)(0.293) \quad [\text{Btu/hr}] \frac{[\text{W}]}{[\text{Btu/hr}]} = [\text{W}]$$

ANSWER

$$\dot{Q} = 2.93 \quad [\text{W}]$$

B. Convert a thermal conductivity (k) of 10 Btu/hr·ft·Δ_1°F to W/m·Δ_1°C.

SOLUTION

EQUATION FOR CONVERTED PARAMETER

$$k = k'F_c \quad [\text{W/m} \cdot \Delta_1 °\text{C}] \qquad [\text{Eq. (1-1)}]$$

PARAMETERS

$$k' = 10 \quad [\text{Btu/hr} \cdot \text{ft} \cdot \Delta_1 °\text{F}] \qquad (\text{given})$$

Equation of Equality

$$1 \quad [\text{Btu/hr} \cdot \text{ft} \cdot \Delta_1 °\text{F}] = 1.73 \quad [\text{W/m} \cdot \Delta_1 °\text{C}] \qquad \text{Appendix C-4}$$

$$F_c = 1.73 \quad \frac{[\text{W/m} \cdot \Delta_1 °\text{C}]}{[\text{Btu/hr} \cdot \text{ft} \cdot \Delta_1 °\text{F}]}$$

SUBSTITUTION

$$k = (10)(1.73) \quad [\text{Btu/hr} \cdot \text{ft} \cdot \Delta_1 °\text{F}] \frac{[\text{W/m} \cdot \Delta_1 °\text{C}]}{[\text{Btu/hr} \cdot \text{ft} \cdot \Delta_1 °\text{F}]} = [\text{W/m} \cdot \Delta_1 °\text{C}]$$

ANSWER

$$k = 17.3 \quad [\text{W/m} \cdot \Delta_1 °\text{C}]$$

C. Convert a film coefficient (h) of 100 Btu/hr·ft^2·Δ_1°F to W/m^2·Δ_1°C.

SOLUTION

EQUATION FOR CONVERTED PARAMETER

$$h = h'F_c \quad [\text{W/m}^2 \cdot \Delta_1 °\text{C}] \qquad [\text{Eq. (1-1)}]$$

PARAMETERS

$$h' = 100 \quad [\text{Btu/hr} \cdot \text{ft}^2 \cdot \Delta_1 °\text{F}] \qquad (\text{given})$$

Equation of Equality

$$1 \; \left[\text{Btu/hr}\cdot\text{ft}^2\cdot\Delta_1{}^\circ\text{F}\right]=5.68 \; \left[\text{W/m}^2\cdot\Delta_1{}^\circ\text{C}\right] \quad \text{Appendix C-4}$$

$$F_c=5.68\frac{\left[\text{W/m}^2\cdot\Delta_1{}^\circ\text{C}\right]}{\left[\text{Btu/hr}\cdot\text{ft}^2\cdot\Delta_1{}^\circ\text{F}\right]}$$

SUBSTITUTION

$$h=(100)(5.68) \; \left[\text{Btu/hr}\cdot\text{ft}^2\cdot\Delta_1{}^\circ\text{F}\right] \; \frac{\left[\text{W/m}^2\cdot\Delta_1{}^\circ\text{C}\right]}{\left[\text{Btu/hr}\cdot\text{ft}^2\cdot\Delta_1{}^\circ\text{F}\right]}$$

$$=\left[\text{W/m}^2\cdot\Delta_1{}^\circ\text{C}\right]$$

ANSWER

$$h=568 \; \left[\text{W/m}^2\cdot\Delta_1{}^\circ\text{C}\right]$$

PROBLEMS

1-1. Give a practical example for each of the three basic modes of heat transfer.

1-2. Explain the basic difference between free convection and forced convection heat transfer.

1-3. In the SI (metric) system of units, specify the appropriate units and symbols for the following:
(a) heat or thermal energy
(b) temperature difference
(c) thermal conductivity
(d) thermal resistance
(e) specific heat
(f) latent heat of vaporization

1-4. Convert the thermal conductivity of aluminum from English units (118 Btu/hr·ft·$\Delta_1{}^\circ$F) to SI units.

1-5. What is the value in SI units for a thermal resistance of 100 $\Delta{}^\circ$F·hr·ft^2/Btu?

1-6. Convert the value of specific heat for water ($c_p=1.0$ cal/g·$\Delta_1{}^\circ$C) to J/kg·Δ_1K.

1-7. Convert the value of specific heat for oil ($c_p=0.5$ Btu/lb·$\Delta_1{}^\circ$F) to the SI units.

1-8. Where are the SI (metric) units for the parameter called heat flux?

1-9. What are the SI (metric) units for the parameter called work?

1-10. Convert the absolute viscosity (μ) of water at 80°F (0.573×10^{-3} lb^m/ft·s) to Pa·s.

1-11. Convert the absolute viscosity of glycol at 120°F (7.0 centipoises) to Pa·s.

1-12. Convert 125°F to °C.

1-13. Convert 80°F to K.

1-14. Convert the thermal conductance of 50 Btu/hr·ft^2·Δ_1°F to W/m^2·Δ_1°C.

1-15. Convert the thermal conductance of 1280 Btu/hr·Δ_1°F to W/Δ_1°C.

1-16. Convert the thermal resistance of 0.2 ft^2·Δ°F/Btu/hr to m^2·Δ°C/W.

1-17. Convert the thermal resistance of 0.001 Δ°F/Btu/hr to Δ°C/W.

Chapter 2
Steady State Conduction

Conduction is fundamental to almost all facets of heat transfer. Consequently, the beginning student should acquire a good understanding of the basic conduction process. Four key terms that may be new to the student are:

- Thermal conductivity (k)
- Thermal conductance ($\mathcal{K}$) or ($\overline{\mathcal{K}}$)
- Thermal resistivity (r)
- Thermal resistance ($\mathcal{R}$) or ($\overline{\mathcal{R}}$)

Thermal conductivity (k) is a measurement of the capacity of a *material* to conduct heat. Thermal conductivity can be considered to be an "index" of the material indicating the amount of heat conducted per unit conduction path area per unit temperature gradient.

$$k = \frac{\dot{Q}}{A\,\Delta T / x}$$

(2-1)

where

$\dot{Q}$ = heat flow rate [W]

A = area of heat transfer path [m²]

$\Delta T/x$ = temperature gradient along path [Δ°C/m]

$$\text{units of } k = \frac{[\text{W}]}{[\text{m}^2][\Delta°\text{C}/\text{m}]} = \frac{[\text{W}]}{[\text{m}][\Delta_1°\text{C}]}$$

Thermal conductance ($\mathcal{K}$) or ($\overline{\mathcal{K}}$) is a measurement of the capacity of a heat transfer path to conduct heat. The thermal conductance ($\mathcal{K}$) is for a unit area of the conduction path.

$$\mathcal{K} = \frac{\dot{Q}}{A\,\Delta T} \qquad\qquad (2\text{-}2)$$

where

$\dot{Q}$ = heat flow rate [W]

A = area of heat transfer path [m²]

ΔT = temperature difference between ends of path [Δ°C]

$$\text{units of } \mathcal{K} = \frac{[\text{W}]}{[\text{m}^2][\Delta_1°\text{C}]}$$

The thermal conductance ($\overline{\mathcal{K}}$) is for the conduction path.

$$\overline{\mathcal{K}} = \frac{\dot{Q}}{\Delta T}$$

$$\text{units of } \overline{\mathcal{K}} = \frac{[\text{W}]}{[\Delta_1°\text{C}]}$$

Thermal resistivity (r) and thermal resistance ($\mathcal{R}$) and ($\overline{\mathcal{R}}$) are the inverse of thermal conductivity (k) and thermal conductance ($\mathcal{K}$) and ($\overline{\mathcal{K}}$).

The rate of heat transfer ($\dot{Q}$) by a conduction path is

$$\dot{Q} = \frac{Ak\,\Delta T}{x} = A\mathcal{K}\,\Delta T = \overline{\mathcal{K}}\,\Delta T = \frac{A\,\Delta T}{rx} = \frac{A\,\Delta T}{\mathcal{R}} = \frac{\Delta T}{\overline{\mathcal{R}}} \quad [\text{W}]$$

The following sections will first describe the thermal conductivity characteristics of various materials, and then describe the heat transfer equations for various conductive heat transfer paths.

2-1 THERMAL CONDUCTIVITY

Thermal conductivity is the property of a material which indicates its ability to conduct heat. A material with a high thermal conductivity is termed a "good conductor," and a material with a low thermal conductivity is termed a "good insulator." Materials have a wide range of thermal conductivities as indicated in Figure 2-1. This is a plot of thermal conductivities versus temperature for an illustrative group of gases, liquids, and solids. In general gases have the lowest conductivities, liquids are in the midrange, and solids have the highest conductivities. Pure silver has the highest thermal conductivity of any material.

While the mechanism of the propagation of molecular motion in heat transfer has similarities to the propagation of sound and to the electrical properties of the material, only limited theoretical correlations have been achieved. Consequently, the evaluation of thermal conductivities has been primarily experimental, and such data is available from many published sources expanding on the information given in Appendix B of this text. Typical values are presented in the following sections discussing the thermal conductivities of gases, liquids, and solids in more detail.

2-1.1 Definition of Thermal Conductivity

The basic equation for the conduction of heat states that the instantaneous rate of heat flow (dQ/dt) is equal to the product of three factors:

1. the area A of the section of the heat flow path, perpendicular to the direction of heat flow;
2. the temperature gradient $(-dT/dx)$, which is the rate of change of temperature T with respect to the length of the path x;
3. and a proportionality factor k, known as the thermal conductivity.

This is mathematically expressed in the following equation.

$$\frac{dQ}{dt} = -kA\frac{dT}{dx} \tag{2-3}$$

This basic relation for heat transfer by conduction was proposed by the French scientist J. B. J. Fourier in 1822, and is known as Fourier's law. This law was used as a fundamental equation in his analytic theory of heat.[1]

Thermal conductivity as defined by the basic conduction heat transfer equation is

$$k = \frac{\dot{Q}x}{A(T_1 - T_2)} \tag{2-4}$$

[1] Fourier, J. B., "Theorie analytique de la chaleur," Oeuvres de Fourier, Gauthier-Villars et Fils, Paris, 1822.

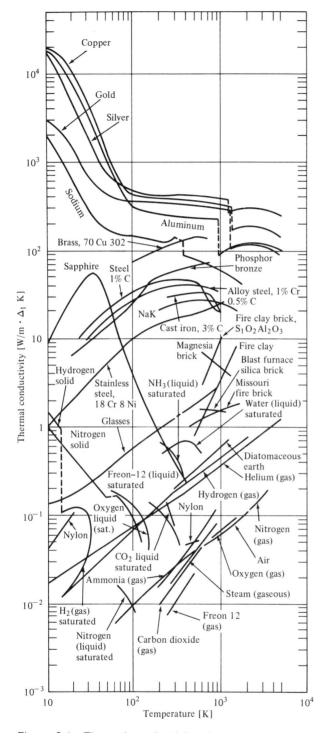

Figure 2-1   Thermal conductivity of gases, liquids, and solids.

where

$\dot{Q}$ = rate of heat flow through heat flow path $[W]$

A = area of path section normal to heat flow $[m^2]$

k = thermal conductivity of heat flow path $[W]/[m][\Delta_1°C]$

x = length of heat flow path $[m]$

T_1 = temperature at the start of heat flow path $[°C]$

T_2 = temperature at the end of heat flow path $[°C]$

When thermal conductivity is used in an equation, the units of the dimensions must be consistent with the units used for the other parameters in the equation. Typical sets of units for thermal conductivity are

SI metric: $[W]/[m^2][\Delta°C/m] = [W]/[m][\Delta_1°C]$

Other sets of units may be used for convenience. For example, in a specific area of application, such as in rocket thrust chambers where heat transfer rates are very high, the heat flow rates may be expressed in $[kW]/[cm][\Delta_1°C]$.

2-1.2 Theoretical Concepts of Thermal Conductivity

For a free flow of heat through a material there must be a temperature gradient, with the heat flowing from the hotter region to the colder region. According to the kinetic theory of heat, the temperature of a molecule is proportional to the kinetic energy of the molecule. The basic concept of thermal conductivity is that of the rapidity of the diffusion of the molecular kinetic energy through the material. The diffusion or transfer of the molecular kinetic energy takes place by molecular collisions and/or by diffusion of faster moving electrons from higher to lower temperature molecules. The theoretical models for thermal conductivities and the magnitude of the values for thermal conductivities differ for gases, liquids, and solids. These are discussed in more detail in the following sections.

2-1.3 Thermal Conductivities of Gases

The general range of the thermal conductivities of various gases is from 0.009 to 0.03 $[W]/[m·\Delta_1°C]$. Gases are regarded as thermal insulators because of these low values of thermal conductivity. The thermal conductivity of a gas is affected by various parameters as indicated in Table 2-1.

Values for thermal conductivities are usually obtained from published experimental values, such as listed in the tables in Appendix B of this text, or from other references. Some published values of thermal conductivities of gases may be in error as much as 10 to 20%. Accurate measurements of

Table 2-1 THE EFFECT OF VARIOUS PARAMETERS UPON THE THERMAL CONDUCTIVITY OF GASES

PARAMETER	EFFECT ON THERMAL CONDUCTIVITY
Temperature	Increases with increasing temperature
Pressure	Not affected by pressure over a wide range; Lower limit of insensitivity:
	Heavy gases (air, etc.) 0.05 mm Hg abs.
	Hydrogen 0.02
	Helium 0.01
	Upper limit of insensitivity: Approaching the critical pressure
Molecular weight	Decreases with increasing molecular weight
Molecular structure	Decreases with increasing number of degrees of freedom of motion within the molecule.

the thermal conductivity of gases are relatively difficult to achieve, as discussed on page 22.

When mixtures of gases are encountered and experimental values of the thermal conductivity of the mixture are not available, a rough estimate can be used. Such an estimate is made by summing the products of the thermal conductivity times the ratio of the partial pressure to the total pressure of the mixture of each constituent. When a resort is made to the use of this rough estimate of the thermal conductivity, it must be appreciated that the estimate is not accurate. This inaccuracy is due to the fact that the diffusion of kinetic energy through a gas mixture with dissimilar molecules is significantly different from the diffusion through a gas of homogeneous molecules.

THEORETICAL MODEL OF HEAT CONDUCTION
Heat conduction in a gas is a process of diffusion of molecular kinetic energy. The two major mechanisms of this diffusion are:

- the migration of molecules (of different average velocities and kinetic energies) from warmer to colder zones, and vice versa;
- the exchange of kinetic energy between molecules by collisions.

• *The Effect of Temperature.* In a gas an increase of temperature (under constant pressure) increases the average kinetic energy of a molecule and decreases the number of molecules in a given volume. This decrease in the number of molecules in a given volume increases the length of the mean free path of the molecule. The combination of a longer mean free path length and a higher kinetic energy of the molecule results in a more rapid diffusion of molecular kinetic energy. Consequently, the thermal conductivity increases.

• *The Effect of Pressure.* In a gas an increase of pressure (at constant temperature) increases the number of molecules in a given volume with no change in the average kinetic energy of the molecule. This increase in the number of molecules in a given volume decreases the length of the mean free path of the molecule. The net overall effect of the increase in the number of molecules and the decrease of the mean free path length result in no change in the diffusion rate of the molecular energy. Consequently, the thermal conductivity remains constant.

As the pressure of a gas approaches a vacuum (on the low side) or the critical pressure (on the high side), significant deviations in the molecular behavior in the gas are encountered. Under these conditions, the thermal conductivity of the gas is no longer independent of pressure.

The low pressure level at which the independence of the thermal conductivity to pressure level ceases varies for different gas molecular structures. Experimental measurements have shown the following pressure limits in Table 2-2, where the thermal conductivity is no longer insensitive to pressure.

The high pressure level at which the independence of the thermal conductivity to pressure level ceases varies for different gases. A pressure effect upon the gas thermal conductivity occurs as the pressure level approaches the critical pressure of the gas.

• *The Effect of Molecular Weight.* The thermal conductivity decreases with increasing molecular weight as the average kinetic energy of a molecule (at a constant pressure level) becomes less as the molecular weight increases. The effect of this is to reduce the rate of the diffusion of the molecular kinetic energy. Consequently, the thermal conductivity decreases.

• *The Effect of Molecular Structure.* The thermal conductivity decreases with increasing complexity (increasing number of degrees of freedom of motion within the molecule) because this more complex interchange of kinetic energy upon collisions between molecules results in a slower diffusion of the molecular kinetic energy.

Efforts to establish relationships between various gas parameters and thermal conductivity have resulted in obtaining good correlations between

Table 2-2 LOW PRESSURE LIMIT OF GAS THERMAL CONDUCTIVITY INSENSITIVITY TO PRESSURE LEVEL

PRESSURE (mm Hg abs.)	GAS
0.05	Heavy gases (air, nitrogen, oxygen, etc.)
0.02	Hydrogen
0.01	Helium

thermal conductivity, viscosity, and specific heat at constant pressure. The equation given below was derived by Maxwell[2] from kinetic theory considerations. This equation has been supported to a remarkable degree by experimental data.

$$k = a\mu c_v \quad [\text{W/m} \cdot \Delta_1 {}^\circ \text{C}] \tag{2-5}$$

where

μ = absolute viscosity $[\text{Pa} \cdot \text{s}]$

c_v = specific heat at constant volume $[\text{J/kg} \cdot \Delta_1 {}^\circ \text{C}]$

a = constant for different molecular structures

The constant "a" is discussed in more detail in the following section.

EVALUATING THERMAL CONDUCTIVITIES

The student and practicing engineer will usually obtain values for the needed thermal conductivities from published tables, such as the tables in Appendix B of this text or from other references. Measurements have been made by numerous investigators since the late 1800s and thermal conductivity data for a wide variety of gases and vapors have been published in technical papers as well as in summary tables in texts and other publications. As the thermal conductivities are insensitive to pressure over a wide pressure range, as outlined on page 20, references to pressures are usually omitted in the tables if the data are for normal pressures.

Tables of thermal conductivities of gases usually present the data in two different ways. One way is to list the thermal conductivity for specific temperatures. The other is to give the values for k_{32}, C, and the applicable temperature range for use in the Eucken[3] formula.

When thermal conductivities are listed for specific temperatures, the extreme temperature values usually represent the experimental range. For extrapolation to other temperatures the given data can be plotted as $\log k$ versus $\log T$, or the assumption made that c_p/k is a constant.

Some tables of thermal conductivities are in the English system of units ($\text{Btu/hr} \cdot \text{ft}^2 \cdot \Delta_1 {}^\circ \text{F/ft}$) give values of k_{32} and C. When these values are given the thermal conductivity for the desired temperature can be calculated using the following equation.

$$k = k_{32} \left(\frac{492 + C}{T + C} \right) \left(\frac{T}{492} \right)^{3/2} \quad [\text{Btu/hr} \cdot \text{ft}^2 \cdot \Delta_1 {}^\circ \text{F/ft}] \tag{2-6}$$

where

T = absolute temperature °R (°F + 460)

[2] Maxwell, J. C., *Collected Works*, Vol. II, p. 1, Cambridge, London, 1890.
[3] Eucken, A., *Physik, Z.*, 14, 324 (1913).

The value of k in English units [Btu/hr·ft·Δ_1°F] is then converted to the SI (metric) units [W/m·Δ_1°C] by multiplying the English value by 1.73.

If the thermal conductivity is not readily available from tables, but the data for use in the Maxwell equation is, the Maxwell equation [Eq. (2-5)] can be used. To do so, value for constant "a" must be selected. As thermal conductivity data are readily available for common monatomic, diatomic, and triatomic gases, the presumed interest in the Maxwell equation would be for more complex gases. As a guide, values for four gases are given in Table 2-3.

To estimate "a" Eucken[3] has suggested the empirical relationship

$$a = 0.25(9\gamma - 5) \tag{2-7}$$

where

γ = ratio of the specific heat at constant pressure to that at constant volume

As a matter of interest, the experimental values of "a" are listed in Table 2-4 for common monatomic, diatomic, and triatomic gases.

MEASUREMENT OF THERMAL CONDUCTIVITY OF GASES

The usual method for measuring the thermal conductivity of gases employs a fine wire stretched along the axis of a capillary tube, or a concentric cylinder and tube with a narrow annulus of the gas to prevent any convection currents in the gas. This configuration originates from Schleiermacher (1888).[4] The determination of k then consists basically of measuring the electrical input to the wire or the heaters of the cylinder necessary to maintain a measured temperature difference across the annulus. Using the geometry of the test apparatus, the k can be calculated using the thick-walled or thin-walled pipe equation [Eq. (2-18) or (2-21)] as appropriate. The resulting value is the average thermal conductivity within the test temperature range. For small temperature differences it is considered to be the thermal conductivity (k) at the average of the two temperatures.

Table 2-3 VALUES OF CONSTANT "a" FOR THE MAXWELL EQUATION [EQ. (2-5)]

GAS	EXPERIMENTAL VALUE AT $0°C$
C_2H_6 (Ethane)	1.57
$CHCl_3$ (Chloroform)	1.48
CCl_4 (Carbon tetrachloride)	1.385
$CH_3 \cdot COO \cdot C_2O_5$ (14 atomic ethyl acetate)	1.19

SOURCE: Eucken, A., *Physic. Zeitschr.* 14, 324; 1913.

[4]Schleiermacher, A., *Wiedemanns Annalen*, 34, 623; 1888.

Table 2-4 EXPERIMENTAL VALUES OF CONSTANT "*a*" FOR THE
MAXWELL EQUATION [EQ. (2-5)]

GAS STRUCTURE		VALUES OF "*a*" AT $0°C$
MONATOMIC SPHERICAL MOLECULES		
Helium	He	
Argon	A	2.5
DIATOMIC MOLECULES		
With negligible energy of oscillation		
Hydrogen	H_2	2.03
Nitrogen	N_2	1.95
Oxygen	O_2	1.915
Air		1.95
With appreciable energy of oscillation		
Carbon monoxide	CO	1.865
Nitric oxide	NO	1.83
Chlorine	Cl_2	1.803
TRIATOMIC MOLECULES		
Carbon dioxide	CO_2	1.67

SOURCE: Adapted from Eucken's data given by Jakob, Max, *Heat Transfer*, Vol. 1, p. 75,
John Wiley & Sons, Inc., New York, 1956.

Compensation must be made for any heat flow through the ends of
the capillary tube or the concentric cylinder and tube. Various methods of
compensation have been used by investigators such as Eucken (1911)[5] and
Sellschopp (1934).[6] Corrections must be made for any radiation transfer
between the hot and cold surfaces of the annulus. In making any experimental measurements, it is very important that equilibrium temperatures
have been reached. Because of the low values of the thermal conductivities
of gases, even slight transient conditions can give results significantly in
error.

2-1.4 Thermal Conductivities of Liquids

For conduction heat transfer considerations, liquids can be separated into
four general groups, namely:

- liquid metals
- water
- aqueous solutions and inorganic liquids
- organic liquids

Liquid metals are excellent thermal conductors. Water has the highest
thermal conductivity of the other three groups, which are characterized by
relatively low thermal conductivities. Typical ranges of the thermal conductivities of these four groups of liquids are shown in Table 2-5.

[5] Eucken, A., *Annalen d. Physik* (4) 34, 185: 1911.
[6] Sellschopp, W., *Forschung a.d. Geb.d. Ingenieurwes.* 5, 162; 1934.

Table 2-5 TYPICAL RANGE IN THERMAL CONDUCTIVITIES OF DIFFERENT TYPES OF LIQUIDS

| | THERMAL CONDUCTIVITY | | |
LIQUID TYPE	$[W/m \cdot \Delta_1 {}^\circ C]$		
Liquid metals	8	to	120
Water (best nonmetallic conductor)	0.55		0.675
Aqueous solutions	0.12		0.70
Inorganic liquids	0.10		0.70
Organic liquids	0.05		0.35

The thermal conductivity of a liquid is affected by various parameters as indicated in Table 2-6.

Values of thermal conductivities are usually obtained from published experimental values, such as listed in the tables in Appendix B of this text, or from other references. Accurate measurements of thermal conductivities of liquids can be made more easily than for gases. Consequently, the published values are more accurate than those for gases. Earlier experimenters often obtained erroneously high values for the thermal conductivities. Subsequent improvement in the experimental techniques have enabled accurate values to be obtained.

When mixtures of liquids are encountered and experimental values or a formula for the thermal conductivity of the mixture are not available, a rough estimate can be used. For both mixtures of mixable liquids and aqueous solutions, the estimate of the thermal conductivity of the mixture is made by summing the products of the thermal conductivity times the ratio of the weight to the total weight of the mixture of each constituent. When a resort is made to the use of this rough estimate of the thermal conductivity, it must be appreciated that the estimate is not accurate.

Heat conduction in a liquid is a process of diffusion of molecular kinetic energy by the exchange of kinetic energy between molecules by collisions, and in the case of liquid metals by the additional effect of the

Table 2-6 THE EFFECT OF VARIOUS PARAMETERS UPON THE THERMAL CONDUCTIVITY OF LIQUIDS

PARAMETER	EFFECT ON THERMAL CONDUCTIVITY
Temperature	decreases for most liquids with increasing temperature
	increases for water and some aqueous solutions with increasing temperature
Pressure	not affected by normal pressures
	increases for some fluids under high pressures:
	15% at 200 MPa
	100% at 1200 MPa
Modulus of compressibility	increases with increasing modulus of compressibility

migration of free electrons. The random translatory motion is small, by comparison with that in gases, and so is the translation of energy due to this motion.

The most promising theories of heat conduction suppose heat transfer by longitudinal vibrations similar to the propagation of sound. The equation proposed by Kardos[7] based on a cubic array of the molecules is

$$k = \rho c_p V L \quad [W/m \cdot \Delta_1 °C] \tag{2-8}$$

where (in consistent units)

k = thermal conductivity $[W/m \cdot \Delta_1 °C]$

ρ = liquid density $[kg/m^3]$

c_p = specific heat at constant pressure $[J/kg \cdot \Delta_1 °C]$

V = velocity of sound in the liquid $[m/s]$

L = distance between the surfaces of adjacent molecules $[m]$

The comparison of computed and observed values for the thermal conductivities of some liquids is indicated in Table 2-7.

EVALUATING THERMAL CONDUCTIVITIES

The student and practicing engineer will usually obtain values for the needed thermal conductivities from published tables, such as the tables in Appendix B of this text, or from other references. Thermal conductivities

Table 2-7 COMPARISON OF COMPUTED AND OBSERVED THERMAL CONDUCTIVITIES OF LIQUIDS AT 30°C

			k_K AND k_0 $[W/m \cdot \Delta_1 °C]$		
LIQUID	CHEMICAL FORMULA	NUMBER OF C ATOMS	V [m/s]	COMPUTED k_K [EQ. (2-8)]	OBSERVED k_0
Water	H_2O	0	1500	0.596	0.610
Methyl alcohol	CH_4O	1	1130	0.202	0.211
Ethyl alcohol	C_2H_6O	2	1140	0.198	0.182
n-Propyl alcohol	C_3H_8O	3	1240	0.207	0.171
Isobutyl alcohol	$C_4H_{10}O$	4	1050	—	—
n-Butyl alcohol	$C_4H_{10}O$	4	—	0.198	0.167
Isoamyl alcohol	$C_5H_{12}O$	5	1240	0.221	0.148
Carbon bisulfide	CS_2	1	1180	0.147	0.159
Acetone	C_3H_6O	3	1140	0.182	0.179
Ether	$C_4H_{10}O$	4	920	0.140	0.137
Ethyl bromide	C_2H_5Br	2	900	0.113	0.120
Ethyl iodide	C_2H_5I	2	780	0.103	0.111

SOURCE: Jakob, M., *Heat Transfer*, Vol. 1, John Wiley & Sons, Inc., New York, 1956. By permission.

[7] Kardos, A., *Forschung a.d. Geb. d. Ingenieurwes.* 5, 14; 1934.

are listed for specific temperatures. A linear variation with temperature can be used. Often the given temperature range defines the temperature limits over which the data are recommended.

If the thermal conductivity is not readily available from tables, but the data for use in the Kardos equation are, the Kardos equation can be used to estimate the thermal conductivity.

MEASUREMENT OF THERMAL CONDUCTIVITIES OF LIQUIDS

The thermal conductivity of liquids is usually measured in much the same manner as the thermal conductivity of gases. Accurate measurements of the thermal conductivities of liquids are higher. In early experiments relatively thick liquid layers and large temperature differences were used. Resulting convective currents in the liquid layer and radiation heat transfer through the liquid layer gave erroneously high measurements for the thermal conductivities. More recent experimenters[8] have reduced the liquid layer thickness to approximately 0.4 mm and have used temperature differences of 0.5 ΔK. Such improved test configurations reduce convective and radiation effects to negligible amounts.

2-1.5 Thermal Conductivities of Solids

For conduction heat transfer considerations, solids can be separated into six general groups, namely:

- metals
- nonmetallic crystalline materials
- amorphous materials
- porous materials
- building materials
- electrical insulation

Metals are excellent thermal conductors. Typical ranges of the thermal conductivities of these six groups of solids are shown in Table 2-8.

Table 2-8 TYPICAL RANGE IN THERMAL CONDUCTIVITIES OF DIFFERENT TYPES OF SOLIDS

SOLID TYPE	THERMAL CONDUCTIVITY $[W/m \cdot \Delta_1 °C]$		
Metals	8	to	415
Nonmetallic crystalline solids	0.85		70.
Amorphous solids	0.17		1.2
Porous solids (insulating materials)	0.025		1.2
Building materials	0.05		4.3
Electrical insulation materials	0.03		0.5

[8]Schmidt, E., and Sellschopp, W., *Forschung a.d. Geb. d. Ingenieures.* 3, 227; 1932.

Table 2-9 THE EFFECT OF VARIOUS PARAMETERS UPON THE
THERMAL CONDUCTIVITY OF METALS

PARAMETER	EFFECT OF THERMAL CONDUCTIVITY
Temperature	decreases for pure metals with increasing temperature
	increases for most alloys with increasing temperature
Pressure	not affected by normal pressures
Crystalline structure	increases (above that of a normal poly-crystalline structure) for a single crystal
	changes with phase change of an alloy (such as occur during heat treatment or precipitation hardening of the metal)
	decreases with the presence of impurities in pure metals
	changes with the direction of the heat transfer relative to the crystal lattice structure

Because of these characteristics, solids range from good thermal conductors
to thermal insulators. Because the thermal conductivities of these types of
solids are affected differently by various parameters, each type of solid will
be discussed in separate following sections.

THERMAL CONDUCTIVITIES OF METALS

Metals are good thermal conductors. Pure silver has the highest thermal
conductivity of all materials, followed by pure copper, and then by pure
aluminum. The general range of the thermal conductivities of various
metals is from 8 to 415 $[W/m \cdot \Delta_1 °C]$. The thermal conductivity of a
metal is affected by various parameters as indicated in Table 2-9.

Values of thermal conductivities are usually obtained from published
experimental data, such as listed in the tables in Appendix B of this text, or
from other references, such as the *Thermophysical Properties of Matter*.[9]
Accurate measurements of the thermal conductivity of metals are relatively
easy to make. The measurement experimental techniques are simplified as
radiation and convective effects are not present.

The thermal conductivity of an alloy cannot be calculated, or even
estimated by the ordinary arithmetic mixing rule. Wide variations are
encountered, as illustrated by Table 2-10.

The effect of impurities upon the thermal conductivity of pure metals
cannot be reliably calculated. Nonlinear variations are encountered, as
illustrated by Table 2-11.

[9]*Thermophysical Properties of Matter*, Vol. 1; *Thermal Conductivity Metallic Elements and
Alloys*, IFI/Plenum, New York, Washington, 1970

Table 2-10   THE THERMAL CONDUCTIVITIES OF SELECTED ALLOYS COMPARED TO THE THERMAL CONDUCTIVITIES OF THE ALLOY CONSTITUENTS [W/m·Δ_1°C]

ALLOY	k OF ALLOY	k_1	k_2
Brass 60% Cu 40% Zn	102.	Cu 393	Zn 112
29% nickel steel	7.8	Fe 93	Ni 83.9
33% Cd 67% Sb	1.25	Cd 102	Sb 15.9

• *Theoretical Model.* Heat conduction in a metal combines the processes of diffusion of molecular kinetic energy by the vibrations of the crystal lattice and the migration of free electrons through the crystal lattice. The thermal conductivity associated with the migration of free electrons is analogous to electrical conductivity, except that the property of metals to lose electrical resistance close to absolute zero does not hold for thermal conductivities.

The diffusion of molecular kinetic energy by vibrations of the crystal lattice is affected by the nature of the lattice structure. Consequently, the thermal conductivity may vary along the different axes of the crystal lattice. In pure metals even a small amount of an impurity will cause discontinuities in the crystal lattice structure, and these discontinuities will impede the diffusion of the molecular kinetic energy by vibrations of the crystal lattice.

Alloys have more complex crystalline structures than pure metals, often being composed of constituents with crystal lattices of different dimensions and basic geometries. The resulting complex structure impedes the diffusion of the kinetic energy.

Precipitation hardening and heat treatment often cause material to migrate to a position in the boundary between crystals where it impedes the diffusion of the molecular kinetic energy across the crystal boundaries from one crystal to another.

The usual polycrystalline metallic structure has discontinuities in the crystal lattice structure at each crystal boundary. This results in a significantly lower thermal conductivity in a polycrystalline structure than in a single crystal of the same material.

Table 2-11   THERMAL CONDUCTIVITIES FOR VARIOUS PURITIES [W/m·Δ_1°C] (Typical Values from Reference 9)

ALUMINUM			COPPER		
PURITY (%)	$25K$	$273K$	PURITY (%)	$20K$	$273K$
99.999	7900	—	99.999	8800	391
99.99	2690	239	99.98	1190	—
99.94	846	—	99.8	—	213
Commercial	290	193	99.4	335	—

Table 2-12 COMPARATIVE THERMAL CONDUCTIVITIES OF METALS
IN THE SOLID (POLYCRYSTALLINE) STATE AND IN THE LIQUID STATE
$[W/m \cdot \Delta_1 {}^\circ C]$

METAL	k_s (SOLID STATE)	k_i (LIQUID STATE)
Aluminum	225.	88.
Lead	31.	15.
Mercury	37.2	10.4
Zinc	93.	60.

The heat conducting capability of the crystal lattice is indicated by the reduction in the thermal conductivity when the metal melts destroying the crystal lattice structure. The comparative thermal conductivities of several metals in the solid (polycrystalline) state and in the liquid state at approximately the same temperature is illustrated in Table 2-12.

Numerous investigators have endeavored to develop theoretical equations for calculating thermal conductivities of metals. Relationships were found which could be applied to limited cases only. While these theoretical studies were in process, many measurements of thermal conductivities of metals and alloys at many temperatures have been made. Currently, the numerical values for use have been supplied by the experimental tests, and the theoretical studies have mainly served to give an insight into the mechanism of heat conduction in metals.

Attempts have been made to correlate the electrical resistance of the metal with its thermal conductivity. Correlations have been obtained and empirical relationships have been developed.[10] However, the application of these relationships is limited, usually being valid only for comparing the thermal conductivities of various lots of the same metal or alloy, or between similar alloys.

• *Evaluating Thermal Conductivities.* The student and practicing engineer should obtain values for needed thermal conductivities from published tables, as contrasted to attempting evaluations from theoretical approaches. Tables of thermal conductivities for metals are in Appendix B of this text and can be found in other published reference material. Thermal conductivities of metals are usually listed for specific temperatures, or a thermal conductivity is given for 0° (k_0) and a temperature constant α is also given. The thermal conductivity at a temperature T is then calculated by the following formula.

$$k = k_0 + (\alpha T) \tag{2-9}$$

If α is positive, the thermal conductivity increases with temperature; if negative, it decreases. Where thermal conductivity values are given for

---

[10] Jakob, M., *Heat Transfer*, Vol. 1, pp. 112–114, John Wiley & Sons, Inc., New York, 1956.

specific temperatures, a linear variation with temperatures can be used to obtain the value of the thermal conductivity at an intermediate temperature.

Some authors give k_0 at 0°F and α per °F; others give k_0 at 0°C and α per °C. Also, mixtures of these may occur, such as k_0 being given for 32°F, and α per °F above or below 32°F.

• *Measurement of Thermal Conductivities of Metals.* Experimental measurements of thermal conductivities of metals have been accomplished by a variety of methods. These methods are based on the measurement of the steady state temperature gradient along a constant area cross section path of the material conducting a known heat flow. Each of these methods has limitations. The selection of one method over another for a particular measurement is governed by whether the metal is a good or poor conductor and by the general temperature level at which the thermal conductivity is to be measured.

For example, if the specimen has a relatively high thermal conductivity, and the temperature level is moderate, the relatively simple Searle method would be suitable. The Searle method is illustrated in Figure 2-2. The specimen is in the form of a rod with a constant cross section. The rod is placed in a bed of insulation. A known heat input is applied to one end, and the other end is cooled. Thermocouples are attached at various points along the length of the test specimen. After a sufficient time has elapsed to achieve steady state, the temperatures along the test specimen are measured, and the linear temperature gradient $(\Delta T/x)$ is determined. The average value of the thermal conductivity is obtained from Eq. (2-10).

$$\bar{k} = \frac{\dot{Q}}{A(\Delta T/x)} \quad [\mathrm{W/m \cdot \Delta_1 {}^\circ C}] \tag{2-10}$$

where

$\dot{Q}$ = heat flow rate [W]

A = heat flow path cross section area
  (perpendicular to the heat flow) [m²]

ΔT = temperature drop across the path length [Δ°C]

x = heat flow path length [m]

The smaller the temperature drop along the test specimen, the more nearly the measured thermal conductivity approaches the true value at the average temperature. However, the correction for heat losses, and so forth, becomes more difficult as the heat losses become relatively large with respect to the total heat flow, resulting in increased errors in the measurement.

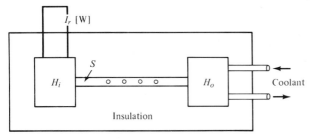

Figure 2-2 Thermal conductivity by Searle's method.

To measure the thermal conductivity of poor conductors, the specimens are usually in the form of a sheet. The heat is passed through the thickness of the sheet, so the heat flow cross-sectional area will be large and the length of the heat flow path will be short. In this method, the heat leakage is the principle source of error, and techniques have been developed to minimize these losses. The "guarded-plate method" has been standardized by the ASTM and is used by such research and testing organizations as the National Bureau of Standards. This method is illustrated schematically in Figure 2-3. In this method, electrically heated thermal guards are placed adjacent to the exposed surfaces of the heat source H_i, the test specimen S, and the heat sink H_o. These thermal guard plates are maintained independently at the same temperature as the adjacent surfaces so that ideally no heat-leaks in or out of the measuring system occur.

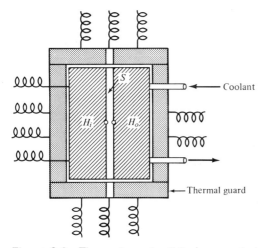

Figure 2-3 Thermal conductivity by guarded-plate method.

Table 2-13 THE EFFECT OF VARIOUS PARAMETERS UPON THE THERMAL CONDUCTIVITY OF NONMETALLIC CRYSTALLINE SOLIDS

PARAMETER	EFFECT ON THERMAL CONDUCTIVITY
Temperature	decreases with increasing temperature
Pressure	not affected by normal pressures
Crystalline structure	decreases (below that of a single crystal) for a polycrystalline structure
	changes with the direction of the heat transfer relative to the crystal lattice structure
	decreases with the presence of impurities
	decreases with increasing complexity of the crystal structure

THERMAL CONDUCTIVITIES OF NONMETALLIC CRYSTALLINE SOLIDS

Nonmetallic crystalline solids rank next to metals as good thermal conductors. The general range of the thermal conductivities of various nonmetallic crystalline solids is from 0.60 to 300 $[W/m \cdot \Delta_1 °C]$. Materials, such as Acheson graphite, have the highest thermal conductivity in the group of nonmetallic crystalline solids. The thermal conductivity of nonmetallic crystalline solids is affected by various parameters as indicated in Table 2-13.

Values of thermal conductivities are usually obtained from published experimental data, such as listed in the tables in Appendix B of this text, or from other references. Accurate measurements of the thermal conductivity of nonmetallic crystalline solids are relatively easy to make.

• *Mechanism of Heat Conduction.* Heat conduction in a crystalline nonmetallic solid is primarily the diffusion of molecular energy by the vibrations of the crystal lattice. Debye[11] developed the following formula:

$$k = \tfrac{1}{2} \gamma \omega \rho c_v \qquad (2\text{-}11)$$

where

γ = average distance of effect of thermoelastic wave

ω = average velocity of propagation of the elastic waves

ρ = density of the crystal

c_v = specific heat of the crystal

[11] Debye, P., *Vortraege ueber die kinetische Theorie de Materie und Elektrizitaet*, Leipzig and Berlin, 1914.

For many crystals the product of the thermal conductivity times the absolute temperature remains essentially constant $(kT = A)$ over various temperature ranges at a value characteristic for the crystal. This relationship tends to be more valid in temperature ranges from $0°$ to $-200°C$. At lower temperatures approaching absolute zero, significant changes in the temperature/thermal conductivity relationship often occur. At elevated temperatures from $0°$ to $1000°C$, the decrease in the thermal conductivity is usually significantly less than that indicated by the $kT = A$ relationship. This results in an increase in the kT product. Table 2-14 illustrates the kT values for several crystals. Although the influence of temperature upon conductivity of nonmetallic crystals is generally the same, the absolute values of k differ widely, as indicated by the variations of the kT values in Table 2-14 for the different crystals at the same temperature.

• *Evaluating Thermal Conductivities.* Values for thermal conductivities of crystals are available in published material. In general, this information will be found to a greater extent in scientific rather than in engineering literature. A limited amount of thermal conductivity data is presented in Appendix B of this text. Limited extrapolations to other temperatures than those given can be made by the use of the $(k)(T)$ product as a constant, where

$$k = \text{thermal conductivity} \quad [W/m\Delta_1{}°C]$$

$$T = \text{temperature (abs.)} \quad [K]$$

• *Measurement of Thermal Conductivities of Crystals.* Experimental measurements of the thermal conductivities of crystals are accomplished in a manner similar to those for measuring the thermal conductivities of metals. Appropriate techniques are applied to adapt to the crystal size and the temperature levels involved.

THERMAL CONDUCTIVITIES OF AMORPHOUS SOLIDS
The general range of the thermal conductivities of homogeneous amorphous solids (glasslike materials) is generally lower than that of nonmetallic crystalline materials and higher than that of nonmetallic liquids.

The thermal conductivity of amorphous solids is affected by various parameters as indicated in Table 2-15.

Values of thermal conductivities are usually obtained from published experimental data, such as listed in tables in Appendix B of this text, or from other references. Accurate measurements of the thermal conductivity of amorphous solids are relatively easy to make.

• *Mechanism of Heat Conduction.* Because a crystal lattice structure is not present in an amorphous solid, the diffusion of the molecular kinetic

Table 2-14 $(k)(T)$ VALUES FOR SOME CRYSTALS [W/m]

TEMP. [°C]	ICE (H_2O)	$CaCO_3$ NORMAL TO AXIS	MULLITE ($3\,Al_2O_3 \cdot 2\,SiO_2$)	SALT (NaCl)	KCl	CORUNDUM (Al_2O_3)	QUARTZ NORMAL TO AXIS	QUARTZ PARALLEL TO AXIS	PERICLASE (MgO)	CARBORUNDUM (SiC)
−200		1297.0		1757.3	2217.5		2905.4	7321.9		
−100	601.4	1129.7		2050.2	2008.3		2992.9	5985.8		
0	609.1	1171.5	1605.2	1882.8	1882.8	2833.7	2839.2	5187.0	11332.2	19355.7
+100		1338.9	1935.9	1799.1	1841.0	3226.4	2573.7	4513.3	11936.0	21596.7
+200			2208.9			3273.2			11872.3	22893.2
+400			2328.6			3957.2			11037.2	24429.9
+600			2566.6			4530.9			9777.6	25666.2
+800			3154.6			5010.9			9281.4	25000.9
+1000			2864.2			5066.5			8147.2	25332.7

SOURCE: Adapted from various data presented by Jakob, M., *Heat Transfer*, Vol. 1, John Wiley & Sons, Inc. New York, 1956.

Table 2-15 THE EFFECT OF VARIOUS PARAMETERS UPON THE
THERMAL CONDUCTIVITY OF AMORPHOUS SOLIDS

PARAMETER	EFFECT ON THERMAL CONDUCTIVITY
Temperature	increases with increasing temperature
Pressure	not affected by normal pressure

energy must be accomplished by collisions between molecules, in much the same manner as in liquids. Because the relationship between molecules is vastly different between that in amorphous solids and in liquids, the characteristics of the diffusion of the molecular energy can be expected to be different, and it is. The thermal conductivity is higher, and the effect of temperature is opposite.

Some materials, such as quartz, can exist in either the crystalline form or in the amorphous form. As a matter of interest, Table 2-16 shows the thermal conductivities over a temperature range for quartz in both the crystalline and amorphous form.

• *Evaluating Thermal Conductivities.* The homogeneous amorphous solids of primary interest in engineering applications are those of glasses. Values for thermal conductivities of these materials are available in published material. A limited amount of thermal conductivity data is presented, in Appendix B of this text. For limited extrapolations to other temperatures linear variation with temperature can be used.

• *Measurement of Thermal Conductivity.* The thermal conductivity of homogeneous amorphous solids is usually measured in much the same manner as for the thermal conductivity of poor conductor metals as described on page 31.

THERMAL CONDUCTIVITIES OF POROUS MATERIALS
Porous materials are characterized by low thermal conductivities, and comprise the bulk of the "insulators." These products are manufactured

Table 2-16 A COMPARISON OF THE THERMAL CONDUCTIVITY OF
QUARTZ IN THE AMORPHOUS AND CRYSTALLINE FORM $k[W/m \cdot \Delta_1{}^\circ C]$

| | | QUARTZ CRYSTAL | |
TEMPERATURE [°C]	QUARTZ GLASS (FUSED SILICA)	NORMAL TO AXIS	PARALLEL TO AXIS
−200	0.93	39.8	100.3
−100	1.56	17.3	34.6
0	1.90	10.4	19.0
+100	2.08	6.9	12.1

SOURCE: Jakob, M., *Heat Transfer*, Vol. 1, John Wiley & Sons, Inc., New York, 1956. By permission.

largely from various raw materials for specific applications. The resulting materials are nonhomogeneous, being composed of a solid component and volume or space component. The volume component, when surrounded by the solid component, will be in the form of pores that may or may not be interconnecting. When the solid component is in the form of fibers or granular materials, the solid material will be surrounded largely by the volume component, with local points of contact between the fibers or granules.

The heat conducted through such a porous material is the sum of the heat flow along thermal paths through both the solid component and the gas in the volume component. The heat transfer mechanisms along these paths include conduction through the solid component; conduction and convection through the gas in the volume component; and radiation across the volume component.

To obtain the best insulation, the volume component is made as large as possible to take advantage of the lower thermal conductivity through the gas in the volume component. To provide a measure of the volume component content, insulating materials are often characterized by their specific weight. Two major factors affecting the optimum amount of volume content are:

- the physical strength requirements of the insulation, as a higher strength and load bearing capacity requires an increased amount of the solid component, resulting in reducing the effectiveness of the insulating material
- increased size of the pores in the volume component increases the convective heat transfer through the volume component, reducing the effectiveness of the insulating material

The thermal conductivities of porous materials usually increase with increasing temperatures because the combined effects of the characteristics of thermal conductivities of the constituent gas, amorphous solids, and radiation are dominant. These effects usually more than offset the opposing characteristics of the thermal conductivities of any crystalline material content.

Insulating materials are generally categorized in accordance with their intended application, as indicated by Table 2-17.

Table 2-17 CATEGORIES OF INSULATING MATERIALS

APPLICATION	CHARACTERISTIC TEMPERATURE RANGE
Low temperature	below $-45°C$
Normal refrigeration	$-45°C$ to room temperature
Building	$-45°C$ to $+65°C$
Heating and process	$+65°C$ to $+150°C$
Power generation	$+150°C$ to $+550°C$
Refractory (furnace lining)	over $+550°C$

In the following paragraphs the general "typical" characteristics of the insulation for each category will be described. Tables of thermal conductivities for these insulating materials will be found in Appendix B.

• *Low Temperature Insulation.* Low temperature insulating material is used to minimize heat flow into low temperature systems. Vacuum jackets with radiation shields are used a great deal for low temperature insulation. Main requirements of the insulation material include:

- suitable physical and mechanical properties of the insulating material at the low temperatures
- complete moisture and pressure sealing from the surrounding atmosphere
- a minimum of gas circulation within the insulation
- a minimum of conduction heat transfer paths through the solid constituent of the insulation
- a minimum of radiation heat transfer through the insulation

When porous insulation is applied to cryogenic temperature systems without a high vacuum environment in the porous insulation, the air or other gas in the pores of the insulation may be subject to liquefaction. This reduces the pressure in the affected pores, and if a complete pressure seal is not present, additional gas migrates into the affected pores. This results in poor insulation characteristics. Also, when air is involved, the presence of liquid air in contact with a combustible material presents a safety hazard. Spontaneous combustion or even explosions may result.

• *Normal Refrigeration.* Normal refrigeration insulation material is used to provide insulation for refrigerators, cold storage rooms, refrigerated trucks, and railroad cars, to prevent freezing or sweating of water pipes, insulate ice-water pipes, and so forth. In the application of these insulating materials, it is of paramount importance to seal the insulation from infiltration of moisture and to keep the moisture away from the cold surface. A buildup of moisture destroys the insulation characteristics of the insulating materials, because the thermal conductivity of water or ice is significantly higher than that of the dry insulating material. Freezing of the water in the insulation may be physically destructive to the insulation structure.

A wide range of insulating materials is available. The selection of a particular insulating material will depend upon the requirements of the application and use. Typical materials include cork, foam plastics, fiber glass and rock wool, granular materials such as vermiculite, hair and fibers, and wool felt. Cork is used in many forms such as sheets, blocks, bulk, or granulated mixed with a binder to form a heavy paintlike substance that can be applied to a surface. Foam plastics are used in the form of sheets, blocks, or bulk, as well as having the capability of being "foamed in place."

• *Building Insulation.* Insulation in buildings is used primarily to reduce heat loss during cold weather and to reduce heat penetration during warm weather. In the application of these insulation materials, moisture penetration should be minimized, but a complete moisture seal, as required for refrigeration insulations, is not necessary. Any moisture that is collected during warm weather when the inside of the building is cooler is driven off during cold weather when the inside of the building is warmer.

A wide range of materials is used, with an emphasis on low cost. This favors the use of cork, rock wool, slag wool, glass wool, vermiculite or other granular materials, and reflective metal foils. These insulators are applied in bulk, blankets, and reinforced batts.

• *Heating and Process Insulation.* This category of insulation is used primarily to reduce the heat loss in space heating systems and in various process systems. In these applications a moisture seal is neither required nor is it desirable. Asbestos paper and asbestos paper structures are used extensively because of their good insulation characteristics, low cost, and convenient application. Layers of asbestos paper with thin air spaces between layers can readily be wrapped around pipes or ducts. Corrugated asbestos paper boards can be readily applied to flat surfaces. Mineral wools and granular materials are also used.

• *Power Generation Insulation.* Insulation is used primarily to reduce the heat loss from power generation and high temperature chemical processing systems. In these applications a moisture seal is neither required nor is it desirable. In exposed installations, protection should be provided to keep rain from penetrating into the installation. The materials normally used must be resistant to mechanical vibrations and thermal shock. A material composed of 85% magnesia and 15% asbestos is often used. This is formed into quite rigid flat slabs or preformed sections to fit around pipes. In applications where thick insulation is required, the preformed insulation is applied in layers, with the outer layers being composed of less expensive lower temperature insulation.

• *Refractory Insulation.* Refractory insulation is used primarily for lining furnaces and high temperature retorts and crucibles for molten metals. The primary functions of the refractory insulation include:

- providing a wear resistant hot radiating surface to assist in obtaining high furnace temperatures
- insulating the furnace, retort, or crucible structure to prevent it from overheating
- restricting the heat loss from the furnace, retort, or crucible

• providing a lining to a retort or crucible that will resist the forces of hot molten metal in contact with or flowing over the refractory insulation

In the selection of refractories for specific applications, a low thermal conductivity is not always a required characteristic. The density of refractories is significantly higher than that of the other types of insulation discussed, typically ranging from 2250 to 4800 kg/m^3. When a refractory lining of some height is required, the bottom bricks are subjected to a significant loading, and must have adequate strength to support the lining.

The usual application of refractory insulation is in the form of bricks. These can be laid on top of one another to form a vertical wall or be laid in an arched ceiling.

• *Evaluating Thermal Conductivities of Porous Materials.* Values of thermal conductivities of porous materials have been obtained by experimental measurements. These values are available in published material. Tables of the thermal conductivities of representative insulating materials are given in Appendix B-4 of this text. Additional data can be obtained from other published reference material and from the manufacturers or suppliers of the insulation material of interest.

• *Measurement of Thermal Conductivities of Porous Materials.* Experimental measurements of thermal conductivities of porous materials have been accomplished by several methods. Large flat samples can be used in the guard plate method developed for poor conducting metals. Preformed sections to insulate pipes can be tested in a manner analogous to the concentric cylinder and tube method used for measuring the thermal conductivities of gases. A spherical apparatus, as illustrated in Figure 2-4, has also been used. It is relatively difficult to fill the hollow sphere homogeneously with a loose material, particularly if a definite and uniform density is to be obtained. Also after filling, sagging of the material cannot be entirely prevented. The advantage of the spherical method is that all edge losses present in the other methods are eliminated. In the evaluation of the thermal conductivities of low conductivity material, even small losses can result in errors in the experimental values obtained.

THERMAL CONDUCTIVITIES OF BUILDING MATERIALS

The materials used in building are varied. The thermal conductivity of the same type of material, such as wood, brick, or concrete, varies from sample to sample. The thermal conductivity of all materials that can absorb moisture is significantly affected by the moisture content, increasing with increased moisture content. Studies have shown that for a particular

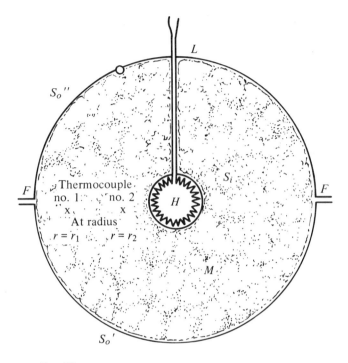

F = Flange
H = Electric heater
L = Lid
M = Material to be tested
S_i = Inner spherical metal shells
$S_o' S_o''$ = Outer hemispherical metal shells

Figure 2-4 Sketch of spherical apparatus for measurement of the conductivity of loose insulating and building materials.

material, a good correlation can be made between the thermal conductivity and the specific weight of the material with a constant moisture content and at the same average temperature. Examples of this type of correlation for wood are shown in Figures 2-5 and 2-6. The relationship between thermal conductivity and the average specific weight for 20 different woods at a mean temperature of 25°C and 12% moisture content is shown in Figure 2-5.

The effect of different moisture content upon the thermal conductivity of Douglas Fir at a mean temperature of 24°C is shown in Figure 2-6.

• *Evaluating Thermal Conductivities of Building Materials.* Values of thermal conductivities of building materials have been obtained by experimental measurements. These values are available in published material. Usually

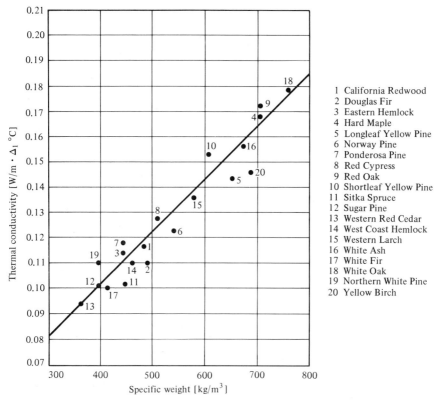

Figure 2-5 Relation between thermal conductivity and average specific weight for 20 different woods at a mean temperature of 24°C and 12% moisture content.

the specific weight of the material is given in conjunction with the value of the thermal conductivity. For materials that can absorb moisture, the moisture content is also given. Tables of the thermal conductivities of representative building materials are given in Appendix B of this text. Additional data can be obtained from other published reference material.

• *Measurement of Thermal Conductivities of Building Materials.* Experimental measurements of thermal conductivities of building materials have been accomplished by methods that have been outlined previously for low conductivity solids.

THERMAL CONDUCTIVITIES OF ELECTRICAL INSULATORS
The design of many electrical devices is limited by the ability to remove heat from the electrical windings. The electrical insulating materials are required to provide sufficient electrical insulating characteristics on one

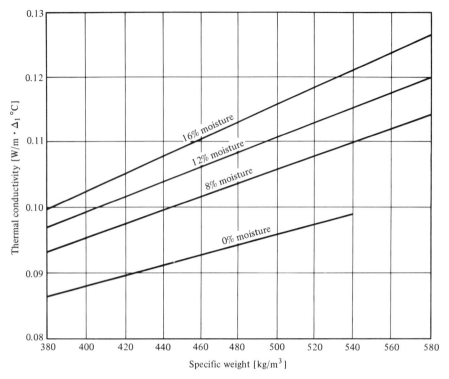

Figure 2-6 The effect of moisture content upon the thermal conductivity of Douglas fir.

hand, and to provide a high thermal conductivity on the other hand, while withstanding elevated temperatures. The early electrical insulating materials were organic materials with low thermal conductivities. To produce electrical insulating materials with significantly higher thermal conductivities, crystalline materials, such as quartz, with higher thermal conductivities as well as good electrical insulating properties have been added to the organic materials.

Values of thermal conductivities of electrical insulators have been obtained by experimental measurements. The data for most of the insulating materials show large variances due to both the difficulties in defining the material accurately and the difficulties of measuring thermal conductivities of thin sheets of material that are used in the electrical device. These values are available in published material. Tables of thermal conductivities of representative insulating materials are given in Appendix B of this text. Additional data can be obtained from other published reference material and from the manufacturers of suppliers of the insulation material of interest.

2-2 CONDUCTION HEAT TRANSFER PATHS

Any heat transfer path can be characterized by a "thermal conductance" and a "thermal resistance," which is the reciprocal of the thermal conductance. The heat flow through the conductive heat transfer path is described by the following equation.

$$\dot{Q} = \overline{\overline{\mathcal{K}}}(T_1 - T_2) = \overline{\overline{\mathcal{K}}}\,\Delta T = \frac{\Delta T}{\overline{\overline{\mathcal{R}}}} \quad [\text{W}] \tag{2-12}$$

where

$\dot{Q}$ = heat flow rate   [W]

$\overline{\overline{\mathcal{K}}}$ = thermal conductance   [W/Δ_1°C]

$\overline{\overline{\mathcal{R}}}$ = thermal resistance   [Δ°C/W]

T_1 = temperature at one end of the path   [°C]

T_2 = temperature at the other end of the path   [°C]

ΔT = temperature drop along the path   [Δ°C]

This equation is similar in form to Ohm's law for the flow of electrical current, which is usually expressed as follows.

$$I = \frac{E}{R_e} \tag{2-13}$$

where

I = current flow rate

E = voltage drop

R_e = electrical resistance

The conduction of heat (energy) can be shown to be analogous to the flow of electrical current, and thus $\dot{Q} = \Delta T/\overline{\overline{\mathcal{R}}}$. A further discussion of an electrical analogy in heat transfer can be found in references such as Kreith.[12]

In most engineering applications, conduction heat transfer paths are varied in form. The conduction heat transfer paths discussed in the following sections include:

- linear conduction paths
- two-dimensional conduction paths
- three-dimensional conduction paths
- extended surfaces (rods and fins)
- bodies with internal heat sources

[12]Kreith, Frank, *Principles of Heat Transfer*, pp. 18–21, International Textbook Company, Scranton, Pa., 1962.

2-2.1 Linear Conduction Heat Transfer Paths

Linear conduction heat transfer path elements can be used singly, or arranged to provide a number of paths with the same temperature drop (which is termed a "parallel" system), or arranged to provide a single path through a number of path elements (which is termed a "series" system). A number of linear conduction path elements can be arranged to provide:

- a number of paths with the same temperature drop forming a "parallel" system
- a single path across the temperature drop forming a "series" system

The characteristics of the path element and the parallel and series systems are discussed in the following sections.

THE LINEAR CONDUCTION HEAT TRANSFER PATH ELEMENT

The "classical" linear conduction heat transfer path element has a constant cross section perpendicular to the heat flow and a constant value of thermal conductivity of its material throughout its length, as illustrated in Figure 2-7.

In the application of Fourier's law to these steady state conditions, the following conditions apply.

1. The rate of heat flow dQ/dt is constant, and can be represented by $\dot{Q}$.
2. The temperature gradient dT/dx is constant, and can be represented by $\Delta T/x$ or $(T_1 - T_2)/x$.
3. The temperature at any point does not vary with time so no heat is stored or released.

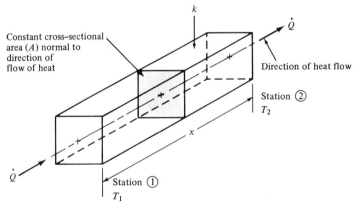

Figure 2-7 The linear conduction heat transfer path element.

Applying these conditions to Eq. (2-3) (Fourier's law) the following equation for simple linear steady state heat conduction heat transfer is obtained.

$$\dot{Q}=A\frac{k}{x}(T_1-T_2)=A\frac{k}{x}\Delta T \quad [W] \tag{2-14}$$

Consistent units must be used, such as:

$\dot{Q}=$ rate of heat flow through heat flow path $[W]$

$A=$ area of path section normal to heat flow $[m^2]$

$k=$ thermal conductivity of heat flow path $[W/m\cdot\Delta_1°C]$

$x=$ length of heat flow path $[m]$

$T_1=$ temperature at the start of heat flow path $[°C]$

$T_2=$ temperature at the end of heat flow path $[°C]$

$\Delta T=$ temperature difference $[\Delta°C]$

Table 2-18   ARRANGEMENTS OF THE BASIC EQUATION FOR SIMPLE LINEAR CONDUCTION

Rate of heat flow	$\dot{Q}=A\dfrac{k}{x}(T_1-T_2)=A\dfrac{k}{x}\Delta T=A\mathcal{K}\Delta T=\overline{\mathcal{K}}\Delta T \quad [W]$
Path cross-sectional area	$A=\dfrac{\dot{Q}x}{k(T_1-T_2)}=\dfrac{\dot{Q}x}{k\,\Delta T}=\dfrac{\dot{Q}}{\mathcal{K}\Delta T} \quad [m^2]$
Heat flow flux	$\dfrac{\dot{Q}}{A}=\mathcal{K}\Delta T=\Delta T/\mathcal{R} \quad [W/m^2]$
Path length	$x=A\dfrac{k}{\dot{Q}}(T_1-T_2)=\dfrac{Ak\,\Delta T}{\dot{Q}} \quad [m]$ $x=\mathcal{R}k \quad [m]$ $x=\dfrac{k}{\mathcal{K}} \quad [m]$
Path material conductivity	$k=\dfrac{\dot{Q}x}{A(T_1-T_2)}=\dfrac{\dot{Q}x}{A\,\Delta T} \quad [W/m\cdot\Delta_1°C]$
Path element conductance	$\overline{\mathcal{K}}=\dfrac{Ak}{x} \quad [W/\Delta_1°C]$ $\mathcal{K}=\dfrac{k}{x} \quad [W/m^2\cdot\Delta_1°C]$
Path element resistance	$\overline{\mathcal{R}}=\dfrac{x}{Ak} \quad [\Delta°C/W]$ $\mathcal{R}=\dfrac{x}{k}=\dfrac{A\,\Delta T}{\dot{Q}} \quad [m^2\cdot\Delta°C/W]$
Path temperature drop	$\Delta T=(T_1-T_2)=\dfrac{\dot{Q}x}{Ak}=\dfrac{\dot{Q}}{\mathcal{K}}=\dot{Q}\overline{\mathcal{R}}=\dfrac{\dot{Q}}{A} \quad [\Delta°C]$
Temperature at one end of path T_1	$T_1=T_2+\dfrac{\dot{Q}x}{Ak}=T_2+\dfrac{\dot{Q}}{\mathcal{K}}=T_2+\dot{Q}\overline{\mathcal{R}} \quad [°C]$
Temperature at other end of path T_2	$T_2=T_1-\dfrac{\dot{Q}x}{Ak}=T_1-\dfrac{\dot{Q}}{\mathcal{K}}=T_1-\dot{Q}\overline{\mathcal{R}} \quad [°C]$

Combining Eqs. (2-12) and (2-14),

$$\dot{Q} = \frac{Ak}{x}\Delta T = \overline{\overline{\mathfrak{K}}}\Delta T = \frac{\Delta T}{\mathfrak{R}} \qquad (2\text{-}15)$$

from which it can be seen that

$$\overline{\overline{\mathfrak{K}}} = \frac{Ak}{x} \quad [\text{W}/\Delta_1{}^\circ\text{C}] \qquad (2\text{-}15\text{a})$$

$$\mathfrak{R} = \frac{x}{Ak} \quad [\Delta{}^\circ\text{C}/\text{W}] \qquad (2\text{-}15\text{b})$$

The thermal conductance ($\overline{\overline{\mathfrak{K}}}$) of a classical linear conduction heat transfer path element is

- directly proportional to the cross section area
- directly proportional to the thermal conductivity of the path material
- inversely proportional to the length of the path

In the usual conduction problem encountered, one of the parameters of Eq. (2-14) is the unknown and the balance of the parameters are given. Equation (2-14) can be rearranged for any of the parameters as the unknown, as listed in Table 2-18.

Example Problem 2-1

What is the thermal conductance of the heat transfer path along the length of a square metal bar 2 cm $\times$ 2 cm $\times$ 1 m long if the thermal conductivity of the metal were 3.70 W/cm $\cdot$ $\Delta_1{}^\circ$C?

SOLUTION

SKETCH

Metal thermal conductivity k

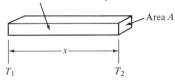

Figure 2-8   Problem category—single one-dimensional conduction path.

EQUATION

$$\overline{\overline{\mathfrak{K}}} = \frac{Ak}{x} \quad [\text{W}/\Delta_1{}^\circ\text{C}] \qquad [\text{Eq. (2-15a)}]$$

PARAMETERS

$$A=2\times2=4 \quad [\text{cm}^2] \qquad \text{(given)}$$
$$k=3.70 \text{ W/cm}\cdot\Delta_1{}^\circ\text{C} \qquad \text{(given)}$$
$$x=1\times100=100 \text{ cm} \quad \text{(given)}$$

SUBSTITUTION

$$\overline{\mathcal{K}}=\frac{4\times3.70}{100} \quad \frac{[\text{cm}^2][\text{W/cm}\cdot\Delta_1{}^\circ\text{C}]}{[\text{cm}]}=[\text{W}/\Delta_1{}^\circ\text{C}]$$

ANSWER

$$\overline{\mathcal{K}}=0.148 \text{ W}/\Delta_1{}^\circ\text{C}$$

Example Problem 2-2

How much heat would be conducted by the metal bar in Example Problem 2-1 if the temperature drop between the ends of the bar were 100 Δ°C?

SOLUTION

SKETCH

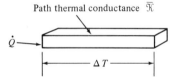

Figure 2-9

EQUATION

$$\dot{Q}=\overline{\mathcal{K}}\Delta T \quad [\text{W}] \qquad [\text{Eq. (2-15)}]$$

PARAMETERS

$$\overline{\mathcal{K}}=0.148 \text{ W}/\Delta_1{}^\circ\text{C} \qquad \text{(from Example Problem 2-1)}$$
$$\Delta T=100\,\Delta^\circ\text{C} \qquad \text{(given)}$$

SUBSTITUTION

$$\dot{Q}=0.148\times100 \quad [\text{W}/\Delta_1{}^\circ\text{C}][\Delta^\circ\text{C}]=[\text{W}]$$

ANSWER

$$\dot{Q}=14.8 \text{ W}$$

Example Problem 2-3

If the metal bar in Example Problem 2-1 were conducting 22.2 W and the hot end were at 175°C, what would be the temperature of the cold end of the bar?

SOLUTION

SKETCH

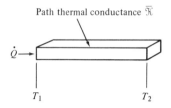

Path thermal conductance $\overline{\mathcal{K}}$

$\dot{Q} \rightarrow$

T_1 T_2

Figure 2-10

EQUATION

$$T_2 = T_1 - \frac{\dot{Q}}{\overline{\mathcal{K}}} \quad [°C] \quad \left[\text{Eq. (2-12x)}\right]$$

PARAMETERS

$\quad T_1 = 175°C \qquad$ (given)

$\quad \dot{Q} = 22.2 \text{ W} \qquad$ (given)

$\quad \overline{\mathcal{K}} = 0.148 \text{ W}/\Delta_1°C \qquad$ (from Example Problem 2-1)

SUBSTITUTION

$$T_2 = 175 - \frac{22.2}{0.148} \quad [°C] - \frac{[\text{W}]}{[\text{W}/\Delta_1°C]} = [°C] - [\Delta°C] = [°C]$$

$$= 175 - 150$$

ANSWER

$\quad T_2 = 25°C$

THE PARALLEL CONDUCTION PATH SYSTEM

When a number of heat transfer path elements are arranged so as to provide multiple heat transfer paths between T_1 and T_2 as illustrated in Figure 2-11, the conduction system is called a "parallel" path system. Heat transfer path conductances are usually used with parallel path heat transfer systems. The parallel conduction path system illustrated in Figure 2-11

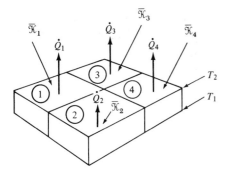

Figure 2-11 The parallel conduction path system.

consists of four separate conduction paths. Each path has the same temperature drop $(T_1 - T_2)$. Each conduction path has its characteristic thermal conductance $(\overline{\mathfrak{K}}_1, \overline{\mathfrak{K}}_2,$ etc.). The total heat flow through the parallel conduction system is the sum of the heat flows of the separate conduction paths.

$$\dot{Q}_t = \dot{Q}_1 + \dot{Q}_2 + \dot{Q}_3 + \dot{Q}_4 \tag{2-16}$$

The heat flow rate through any section is given by Eq. (2-12)

$$\dot{Q}_i = \overline{\mathfrak{K}}_i \Delta T$$

Substituting the values given by Eq. (2-12)

$$\dot{Q}_t = \overline{\mathfrak{K}}_1 \Delta T + \overline{\mathfrak{K}}_2 \Delta T + \overline{\mathfrak{K}}_3 \Delta T + \cdots + \overline{\mathfrak{K}}_n \Delta T$$

as ΔT is the same for all paths in a parallel conduction path system

$$\dot{Q}_t = \left(\overline{\mathfrak{K}}_1 + \overline{\mathfrak{K}}_2 + \overline{\mathfrak{K}}_3 + \cdots + \overline{\mathfrak{K}}_n \right) \Delta T$$

as

$$\dot{Q}_t = \overline{\mathfrak{K}}_t \Delta T \qquad \left[\text{Eq. (2-12)} \right]$$
$$\overline{\mathfrak{K}}_t = \overline{\mathfrak{K}}_1 + \overline{\mathfrak{K}}_2 + \overline{\mathfrak{K}}_3 + \cdots + \overline{\mathfrak{K}}_n \quad [\text{W}/\Delta_1{}^\circ\text{C}] \tag{2-17}$$

The basic equations for parallel conduction path systems are listed in Table 2-19.

Table 2-19 BASIC EQUATIONS FOR PARALLEL CONDUCTION PATH SYSTEMS

Rate of heat flow	$\dot{Q} = \overline{\mathfrak{K}}_t \Delta T$	[W]
Heat flux	$\dfrac{\dot{Q}}{A} = \mathfrak{K}_t \Delta T$	[W/m²]
Total thermal conductance	$\overline{\mathfrak{K}}_t = \overline{\mathfrak{K}}_1 + \overline{\mathfrak{K}}_2 + \overline{\mathfrak{K}}_3 + \cdots + \overline{\mathfrak{K}}_n$	[W/Δ_1°C]
Temperature drop	$\Delta T = \dfrac{\dot{Q}}{\overline{\mathfrak{K}}_t}$	[°C]

In a parallel conduction path system, a single high thermal conductance path exercises a dominant control of the amount of heat flow. By way of an illustration, if the thermal conductance of one path is 80% of the total parallel system conductance, relatively large changes in the thermal conductances of another path, such as doubling or halving its thermal conductance, would have relatively little effect upon the thermal conductance of the total path system.

Example Problem 2-4

In the construction of a solid wooden (fir) wall 5 cm thick, nails (mild steel) were driven through. The total cross-sectional area of the nails equalled $\frac{1}{2}\%$ of the total wall area. At 25°C, what is:

    a. The thermal conductance of the wooden wall per square meter of wall?

    b. The thermal conductance of the nails per square meter of wall?

    c. The thermal conductance of the combination of the wooden wall with the nails per square meter of wall?

SKETCH

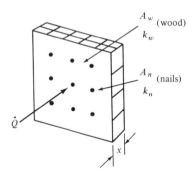

Figure 2-12

SOLUTION

a. For the wooden wall:

EQUATION

$$\overline{\mathcal{K}}_w = \frac{k_w A_w}{x_w} \quad [\text{W}/\Delta_1{}^\circ\text{C}] \quad \left[\text{Eq. (2-15a)}\right]$$

PARAMETERS

$k_w = 0.11$ W/m·Δ_1°C   (from Appendix B, Table B-5)

$A_w = 1 \times 0.995 = 0.995$ m^2 (given)

$x_w = 0.05$ m   (given)

SUBSTITUTION

$$\overline{\mathcal{K}}_w = \frac{0.11 \times 0.995}{0.05} \quad \frac{[W/m \cdot \Delta_1 °C][m^2]}{[m]} = [W/\Delta_1 °C]$$

ANSWER

$\overline{\mathcal{K}}_w = 2.19$ W/Δ_1°C

b. For the nails:

EQUATION

$$\overline{\mathcal{K}}_n = \frac{k_n A_n}{x_n} \quad [W/\Delta_1 °C] \quad [Eq. (2\text{-}15a)]$$

PARAMETERS

$k_n = 59$ W/m·Δ_1°C   (for wrought iron at 25°C from Appendix B, Table B-5)

$A_n = 1 \times 0.005 = 0.005$ m^2 (given)

$x_n = 0.05$ m   (given)

SUBSTITUTION

$$\overline{\mathcal{K}}_n = \frac{59 \times 0.005}{0.05} \quad \frac{[W/m \cdot \Delta_1 °C][m^2]}{[m]} = [W/\Delta_1 °C]$$

ANSWER

$\overline{\mathcal{K}}_n = 5.90$ W/Δ_1°C

c. For the combination:

EQUATION

$$\overline{\mathcal{K}}_t = \overline{\mathcal{K}}_w + \overline{\mathcal{K}}_n \quad [W/\Delta_1 °C] \quad [Eq. (2\text{-}17)]$$

PARAMETERS

$$\overline{\mathcal{K}}_w = 2.19 \ \text{W}/\Delta_1{}^\circ\text{C} \qquad \text{(from Example Problem 2-4a)}$$

$$\overline{\mathcal{K}}_n = 5.90 \ \text{W}/\Delta_1{}^\circ\text{C} \qquad \text{(from Example Problem 2-4b)}$$

SUBSTITUTION

$$\overline{\mathcal{K}}_t = 2.19 + 5.90 \quad [\text{W}/\Delta_1{}^\circ\text{C}] + [\text{W}/\Delta_1{}^\circ\text{C}] = [\text{W}/\Delta_1{}^\circ\text{C}]$$

ANSWER

$$\overline{\mathcal{K}}_t = 8.09 \ \text{W}/\Delta_1{}^\circ\text{C}$$

Example Problem 2-5

If the temperature drop across the wall is 25°C, what is the heat flow rate per square meter of the wall surface through the wooden wall with nails as described in Example Problem 2-4?

SOLUTION

SKETCH

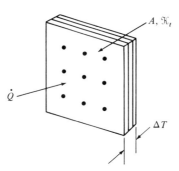

Figure 2-13

EQUATION

$$\frac{\dot{Q}}{A} = \mathcal{K}_t \Delta T \quad [\text{W}/\text{m}^2] \qquad [\text{Eq. (2-12a)}]$$

PARAMETERS

$$\mathcal{K}_t = \frac{\overline{\overline{\mathcal{K}}}_t}{A} \quad [\text{W}/\text{m}^2 \cdot \Delta_1{}^\circ\text{C}]$$

$$\overline{\mathcal{K}}_t = 8.09 \ \text{W}/\Delta_1{}^\circ\text{C} \qquad \text{(from Example Problem 2-4c)}$$

$A = 1 \text{ m}^2$ (given)

$\mathcal{K}_t = 8.09/1$ $[\text{W}/\Delta_1\text{°C}]/[\text{m}^2] = [\text{W}/\text{m}^2 \cdot \Delta_1\text{°C}]$

$T = 25\text{°C}$ (given)

SUBSTITUTION

$$\dot{Q}/A = 8.09 \times 25 \quad [\text{W}/\text{m}^2 \cdot \Delta_1\text{°C}][\text{°C}] = [\text{W}/\text{m}^2]$$

ANSWER

$\dot{Q}/A = 202.25 \text{ W}/\text{m}^2$ of wall surface

Example Problem 2-6

What percentage of the total heat flow in Example Problem 2-5 passes through the nails?

SOLUTION

SKETCH

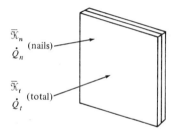

Figure 2-14

EQUATION

$$P = \frac{\dot{Q}_n}{\dot{Q}_t} \times 100 = \frac{\overline{\mathcal{K}}_n}{\overline{\mathcal{K}}_t} \times 100 \quad (\text{ratio}) \quad [\%]$$

PARAMETERS

$\overline{\mathcal{K}}_n = 5.90 \text{ W}/\Delta_1\text{°C}$ (from Example Problem 2-4b)

$\overline{\mathcal{K}}_t = 8.09 \text{ W}/\Delta_1\text{°C}$ (from Example Problem 2-4c)

SUBSTITUTION

$$P = \frac{5.90}{8.09} \times 100 \quad \frac{[\text{W}/\Delta_1\text{°C}]}{[\text{W}/\Delta_1\text{°C}]} \times 100 = \%$$

ANSWER

$P = 72.9\%$

THE SERIES CONDUCTION PATH SYSTEM

When different conduction path elements are connected so the same quantity of heat flows through each conduction path element sequentially, the configuration is called a series conduction path, and is illustrated in Figure 2-15. The heat transfer path element resistances are usually used with series heat transfer path systems instead of the conductances of the path elements. This figure shows a number of conduction path elements, each with its characteristic thermal resistance ($\mathfrak{R}_i$) connected sequentially with the same quantity of heat flow ($\dot{Q}$) passing through each conduction path element. The temperature drop across each element is

$$\Delta T_i = \frac{\dot{Q}}{\mathfrak{K}_{t_i}} = \dot{Q} \overline{\mathfrak{R}}_i$$

The total temperature drop across the entire series path is the sum of the individual temperature drops of each of the sections.

$$\Delta T_t = \Delta T_1 + \Delta T_2 + \Delta T_3 + \cdots + \Delta T_n$$
$$\Delta T_t = \dot{Q}\overline{\mathfrak{R}}_1 + \dot{Q}\overline{\mathfrak{R}}_2 + \dot{Q}\overline{\mathfrak{R}}_3 + \cdots + \dot{Q}\overline{\mathfrak{R}}_n$$
$$\Delta T_t = \dot{Q}\left(\overline{\mathfrak{R}}_1 + \overline{\mathfrak{R}}_2 + \overline{\mathfrak{R}}_3 + \cdots + \overline{\mathfrak{R}}_n\right)$$
$$\Delta T_t = \dot{Q}\overline{\mathfrak{R}}_t$$
$$\overline{\mathfrak{R}}_t = \overline{\mathfrak{R}}_1 + \overline{\mathfrak{R}}_2 + \overline{\mathfrak{R}}_3 + \cdots + \overline{\mathfrak{R}}_n \quad [\Delta°C/W] \tag{2-18}$$

In heat exchange configurations, such as building walls and heat exchanger surfaces, the heat transfer path has a constant cross-sectional area through all of the thermal resistance series elements. In such a heat transfer configuration, the term $\dot{Q}/A$ [W/m^2] is usually used in the solution of problems. When $\dot{Q}/A$ is involved, the thermal resistance per

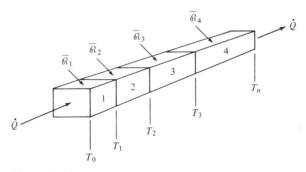

Figure 2-15

unit path area ($\mathcal{R}$) with units $[\text{m}^2 \cdot \Delta°\text{C}/\text{W}]$ is used instead of the path element total thermal resistance $\overline{\overline{\mathcal{R}}}_t [\Delta°\text{C}/\text{W}]$.

The mathematics are indicated below:

$$\dot{Q} = \frac{\Delta T}{\overline{\overline{\mathcal{R}}}_t} \quad [\text{W}]$$

Dividing both sides of the equation by A results in

$$\frac{\dot{Q}}{A} = \frac{\Delta T}{A \overline{\overline{\mathcal{R}}}_t} \quad [\text{W}/\text{m}^2]$$

As $A\overline{\overline{\mathcal{R}}}_t = \mathcal{R}_t \quad [\text{m}^2 \cdot \Delta°\text{C}/\text{W}]$,

$$\frac{\dot{Q}}{A} = \frac{\Delta T}{\mathcal{R}_t} \quad [\text{W}/\text{m}^2]$$

$\mathcal{R}_t$ is the sum of the individual path segments $\mathcal{R}_i$. This can be demonstrated in a manner similar to the development of Eq. (2-18).

$$\Delta T = \mathcal{R}(\dot{Q}/A) \quad [\Delta°\text{C}]$$

Then in a series path comprising n elements,

$$\Delta T_t = \Delta T_1 + \Delta T_2 + \Delta T_3 + \cdots + \Delta T_n \quad [\Delta°\text{C}]$$

Substituting for ΔT_i,

$$\Delta T_t = \mathcal{R}_1(\dot{Q}_1/A_1) + \mathcal{R}_2(\dot{Q}_2/A_2)$$
$$+ \mathcal{R}_3(\dot{Q}_3/A_3) + \cdots + \mathcal{R}_n(\dot{Q}_n/A_n)$$

As

$$\dot{Q}_1/A_1 = \dot{Q}_2/A_2 = \dot{Q}_3/A_3 = \dot{Q}_n/A_n = \dot{Q}/A$$
$$\Delta T_t = (\dot{Q}/A)(\mathcal{R}_1 + \mathcal{R}_2 + \mathcal{R}_3 + \cdots + \mathcal{R}_n)$$

as

$$\Delta T_t = (\dot{Q}/A)\mathcal{R}_t$$

Therefore

$$\mathcal{R}_t = (\mathcal{R}_1 + \mathcal{R}_2 + \mathcal{R}_3 + \cdots + \mathcal{R}_n) \quad [\text{m}^2 \cdot \Delta°\text{C}/\text{W}] \qquad (2\text{-}18\text{a})$$

The temperature drop to the interface between any two sections can be determined by using the sum of the thermal resistances to the selected interface. For example, if the temperature T_2 (at the interface between path elements 2 and 3) of the series linear conduction path illustrated in Figure 2-14 were desired, Eq. (2-14x) could be used. Substituting the thermal resistance $\overline{\mathcal{R}}$ for x/Ak, the equation becomes

$$T_2 = T_o - \dot{Q}(\overline{\overline{\mathcal{R}}}_1 + \overline{\overline{\mathcal{R}}}_2) \quad [°\text{C}] \qquad (2\text{-}19)$$

Table 2-20 ARRANGEMENT OF THE BASIC EQUATION
OF SERIES CONDUCTION PATH

Rate of heat flow	$\dot{Q} = \dfrac{\Delta T}{\mathcal{R}}$	[W]
	$\dfrac{\dot{Q}}{A} = \dfrac{\Delta T}{A\overline{\mathcal{R}}} = \dfrac{\Delta T}{\overline{\mathcal{R}}}$	[W/m²]
Temperature drop across total path	$\Delta T = \dot{Q}\overline{\mathcal{R}}$	[Δ°C]
Total thermal resistance	$\overline{\mathcal{R}}_t = \overline{\mathcal{R}}_1 + \overline{\mathcal{R}}_2 + \overline{\mathcal{R}}_3 + \cdots + \overline{\mathcal{R}}_n$	[Δ°C/W]
Temperature at interface between sections n and $n+1$	$T_n = T_o - \dot{Q}(\overline{\mathcal{R}}_1 + \overline{\mathcal{R}}_2 + \cdots + \overline{\mathcal{R}}_n)$	[°C]

Various arrangements of the basic equation for heat flow in a series conduction path are given in Table 2-20.

Example Problem 2-7

How much heat will be conducted per square meter of wall surface through a wall of a building with a 22°C temperature drop between the inside and outside surface? The wall is constructed with a 20 cm thick concrete (1-2-4 mix) core covered on the inside with a 2 cm thick layer of gypsum plaster and an external layer of face brick (10 cm thick) attached to the concrete wall with a 1 cm thick layer of cement mortar.

SOLUTION

SKETCH

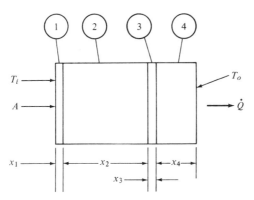

Figure 2-16 1. Gypsum plaster. 2. Concrete (1-2-4 mix). 3. Cement mortar. 4. Face brick.

EQUATION

$$\frac{\dot{Q}}{A} = \frac{\Delta T}{A\overline{\mathcal{R}}_t} \quad [\text{W}/\text{m}^2] \qquad \left[\text{Eq. (2-12a)}\right]$$

PARAMETERS

$\Delta T = 22\,\Delta°\text{C}$ (given)

$A = 1\ \text{m}^2$

$\overline{\mathcal{R}}_t = \overline{\mathcal{R}}_1 + \overline{\mathcal{R}}_2 + \overline{\mathcal{R}}_3 + \overline{\mathcal{R}}_4 \quad [\Delta°\text{C}/\text{W}] \qquad \left[\text{Eq. (2-18)}\right]$

$\overline{\mathcal{R}}_1 = \dfrac{x_1}{A_1 k_1} \quad [\Delta°\text{C}/\text{W}] \qquad \left[\text{Eq. (2-15b)}\right]$

$x_1 = 2\ \text{cm} = 0.02\ \text{m}$ (given)

$k = 0.48\ \text{W}/\text{m}\cdot\Delta_1°\text{C}$ (Appendix B, Table B-5)

$A_1 = 1\ \text{m}^2$ (given)

$\overline{\mathcal{R}}_1 = \dfrac{0.02}{1 \times 0.48} \dfrac{[\text{m}]}{[\text{m}^2][\text{W}/\text{m}\cdot\Delta_1°\text{C}]} = [\Delta°\text{C}/\text{W}]$

$\overline{\mathcal{R}}_1 = 0.042\ \Delta°\text{C}/\text{W}$

$\overline{\mathcal{R}}_2 = \dfrac{x_2}{A_2 k_2} \quad [\Delta°\text{C}/\text{W}]$ (Eq. 2-15b)

$x_2 = 20\ \text{cm} = 0.20\ \text{m}$ (given)

$k_2 = 1.36\ \text{W}/\text{m}\cdot\Delta_1°\text{C}$ (Appendix B, Table B-5)

$A_2 = 1\ \text{m}^2$

$\overline{\mathcal{R}}_2 = \dfrac{0.20}{1 \times 1.36} \dfrac{[\text{m}]}{[\text{m}^2][\text{W}/\text{m}\cdot\Delta_1°\text{C}]} = [\Delta°\text{C}/\text{W}]$

$\overline{\mathcal{R}}_2 = 0.147\ \Delta°\text{C}/\text{W}$

$\overline{\mathcal{R}}_3 = \dfrac{x_3}{A_3 k_3} \quad [\Delta°\text{C}/\text{W}] \qquad \left[\text{Eq. (2-15b)}\right]$

$x_3 = 1\ \text{cm} = 0.01\ \text{m}$ (given)

$k_3 = 1.15\ \text{W}/\text{m}\cdot\Delta_1°\text{C}$ (Appendix B, Table B-5)

$A_3 = 1\ \text{m}^2$ (given)

$\overline{\mathcal{R}}_3 = \dfrac{0.01}{1 \times 1.15} \dfrac{[\text{m}]}{[\text{m}^2][\text{W}/\text{m}\cdot\Delta_1°\text{C}]} = [\Delta°\text{C}/\text{W}]$

$\overline{\mathcal{R}}_3 = 0.009\ \Delta°\text{C}/\text{W}$

$$\overline{\mathcal{R}}_4 = \frac{x_4}{A_4 k_4} \quad [\Delta°C/W] \qquad [\text{Eq. (2-15b)}]$$

$x_4 = 10 \text{ cm} = 0.1 \text{ m} \qquad \text{(given)}$

$k_4 = 1.32 \text{ W/m} \cdot \Delta_1°C \qquad \text{(Appendix B, Table B-5)}$

$A_4 = 1 \text{ m}^2 \qquad \text{(given)}$

$$\overline{\mathcal{R}}_4 = \frac{0.1}{1 \times 1.32} \frac{[\text{m}]}{[\text{m}^2][\text{W/m} \cdot \Delta_1°C]} = [\Delta°C/W]$$

$$\overline{\mathcal{R}}_4 = 0.076 \, \Delta°C/W$$

$$\overline{\mathcal{R}}_t = 0.042 + 0.147 + 0.009 + 0.076 \quad [\Delta°C/W]$$

$$+ [\Delta°C/W] + [\Delta°C/W] + [\Delta°C/W]$$

$$\overline{\mathcal{R}}_t = 0.274 \, \Delta°C/W$$

SUBSTITUTION

$$\frac{\dot{Q}}{A} = \frac{22}{1 \times 0.274} \frac{[\Delta°C]}{[\text{m}^2][\Delta°C/W]} = [\text{W/m}^2]$$

ANSWER

$$\dot{Q}/A = 80 \text{ W/m}^2$$

Example Problem 2-8

In the wall of Example Problem 2-7, (a) what is the section with the largest thermal resistance, (b) what percentage of the total thermal resistance is this section?

SOLUTION

The concrete section $\overline{\mathcal{R}} = 0.147 \, \Delta°C/W$ from Problem 2-7

EQUATION

$$P = \frac{\overline{\mathcal{R}}_2}{\overline{\mathcal{R}}_t} \times 100 \quad [\%]$$

PARAMETERS

$$\overline{\mathfrak{R}}_2 = 0.147 \ \Delta°C/W$$

$$\overline{\mathfrak{R}}_t = 0.274 \ \Delta°C/W$$

SUBSTITUTION

$$P = \frac{0.147}{0.274} \times 100 \quad \frac{[\Delta°C/W]}{[\Delta°C/W]} \times 100 = [\%]$$

ANSWER

$$P = 54\%$$

Example Problem 2-9

What is the temperature drop across the concrete section of the wall?

SOLUTION

EQUATION

$$\Delta T = \dot{Q} \overline{\mathfrak{R}}_2 \quad [\Delta°C]$$

PARAMETERS

$\dot{Q} = 80 \ W$ (from Example Problem 2-6)

$\overline{\mathfrak{R}}_2 = 0.147 \ \Delta°C/W$ (from Example Problem 2-6)

SUBSTITUTION

$$\Delta T = 80 \times 0.147 \quad [W][\Delta°C/W] = [\Delta°C]$$

ANSWER

$$\Delta T = 11.8 \ \Delta°C$$

• *The Controlling Thermal Resistance.* In a series heat transfer path a single high resistance section exercises a dominant control of the amount of heat flow. The consideration of the "controlling" thermal resistance section is a useful tool in a cursory analysis of a series heat transfer path both as to an evaluation of the amount of heat flow and the identification of modifications to increase or decrease the rate of heat flow. By way of an illustration, if a series heat flow path had one section with 80% of the total thermal resistance, relatively large changes in the thermal resistances of another section, such as doubling or halving its thermal resistance, would have relatively little effect upon the total series system thermal resistance.

• *Contact Thermal Resistance.* In many applications, including space vehicles, thermal control is achieved by the use of heat transfer conduction paths. Such conduction paths often include mechanical joints where adjoining pieces are clamped together. Such joints have a gap between the two mating surfaces of a greater or lesser width. The average, or effective, width of this gap and the thermal conductivity of the medium filling the gap determines the thermal resistance of the joint. This thermal resistance is often called "contact" thermal resistance, and can have large magnitudes even though the mating surfaces are smooth and flat by usual mechanical standards. In the design of any conduction thermal path system, the contact resistance of all joints should not be overlooked.

By analyzing a mechanical joint as a parallel heat flow path with some heat flowing through points of direct contact between the two joint surfaces and the balance of the heat being conducted through an effective gap width, the student can ascertain that the heat is effectively all transferred through the gap. Consequently, contact thermal resistance can be treated as an effective gap width filled with a thermally conductive media with a thermal conductivity k. The magnitude of the contact thermal resistance for an air filled joint is illustrated by Example Problems 2-10 and 2-11. As indicated by Eq. (2-13d), the higher the value of k, or the narrower the gap width, the less the contact thermal resistance. If there is no thermally conductive media in the gap, as may be the case in a space vehicle application, the heat can only be transferred by radiation from one joint surface, through the thin layer of vacuum, to the other joint surface. The contact resistance of a dry joint in a vacuum becomes high, as illustrated in Example Problems 2-12 and 2-13. Therefore, in the design of a heat transfer conduction path for a vacuum or space application, some type of a conductive media should be used in all mechanical joints. Such a conductive media could be a low outgassing thermal grease or a soft deformable metal foil.

The gap between the two surfaces of a clamped joint results from:

• the degree of matching of the general surface contour
• the degree of surface roughness

The general surface contour of each of the joint surfaces may be:

• flat
• concave
• convex

as illustrated by (but not limited to) the three typical combinations shown in Figure 2-17. Upon assembly as the two surfaces of the joint are clamped together, the degree of clamping pressure may introduce increased mismatching of the two surfaces, or it can decrease the mismatch depending

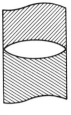 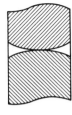 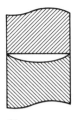

Concave–concave   Convex–convex   Plano–concave

Figure 2-17 · Gaps from general surface contour mismatch.

upon the details of the joint design. A slight mismatch in contour can readily produce gap widths in the 10 to 100 μm range.

The degree of surface roughness is a result of the fabrication or machining processes involved in the making of the surface. The surface roughness can be quantized by using the root mean square of measurements of the variation of the surface height across the surface area. This surface roughness indicator is usually called the RMS value of the surface roughness, and sometimes simply as the surface roughness. Typical surface roughness RMS values are given in Table 2-21. Prior to the adoption of the SI metric system of measurements, the RMS values were usually given in millionths of inches.

The effective gap width is much larger than the sum of the RMS values of the two mating surfaces because there are individual peaks of a much greater height than the RMS value. The surfaces are held apart by these peaks, which act as spacers. Increased clamping pressure reduces the gap width by compressing such peak (spacers). Consequently materials with lower modulus of elasticity, lower compressive yield strengths, and so forth, will show a greater reduction in the joint contact resistance with increased clamping pressure than other materials.

As details of the joint design can significantly affect the matching of the general surface contour (with respect to the very small gap width dimensions involved), it is difficult to directly apply measurements of joint contact resistances, without a full disclosure of the joint design details, to the joints being considered. Numerous investigators have made measurements of contact resistances. As an example, Brunot et al.[13] measured contact resistances for various typical machined surfaces and developed correlations as a function of contact pressure as presented in Figure 2-18. In this figure curves of the contact thermal resistance per unit area as a function of the contact pressure are given for several sets of typical machined surfaces of different finishes (roughnesses). In Figure 2-19(a) the effect of pressure on the contact resistance of a smooth aluminum joint is

[13] Brunot, O. W. and Buckland, Florence F., "Thermal Contact Resistance of Laminated and Machined Joints," *Trans. ASME*, Vol. 71, No. 3, pp. 253–257, April 1949.

Table 2–21

Surface Texture Versus Process

Surface Roughness — Approximate values—will vary with material and equipment used. For specific values check with fabrication engineer.

Left of heavy line = Practical finishes at commercial costs
Right of heavy line = Obtainable finishes at increased costs

NATURAL SURFACE	Surface Roughness (Micrometers)							
	25	6.3	1.6	0.4	0.1	.025	0.005	
	50	12.5	3.2	0.8	0.2	.05	0.012	0.0025
Cast								
Die								
Permanent Mold								
Precision								
Sand								
Shell Mold								
Coin								
Cold Press (Upset)								
Draw (Cold)								
Extrude								
Forge								
Hone (Liquid)								
Hot Press (Upset)								
Peen (Shot)								
Powder Metallurgy								
Roll (Cold)								
Roll (Hot)								
Swage								
Thread Roll								
Protective and Mechanical Finishes								
Galvanize[2]								
Oxide–Black Coat[3]								
Phosphate Coat								
Plate (.0025 Dep)[2]								
Plate (.0005 Dep)[2]								
Sheridize								
Mechanical Barrel Finish								
+ or − (mm)	1.1	0.8 0.4	0.05	0.025 0.013	0.0040	0.0020		
Normal–Practice Tolerance	0.8	0.4 0.12	0.025	0.013 0.0025	0.0025	0.0012		

[1] Dependent on previous finishes, grit and grade of abrasive.

[2] Roughness increases with thickness of deposit.

[3] Surface on which applied does not change.

Source: AESC Drafting Room Handbook, April 1977, by permission.

Table 2-21 (*Continued*)

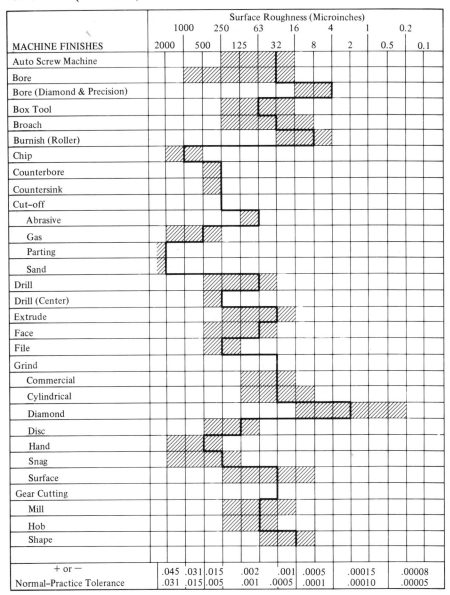

MACHINE FINISHES	Surface Roughness (Microinches)							
	1000	250	63	16	4	1	0.2	
	2000	500	125	32	8	2	0.5	0.1
Auto Screw Machine								
Bore								
Bore (Diamond & Precision)								
Box Tool								
Broach								
Burnish (Roller)								
Chip								
Counterbore								
Countersink								
Cut–off								
Abrasive								
Gas								
Parting								
Sand								
Drill								
Drill (Center)								
Extrude								
Face								
File								
Grind								
Commercial								
Cylindrical								
Diamond								
Disc								
Hand								
Snag								
Surface								
Gear Cutting								
Mill								
Hob								
Shape								
+ or − Normal–Practice Tolerance	.045 .031	.031 .015 .015 .005	.002 .001	.001 .0005	.0005 .0001	.00015 .00010	.00008 .00005	

Table 2-21 (*Continued*)

MACHINE FINISHES (cont)	Surface Roughness (Micrometers)							
	25 / 50	6.3 / 12.5	1.6 / 3.2	0.4 / 0.8	0.1 / 0.2	.025 / .05	0.005 / 0.012	0.0025
Gear Finishing								
Burnish								
Grind								
Lap								
Shave								
Hone								
Cylindrical								
Flat								
Internal								
Micro								
Lap								
Mill								
Finish								
Hollow								
Rough								
Nibble								
Plane								
Planish								
Polish (Buff)[1]								
Profile								
Punch								
Ream								
Saw								
Scrape								
Shape								
Shear								
Slot								
Spin								
Spot Face								
Superfinish								
Cylindrical								
Flat								
Turn								
Smooth								
Rough								
Diamond								
+ or − (mm) Normal–Practice Tolerance	1.1 / 0.8	0.8 0.4 / 0.4 0.12	0.05 / 0.025	0.025 / 0.013	0.013 / 0.0025	0.0040 / 0.0025	0.0020 / 0.0012	

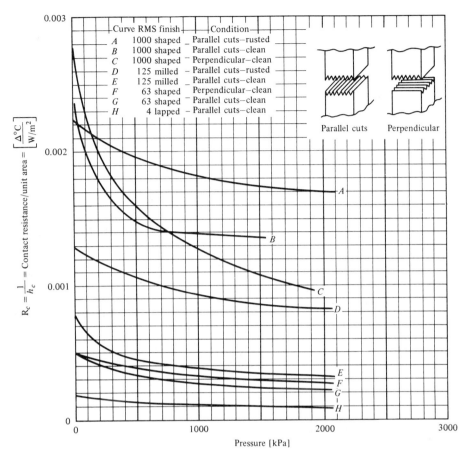

$R_c = \frac{1}{h_c}$ = Contact resistance/unit area = $\left[\frac{\Delta°C}{W/m^2}\right]$

Pressure [kPa]

Figure 2-18 Contact resistance as a function of contact pressure.

shown. In Figure 2-19(b) a comparison is made between rough aluminum and rough steel. Table 2-22 gives the contact resistance at the same pressure for various combinations of dry (air) and oil filled joints.

For extrapolation to other joint surface roughnesses and conductive media (without change in the degree of matching of the general surface contour) this data can be correlated on the basis of a factor times the sum of the RMS values of the two matching surfaces to obtain effective gap width. In this manner the effect of modifying the surface roughness or introducing a thermal grease or other conductive media in the joint can be estimated.

Example Problem 2-10

What is the contact resistance of a joint with an effective gap width (x_e) of 10 μm when the gap is filled with air at 100°C in units of [m²·Δ°C/W]?

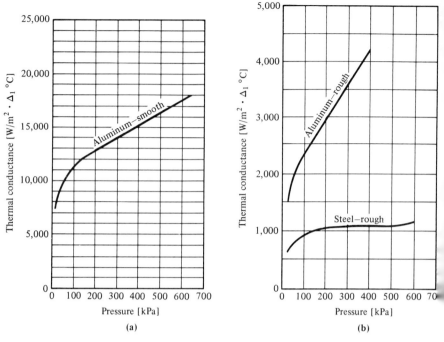

Figure 2-19(a) and (b)   Thermal conductivity as a function of contact pressure.

SOLUTION

SKETCH

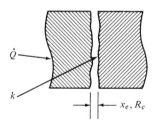

Figure 2-20

EQUATION

$$\mathcal{R}_c = \frac{x_e}{k} \quad [\text{m} \cdot \Delta°\text{C}/\text{W}] \qquad [\text{Eq. from Table 2-18}]$$

PARAMETERS

$$x_e = 10 \times 10^{-6} \text{ m} \qquad (\text{given})$$
$$k = 0.0304 \text{ W}/\text{m} \cdot \Delta_1°\text{C} \qquad (\text{Appendix B, Table B-1})$$

Table 2-22 MEASUREMENTS OF CONTACT THERMAL RESISTANCE ($\mathcal{R}_c$) WITH A JOINT PRESSURE OF 0.079 MPa[a]

| IDENT. | JOINT DESCRIPTION | | SURFACE ROUGHNESS (μm RMS) | | | THERMAL RESISTANCE($\mathcal{R}_c$) $[\text{m}^2 \cdot \Delta°\text{C/W}]$ | | |
	MATERIAL	NATURE	#1	#2	MEAN	150°C DRY	260°C DRY	150°C OIL[b]
#1 and 2	Steel[c]	Smooth vs smooth	0.08	0.08	0.08	0.80×10^{-4}	0.49×10^{-4}	—
	Steel	Rough vs rough	1.76	2.13	1.96	4.40	2.2	1.30×10^{-4}
#1 and 2	Alum[d]	Smooth vs smooth	0.40	0.42	0.40	0.98	0.50	—
		Rough vs rough	1.51	1.51	1.51	1.35	1.17	0.88
#1 and 2	Alum[e]	Smooth vs smooth	0.38	0.25	0.32	0.93	0.70	—
		Rough vs rough	1.26	1.26	1.26	3.52	2.71	1.10
#1 and 2	Bronze[f]	Rough vs rough	1.76	2.01	1.89	2.2	1.47	1.47
#1	Alum[d]	Smooth	0.38		1.66	2.2		1.10
#2	Steel	Rough		2.26			—	

source: Brunot, O. W. and Buckland, Florence F., "Thermal Contact Resistance of Laminated and Machined Joints," *Trans. ASME*, Vol. 71, No. 3, pp. 253–257; April 1949. By permission.

[a] 0.079 MPa = 10 psi.
[b] The thermal conductivity of the oil (k) = 0.113 to 0.166 W/m·Δ_1°C.
[c] Steel—SAE 4140.
[d] Alum—Alcoa No. A-51-S.
[e] Alum—Alcoa No. 18-3.
[f] Bronze—AMS 4846.

SUBSTITUTION

$$\mathcal{R}_c = \frac{10 \times 10^{-6}}{0.0304} \quad \frac{[m]}{[W/m \cdot \Delta_1 {}^\circ C]} = [m^2 \cdot \Delta {}^\circ C / W]$$

ANSWER

$$\mathcal{R}_c = 3.29 \times 10^{-4} \ m^2 \cdot \Delta {}^\circ C / W$$

Example Problem 2-11

What is the length of (a) copper, (b) aluminum, and (c) steel that would have the same thermal path resistance at 100°C as the joint contact resistance of Example Problem 2-10?

SOLUTION

SKETCH

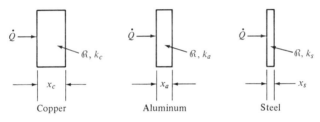

Copper   Aluminum   Steel

Figure 2-21

EQUATION

$$x = \mathcal{R}_c k \quad [m] \qquad [\text{Eq. from Table 2-18}]$$

PARAMETERS

$$\mathcal{R}_c = 3.29 \times 10^{-4} \quad [m^2 \cdot \Delta {}^\circ C / W] \qquad \text{(from Example Problem 2-10)}$$

$$k_c = 384.1 \quad [W/m \cdot \Delta_1 {}^\circ C] \qquad \text{(Appendix B, Table B-3)}$$

$$k_a = 205.9 \quad [W/m \cdot \Delta_1 {}^\circ C] \qquad \text{(Appendix B, Table B-3)}$$

$$k_s = 42.9 \quad [W/m \cdot \Delta_1 {}^\circ C] \qquad \text{(Appendix B, Table B-3)}$$

SUBSTITUTION

$$x_c = (3.29 \times 10^{-4})(384.1) \quad [m^2 \cdot \Delta {}^\circ C / W][W/m \cdot \Delta_1 {}^\circ C] = [m]$$

$$x_a = (3.29 \times 10^{-4})(205.9)$$

$$x_s = (3.29 \times 10^{-4})(42.9)$$

ANSWER

$$x_c = 0.126 \text{ m} \qquad (\text{copper})$$
$$x_a = 0.068 \text{ m} \qquad (\text{aluminum})$$
$$x_s = 0.014 \text{ m} \qquad (\text{steel})$$

Example Problem 2-12

What would the contact resistance of the joint described in Example Problem 2-10 be if it were on a space vehicle and the gap in the joint filled with space vacuum?

SOLUTION

SKETCH

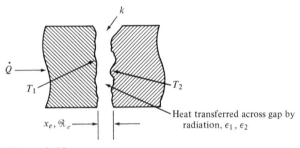

Figure 2-22

EQUATION

$$\mathcal{R}_c = \frac{A \, \Delta T}{\dot{Q}} = \frac{\Delta T}{\dot{Q}/A} \qquad [\text{m}^2 \cdot \Delta°\text{C}/\text{W}]$$

PARAMETERS

$$\Delta T = T_1 - T_2$$
$$T_1 = 105°\text{C} + 273 = 378 \text{ K} \qquad (\text{assumed})$$
$$T_2 = 95°\text{C} + 273 = 368 \text{ K} \qquad (\text{assumed})$$
$$\Delta T = 378 - 368 \quad [\text{K}] - [\text{K}] = [\Delta \text{K}]$$
$$\Delta T = 10 \, \Delta \text{K}$$
$$\dot{Q}/A = F_A F_e \sigma (T_1^4 - T_2^4) \quad [\text{W}/\text{m}^2] \qquad [\text{Eq. (7-31)}]$$
$$F_A = 1 \quad [-]$$
$$F_e = \frac{1}{1/\varepsilon_1 + 1/\varepsilon_2 - 1} \quad [-]$$

$$\varepsilon_1 = 1 \quad [-] \quad \text{(assumed)}$$

$$\varepsilon_2 = 1 \quad [-] \quad \text{(assumed)}$$

$$F_\varepsilon = 1 \quad [-]$$

$$\sigma = 5.670 \times 10^{-8} \quad [W/m^2 \cdot K^4]$$

$$T_1^4 = 204.16 \times 10^8 \quad [K^4] \quad \text{(from } T_1\text{)}$$

$$T_2^4 = 183.40 \times 10^8 \quad [K^4] \quad \text{(from } T_2\text{)}$$

$$\dot{Q}/A = (1)(1)(5.670 \times 10^{-8})\left[(204.16 - 183.40) \times 10^8\right]$$

$$[-][-]\left[W/m^2 \cdot K^4\right][K^4] = \left[W/m^2\right]$$

$$\dot{Q}/A = 117.7 \ W/m^2$$

SUBSTITUTION

$$\mathcal{R}_c = \frac{10}{117.7} \ \frac{[\Delta°C]}{[W/m^2]} = \left[m^2 \cdot \Delta°C/W\right]$$

ANSWER

$$\mathcal{R}_c = 849.6 \times 10^{-4} \quad \left[m^2 \cdot \Delta°C/W\right]$$

Example Problem 2-13

What is the length of copper which would have the same thermal path resistance as the joint contact resistance of Example Problem 2-12?

SOLUTION

SKETCH
See Figure 2-19.

EQUATION

$$x = \mathcal{R}_c k \quad [m]$$

PARAMETERS

$$\mathcal{R}_c = 849.6 \times 10^{-4} \quad \left[m^2 \cdot \Delta°C/W\right] \quad \text{(from Example Problem 2-11)}$$

$$k = 384.1 \quad [W/m \cdot \Delta_1°C] \quad \text{(from Example Problem 2-11)}$$

SUBSTITUTION

$$x = (849.6 \times 10^{-4})(384.1) \quad \left[m^2 \cdot \Delta°C/W\right][W/m \cdot \Delta_1°C] = [m]$$

ANSWER

$x = 32.6$ m

Example Problem 2-14

What is the equivalent gap width of the no. 1 aluminum rough on rough surface joint based upon the contact resistance ($\mathcal{R}_c$) for a dry joint at 150°C in Table 2-22?

SOLUTION

EQUATION

$X_e = \mathcal{R}_c k$ [m]

PARAMETERS

$\mathcal{R}_c = 1.35 \times 10^{-4}$ [m²·Δ°C/W] (from Table 2-22)

$k = 0.334$ W/m·Δ_1°C (air at 149°C) (Appendix B, Table B-2)

SUBSTITUTION

$X_e = 1.35 \times 10^{-4}(0.334)$ [m²·Δ°C/W][W/m·Δ_1°C] = [m]

ANSWER

$X_e = 45 \times 10^{-6}$ m

Example Problem 2-15

How does the effective gap width compare (multiplication factor) with the sum of the RMS roughness of the two surfaces?

SOLUTION

EQUATION

$F = \dfrac{X_e}{RMS_1 + RMS_2}$ [—]

PARAMETERS

$X_e = 45 \times 10^{-6}$ [m] (from Example Problem 2-14)

$RMS_1 = 1.51 \times 10^{-6}$ [m] (from Table 2-26)

$RMS_2 = 1.51 \times 10^{-6}$ [m] (from Table 2-26)

SUBSTITUTION

$$F = \frac{45 \times 10^{-6}}{(1.51 + 1.51) \times 10^{-6}} \quad \frac{[m]}{[m] + [m]} \quad [-]$$

ANSWER

$F = 15$ times sum of RMS surface roughness

Example Problem 2-16

If the joint was filled with a thermal grease with a thermal conductivity equal to 0.25 that of pure aluminum, what would be the joint contact resistance at 149°C?

SOLUTION

EQUATION

$$\mathcal{R}_c = \frac{X_\varepsilon}{k} \quad [m^2 \cdot \Delta°C/W]$$

PARAMETERS

$$X_\varepsilon = 45 \times 10^{-6} \quad [m] \quad \text{(from Problem 2-14)}$$
$$k = 0.25 \, k_{149°C}$$
$$k_{149°C} = 207 \, W/m \cdot \Delta_1°C \quad \text{(extrapolation from Appendix B, Table B-3)}$$

SUBSTITUTION

$$\mathcal{R}_c = \frac{45 \times 10^{-6}}{(0.25)(207)} \quad \frac{[m]}{[W/m \cdot \Delta_1°C]} = [m^2 \cdot \Delta°C/W]$$

ANSWER

$$\mathcal{R}_c = 0.870 \times 10^{-6} \quad [m^2 \cdot \Delta°C/W]$$

THE TWO-DIMENSIONAL CONDUCTION PATH

The classic two-dimensional conduction path element is that of heat flow through the walls of a thick walled pipe or a thick layer of insulation around the pipe. In the following sections the two-dimensional conduction path element will be discussed, followed by discussion of parallel and series two-dimensional path systems.

The classic two-dimensional conduction path element is illustrated in Figure 2-23. The two equations commonly applied to the two-dimensional conduction path element are the:

- thick wall pipe equation
- thin wall pipe equation

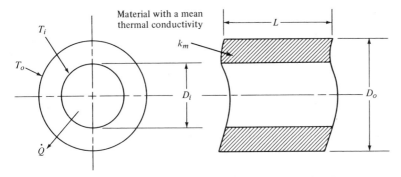

Figure 2-23   Two-dimensional conduction path element.

• *The Thick Wall Pipe Equation* The thick wall pipe equation is an "exact" equation. It has been derived by integrating, in a classical manner, the thermal resistance of the conduction path between the inside and outside diameters of the pipe wall. The thick wall pipe equation is given below. Because this equation is used most often with pipe, the heat flow parameter of greatest utility is the heat flow per unit length of pipe. Consequently, the equation is written for Q/L. Note that Q/L is not affected by the size of the pipe; only the ratio of diameters (D_o/D_i) is involved.

$$\frac{\dot{Q}}{L} = \frac{2\pi k_m (T_i - T_o)}{\ln(D_o/D_i)} \quad [\text{W/m}] \tag{2-20}$$

Consistent units must be used, such as:

$\dot{Q}/L =$ heat flow/unit length $[\text{W/m}]$

$k_m =$ mean thermal conductivity $[\text{W/m} \cdot \Delta_1 {}^\circ\text{C}]$

$T_i =$ temperature of inside wall $[{}^\circ\text{C}]$

$T_o =$ temperature of outside wall $[{}^\circ\text{C}]$

$D_o =$ outside diameter $[\text{m}]$

$D_i =$ inside diameter $[\text{m}]$

$\ln(D_o/D_i) =$ natural (Napierian) log of the diameter ratio

When Eq. (2-20) is written in the form

$$\dot{Q}/L = \mathcal{K}_t \Delta T \quad [\text{W/m}] \tag{2-20a}$$

the thermal conductance per unit length of pipe is

$$\mathcal{K}_t = \frac{2\pi k_m}{\ln(D_o/D_i)} \quad [\text{W/m} \cdot \Delta_1 {}^\circ\text{C}] \tag{2-21}$$

When Eq. (2-20) is written in the form

$$\frac{\dot{Q}}{L} = \frac{1}{\mathcal{R}_t} \Delta T \quad [\text{W/m}] \tag{2-20b}$$

the thermal resistance per unit length of pipe is

$$\mathcal{R}_t = \frac{\ln(D_o/D_i)}{2\pi k_m} \quad [\text{m} \cdot \Delta°\text{C}/\text{W}] \tag{2-22}$$

For convenience in using these equations, a table of natural logarithms is given in Appendix E-1.

Equations 2-21 and 2-22 indicate the thermal conductance through the wall of a pipe is a function of the ratio of the outside to the inside diameters only, and therefore is independent of the magnitude of the pipe diameter.

• *The Thin Wall Pipe Equation* When natural logarithms are not conveniently available, the thin wall pipe equation can be used. This equation is an approximation which gives an erroneously higher thermal conductance. The thin wall pipe equation treats the pipe wall as a linear conduction path with a constant heat flow path area throughout the conduction path length. The heat flow path area is the average of the inside and outside circumference of the pipe. The heat flow path length is equal to the thickness of the pipe wall. Referring to Figure 2-23, the thin wall pipe equation is

$$\frac{\dot{Q}}{L} = \frac{\pi(D_o + D_i)}{(D_o - D_i)} k_m(T_i - T_o) \quad [\text{W}/\text{m}] \tag{2-23}$$

the thermal conductance $\mathcal{K}$ being

$$\mathcal{K} = \frac{\pi(D_o + D_i)}{(D_o - D_i)} k_m \quad [\text{W}/\text{m} \cdot \Delta_1°\text{C}]$$

Consistent units must be used, as illustrated for Eq. (2-20).

When the thin wall pipe equation is used, the resulting heat flow rate $(\dot{Q}/L)$ or the thermal conductance $(\mathcal{K})$ should be corrected by dividing the values obtained with the thin wall pipe equation by the error ratio indicated in Figure 2-21. As the ratio of D_o/D_i increases, the magnitude of the error ratio increases. For the usual schedules of standard thick walled steel pipes (dimensions given in Appendix E-4), the error incurred by using the thin wall pipe equation without correction is less than $2\frac{1}{2}\%$.

However when insulation is applied to the pipe, the insulation diameter ratio usually becomes large enough so the error incurred by the use of the thin walled pipe equation is significant. If the thin walled pipe equation has been used for such configurations, a correction for the error, as indicated by Figure 2-25, should be applied.

Example Problem 2-17

How much heat will be lost per meter length from a 6-in. standard pipe covered with 10-cm radial thickness of 85% magnesia when the outside temperature of the pipe is 450°C and the temperature of the outside surface of the insulation is 50°C?

SOLUTION

SKETCH

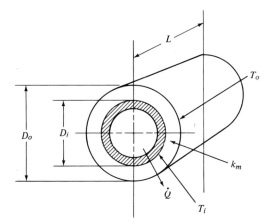

Figure 2-24   Problem category—single two-dimensional conduction path.

EQUATION (THICK WALL PIPE EQUATION)

$$\frac{\dot{Q}}{L} = \frac{2\pi k_m (T_i - T_o)}{\ln(D_o/D_i)} \quad [\mathrm{W/m}] \qquad [\mathrm{Eq.\ (2\text{-}21)}]$$

PARAMETERS

$$k_m = \text{at } 250°\mathrm{C} = 0.0847 \ \mathrm{W/m \cdot \Delta_1 °C}$$

(extrapolated from data in Appendix B-4)

$$T_i = 450°\mathrm{C} \quad \text{(given)}$$

$$T_o = 50°\mathrm{C} \quad \text{(given)}$$

$$D_i = 6.625 \text{ in.} \times 2.54 \text{ cm/in.} = 16.83 \text{ cm}$$

$$D_o = 16.83 + 2(10.00) = 36.83 \text{ cm}$$

$$D_o/D_i = \frac{36.83}{16.83} = 2.188 \quad [-]$$

$$\ln(D_o/D_i) = 0.7830 \quad [-]$$

SUBSTITUTION

$$\frac{\dot{Q}}{L} = \frac{2\pi 0.0847(450 - 50)}{0.7830} \quad \frac{[\mathrm{W/m \cdot \Delta_1 °C}][°\mathrm{C}]}{[-]} = [\mathrm{W/m}]$$

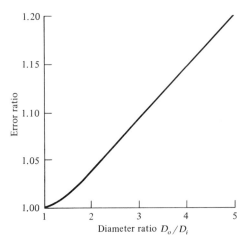

Figure 2-25   Thin walled pipe equation error.

ANSWER

$$\dot{Q}/L = 270 \text{ W/m}$$

To obtain correct heat flow rates divide values calculated using the thin walled pipe equation by the indicated error ratio.

Example Problem 2-18

Solve the example problem for the thick wall pipe equation using the thin wall pipe equation.

SOLUTION

EQUATION (THIN WALL PIPE EQUATION)

$$\frac{\dot{Q}}{L} = \frac{\pi(D_o + D_i)k_m(T_i - T_o)}{(D_o - D_i)} \quad [\text{W/m}] \quad \left[\text{Eq. (2-23)}\right]$$

PARAMETERS

$$k_m = \text{at } 250°C = 0.0847 \text{ W/m} \cdot \Delta_1°C$$

$$\text{(extrapolated from data in Appendix B-4)}$$

$$T_i = 450°C \quad \text{(given)}$$

$$T_o = 50°C \quad \text{(given)}$$

$$D_i = 6.625 \text{ in.} \times 2.54 \text{ cm/in.} = 16.83 \text{ cm}$$

$$D_o = 16.83 + 2(10.00) = 36.83 \text{ cm}$$

SOLUTION

$$\frac{\dot{Q}}{L} = \frac{\pi(36.83+16.83)0.0847(450-50)}{(36.83-16.83)}$$

$$= \frac{\pi(53.66)0.0847(400)}{20.00} \quad \frac{[\text{cm}][\text{W/m}\cdot\Delta_1{}^\circ\text{C}][{}^\circ\text{C}]}{[\text{cm}]} = [\text{W/m}]$$

ANSWER

$$\dot{Q}/L = 285 \text{ W/m}$$

THIN WALL EQUATION ERROR CORRECTION

$$D_o/D_i = \frac{36.83}{16.83} = 2.188 \quad [-]$$

error factor from Figure 2-21 $= 1.048 \quad [-]$

corrected $\dot{Q}/L = 285/(1.048) \; [\text{W/m}]/[-] = [\text{W/m}]$

ANSWER

$$\dot{Q}/L = 271 \text{ W/m}$$

• *Parallel Path Two-Dimensional Conduction Systems* Two-dimensional conduction systems may comprise two or more two-dimensional conduction path elements in parallel. The thermal conductance of such a system is the sum of the thermal conductances of the separate conduction paths as discussed on pages 48–50.

$$\dot{Q} = \mathfrak{K}_t \Delta T \quad [\text{W}] \qquad [\text{Eq. (2-12)}]$$
$$\mathfrak{K}_t = \mathfrak{K}_1 + \mathfrak{K}_2 + \mathfrak{K}_3 + \cdots + \mathfrak{K}_n \quad [\text{W}/\Delta_1{}^\circ\text{C}]$$

An example of a parallel path two-dimensional conduction system is given in the following example problem.

Example Problem 2-19

An electronic component 5 mm in diameter is cooled by conduction through a steel disk 1 mm in thickness. The heat is removed from the steel disk by convection along the exposed perimeter of the disk. To reduce the temperature of the electronic component, the steel disk is plated on one side with copper to the thickness of 0.1 mm. If the outside diameter of the disk is 2 cm, the heat dissipation of the electronic component is 5 W, and the temperature of the periphery of the disk is 35°C, what would be the reduction in temperature of the electronic component achieved by copper plating the disk?

SOLUTION

SKETCH

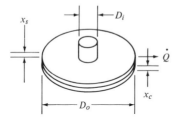

Figure 2-26

$\overline{\mathcal{K}}_s$ = thermal conductance of steel disk

$\overline{\mathcal{K}}_t$ = thermal conductance of copper-plated steel disk

$\overline{\mathcal{K}}_c$ = thermal conductance of the copper plate

EQUATION

$$\Delta T_i = \frac{\dot{Q}_i}{\overline{\mathcal{K}}_i} \quad [\Delta°C] \qquad \left[\text{Eq. (2-12b)}\right]$$

so

$$\Delta T_s = \frac{\dot{Q}_s}{\overline{\mathcal{K}}_s} \quad \text{and} \quad \Delta T_t = \frac{\dot{Q}_t}{\overline{\mathcal{K}}_t}$$

then

$$\Delta T_s - \Delta T_t = \frac{\dot{Q}_s}{\overline{\mathcal{K}}_s} - \frac{\dot{Q}_t}{\overline{\mathcal{K}}_t}$$

as

$$\dot{Q}_s = \dot{Q}_t = \dot{Q}$$

Therefore

$$\Delta T_s - \Delta T_t = \dot{Q} \left[\frac{1}{\overline{\mathcal{K}}_s} - \frac{1}{\overline{\mathcal{K}}_t} \right] \quad [\Delta°C]$$

PARAMETERS

$$\cdot \dot{Q}=5 \text{ W} \qquad \text{(given)}$$

$$\overline{\mathcal{K}}_s = \frac{2\pi k_s}{\ln(D_o/D_i)} x_s \quad [\text{W}/\Delta_1 °\text{C}]$$

$$k_s = 59 \text{ W/m} \cdot \Delta_1 °\text{C} \qquad \text{(Appendix B-4)}$$

$$D_o = 2.0 \text{ cm} \qquad \text{(given)}$$

$$D_i = 0.5 \text{ cm} \qquad \text{(given)}$$

$$D_o/D_i = 4.00 \quad [-]$$

$$\ln(D_o/D_i) = 1.3863 \qquad \text{(Tables, Appendix E-1)}$$

$$x_s = 0.05 \text{ cm} = 5 \times 10^{-4} \text{ m} \qquad \text{(given)}$$

$$\overline{\mathcal{K}}_s = \frac{2\pi(59)}{1.3863} (5 \times 10^{-4}) \quad \frac{[\text{W/m} \cdot \Delta_1 °\text{C}]}{[-]} [\text{m}] = [\text{W}/\Delta_1 °\text{C}]$$

$$\cdot \overline{\mathcal{K}}_s = 0.133 \text{ W}/\Delta_1 °\text{C}$$

$$\overline{\mathcal{K}}_t = \overline{\mathcal{K}}_s + \overline{\mathcal{K}}_c \quad [\text{W}/\Delta_1 °\text{C}]$$

$$\overline{\mathcal{K}}_c = \frac{2\pi k_c}{\ln(D_o/D_i)} x_c \quad [\text{W}/\Delta_1 °\text{C}]$$

$$k_c = 388 \text{ W/m} \cdot \Delta_1 °\text{C} \qquad \text{(Appendix B-4)}$$

$$x_c = 0.01 \text{ cm} \rightarrow 1 \times 10^{-4} \text{ m} \qquad \text{(given)}$$

$$\overline{\mathcal{K}}_c = \frac{2\pi(388)}{1.3863} (1 \times 10^{-4}) \quad \frac{[\text{W/m} \cdot \Delta_1 °\text{C}]}{[-]} [\text{m}] = [\text{W}/\Delta_1 °\text{C}]$$

$$\overline{\mathcal{K}}_c = 0.175 \text{ W}/\Delta_1 °\text{C}$$

$$\overline{\mathcal{K}}_t = 0.133 + 0.175 \quad [\text{W}/\Delta_1 °\text{C}] + [\text{W}/\Delta_1 °\text{C}] = [\text{W}/\Delta_1 °\text{C}]$$

$$\cdot \overline{\mathcal{K}}_t = 0.308 \text{ W}/\Delta_1 °\text{C}$$

SUBSTITUTION

$$\Delta T = 5 \left(\frac{1}{0.133} - \frac{1}{0.308} \right) = 5(7.50 - 3.24)$$

$$[\text{W}] \, [[\Delta °\text{C/W}] - [\Delta °\text{C/W}]] = [\Delta °\text{C}]$$

ANSWER   reduction in temperature $\Delta T = 21.3 \Delta °\text{C}$

• *Series Path Two-Dimensional Conduction Systems.* Two-dimensional conduction systems may comprise two or more two-dimensional path

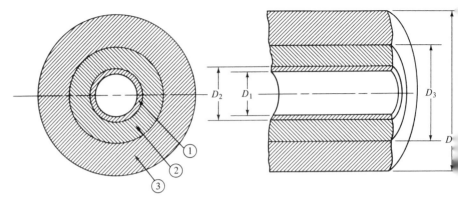

Figure 2-27   Insulated pipe with two layers of insulation.

elements in series. The thermal resistance of such a system is the sum of the thermal resistances of the separate conduction path elements as discussed on pages 54–56.

A common example of such a series conduction system is that of an insulated pipe, which consists of the pipe wall and one or more layers of insulation. This configuration is illustrated in Figure 2-27.

$$\frac{\dot{Q}}{L} = \frac{1}{\mathcal{R}_t}\Delta T \quad [\text{W/m}] \qquad \left[\text{Eq. (2-20b)}\right]$$

$$\mathcal{R}_t = \mathcal{R}_1 + \mathcal{R}_2 + \mathcal{R}_3 \quad [\text{m} \cdot \Delta°\text{C/W}] \qquad \left[\text{Eq. (2-18)}\right]$$

$$\mathcal{R}_i = \frac{\ln(D_{o_i}/D_{i_i})}{2\pi k_{m_i}} \quad [\text{m} \cdot \Delta°\text{C/W}] \qquad \left[\text{Eq. (2-22)}\right]$$

$$\mathcal{R}_t = \frac{1}{2\pi}\left[\frac{\ln(D_2/D_1)}{k_{m_1}} + \frac{\ln(D_3/D_2)}{k_{m_2}} + \frac{\ln(D_4/D_3)}{k_{m_3}}\right] \quad [\text{m} \cdot \Delta°\text{C/W}]$$

Two-dimensional conduction heat transfer problems may be encountered in configurations that cannot be readily solved by applying the "thick wall" pipe equations. Solutions for such problems can be obtained by the application of graphical methods and analogies such as fluid flow and electrical conduction. For descriptions of these methods and their application, the reader is referred to authors such as Jakob[14] and Schneider.[15]

Three-dimensional steady state conduction problems do not usually lend themselves to direct mathematical analysis for equation development. Solutions to these problems are usually obtained by graphical and electrical analog methods along with empirical formulae developed from experimen-

[14] Jakob, M., *Heat Transfer*, Vol. 1, pp. 380–409, John Wiley & Sons, Inc., New York, 1956.
[15] Schneider, P. J., *Conduction Heat Transfer*, pp. 138–170, Addison-Wesley Publishing Company, Inc., Reading, Mass., 1957.

tal data. In the early 1900s Langmuir[16] developed a set of empirical equations that were used for some time. If the student has a three-dimensional heat transfer problem, reference should be made to appropriate technical papers or texts, such as Kreith and Black, *Basic Heat Transfer* (Harper & Row, Publishers, New York, 1980).

2-3 EXTENDED SURFACES

Extended surfaces in the form of rods or fins are often applied to direct heat transfer surfaces to increase the amount of heat flow into the fluid passing over the heat transfer surface. The application of extended surfaces adds indirect heat transfer surface, but at the same time reduces the area of the direct heat transfer surface by the amount covered by the attached extended surfaces. When the effective gain from the added extended heat transfer surface is greater than the effective loss from the covering of some of the direct heat transfer surface by the attachment of the extended surface, a net increase in the amount of heat transferred will result. However, the application of extended surfaces will not always increase the heat transfer.

The addition of extended surfaces will increase the heat transfer only if

$$hA/Pk \leqq 1 \qquad\qquad (2\text{-}24)$$

where

h = convective heat transfer film coefficient (surface to fluid)

A = cross-sectional area of extended surface (normal to heat flow)

P = perimeter of extended surface (on plane of fluid flow)

k = thermal conductivity of the extended surface material

In practice the addition of extended surface is seldom justified unless

$$h < Pk/4A \qquad\qquad (2\text{-}25)$$

Extended surfaces effectively increase the heat transfer to or from a gas, are less effective when the fluid is a liquid in forced convection, and offer no advantage in heat transfer to boiling liquids or from condensing vapors. Extended surfaces are applied to electrical apparatus in which relatively large amounts of generated heat must be dissipated, to cylinders of air

[16] Langmuir, I., Adams, E. Q., and Meikle, G. S., *Trans. Am. Electrochem. Soc.*, 24, 53, 1913.

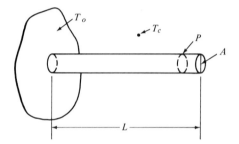

Figure 2-28   Rod extending from a wall.

cooled internal combustion engines, and so on. When extended surfaces are applied to heat exchangers they should be applied to the side of the heat exchange surface where the heat transfer film coefficient (surface to fluid) is the smaller.

2-3.1 Rods

The discussion in this text of extended surfaces begins with a simple rod, followed by the more complex subject of fins.

The general geometric configuration of a rod extended surface is shown in Figure 2-28. The rod has a uniform cross section (A) normal to its length, and a perimeter P, extending a length L from the base. The "wetted surface" of the rod is then PL, plus A, if the end of the rod is exposed to the fluid. The temperature of the base of the rod is T_o and the temperature of the ambient fluid into which the rod extends is considered to be constant at a temperature T_c. Any temperature differences in any cross-sectional area are considered to be sufficiently small so they can be neglected. The values of h (the convective heat transfer film coefficient) and k (the thermal conductivity of the rod material) are assumed to be constant.

The three configurations (or models) usually used are described in Table 2-23. Models I and II readily lend themselves to mathematical analysis, such as presented by authors.[17,18] However, these two configurations are not usually encountered in ordinary applications.

Model III is the configuration most commonly encountered in applications. This configuration does not lend itself readily to precise mathematical analysis. The usual approximation is to treat the model III configuration as a model II configuration with the length of the extended surface increased so the "wetted surface" (PL) for the model II analysis

[17] Jakob, M., and Hawkins, G. A., *Elements of Heat Transfer*, pp. 157–164, John Wiley & Sons, Inc., New York, 1957.
[18] Holman, J. P., *Heat Transfer*, pp. 30–39, McGraw-Hill Book Company, New York, 1972.

Table 2-23 EXTENDED SURFACE CONFIGURATIONS

MODEL	CONFIGURATION	COMMENTS
I	Sufficiently long so end temperature is the same as the fluid temperature.	Seldom used in practice. The length is not independent, being a function of h, k, P, and A.
II	Any length with the end insulated so no heat is transferred from the end.	Usually used for calculations. L is an independent variable.
III	Any length with the end exposed for heat transfer.	Most commonly used in practice. L is an independent variable. Exact equation complex. Model II or simplifying approximations usually used for calculations.

would equal the "wetted surface" ($PL + A$) of the actual extended surface. The resulting error from this assumption is usually small, being dependent upon the extended surface design.

In the model I configuration the comparative relationship between the length of the rod to reach the temperature of the surrounding fluid and the thermal conductivity of the rod material is illustrated in Figure 2-29. The illustrative example shown is for a cylindrical rod with a base temperature of 100°C extending into a fluid with a temperature of 25°C. The temperature of the rod is shown as a function of the distance along the rod from its base (x) for three different thermal conductivities of the rod material, namely:

glass—low thermal conductivity
iron—medium thermal conductivity
copper—high thermal conductivity

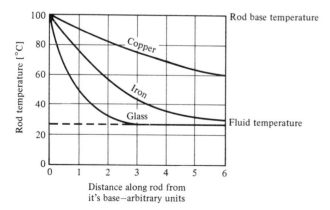

Figure 2-29 The variation of the surface temperature of a rod along its length.

For a rod with a relatively low thermal conductivity, such as glass, the flow of heat along the rod is restricted, and the rod reaches the fluid temperature in a relatively short distance from the base. For the case shown, the temperature of the glass rod reaches the fluid temperature (25°C) at a distance of 3 [arbitrary units] from its base where the rod is at 100°C. An iron rod, with a significantly higher thermal conductivity, offers less restriction to the flow of heat along the rod. Consequently, the temperature of the iron rod reaches the fluid temperature at a distance further from the rod base; in the case shown approximately 5.3 [arbitrary units]. The copper rod, with a much higher thermal conductivity, offers much less restriction to the flow of heat along the rod. The copper rod reaches the fluid temperature at a much greater distance from the rod base, this distance being off the scale of Figure 2-29. Likewise, for any other configuration of an extended surface, there will be a characteristic unique length which is dependent upon the configuration parameters of h, k, P, and A.

In models II and III the length is an independent variable. Any length can be selected at will.

EFFECTIVENESS OF THE ROD
The two key performance factors to consider in the evaluation of the rod as an extended surface are:

- The heat transfer performance of the rod itself.
- The heat transfer performance of a direct surface to which rods have been attached to provide extended heat transfer surface.

The heat transfer performance of the rod itself is usually described by a term called *extended surface efficiency*. The extended surface efficiency η_{es} [—] is defined as the ratio of the actual heat transfer by the rod to the heat transfer that would be obtained if the entire surface of the rod (extended surface) were at the base temperature of the rod.

The heat transfer performance of a direct surface to which rods have been attached to provide extended heat transfer surface is usually described by a term called *heat transfer improvement*. The heat transfer improvement R_{es} [—] is defined as the ratio of the actual heat transfer with the added rods (extended surface) to the heat transfer of the direct surface alone without the extended surface attached.

The heat transfer improvement and the extended surface efficiency are related to a degree. The extended surface efficiency is a measure of the efficiency of the extended surface element alone, and the heat transfer improvement is a measure of the effectiveness of the application of the extended surface to the direct heat transfer surface. Heat transfer improvement is the prime objective of the application of extended surfaces.

The extended surface efficiency is a measure of the efficiency of the extended surface design. This parameter is defined as

extended surface efficiency

$$\eta_{es} = \frac{\text{actual heat transferred}}{\begin{array}{l}\text{heat that would be transferred}\\ \text{if the entire extended surface area were at the base temperature}\end{array}}$$

As stated earlier, extended surface efficiency is only one consideration in the application of extended surfaces. It is not a measure of how much heat transfer is being added by the extended surfaces. It is a measure of how effectively the extended surface material is being utilized.

The material properties have a significant effect upon the extended surface design as indicated in Table 2-24. The relative volume and weight of aluminum and iron fins, with respect to copper, are shown for uniform cross-sectional fins with the same heat transfer performance.

From purely heat transfer considerations, thin, slender, and closely spaced extended surfaces are superior to fewer and thicker extended surfaces. However, for a practical design, consideration must be given to cleaning the passages between the extended surfaces, as well as fabrication methods of producing extended surfaces, and the cost of the extended surfaces. Such considerations often dictate significant compromises with the optimum design based solely upon heat transfer considerations.

The model I configuration extended surface efficiency has little, if any, interest. It is seldom used. The extended surface efficiency is very low.

The equation for the extended surface efficiency for model I is

$$\eta_{es} = \frac{1}{mL} \quad [-] \tag{2-26}$$

where

$$m = \sqrt{hP/kA} \quad [\text{m}^{-1}]$$

$$L = \text{length of rod} \quad [\text{m}] \tag{2-27}$$

The value of L is not an independent variable, being an explicit function of the rod design.

Table 2-24 INFLUENCE OF THE MATERIAL UPON THE VOLUME $\mathcal{V}$ AND WEIGHT W REQUIRED FOR FINS AT THE SAME VALUE OF $\dot{Q}_o/(T_o - T_c)$

METAL	ρ [kg/m^3]	k [W/m$\cdot\Delta_1$°C]	$\mathcal{V}/\mathcal{V}_{cu}$	W/W_{cu}
Copper	8907	384	1.	1.
Aluminum	2707	209	1.83	0.556
Iron	7802	52	7.33	6.43

The model II configuration extended surface efficiency is extensively used in the design of extended surfaces, including the selection of the length L. An iterative design selection and evaluation is often used to reach an "optimum" design.

For model II the extended surface efficiency is

$$\eta_{es} = \frac{\tanh(mL)}{mL} \quad [-] \tag{2-28}$$

In Eq. (2-28), $\tanh(mL)$ is the hyperbolic tangent of (mL). A table of these hyperbolic functions is given in Section E-2 of Appendix E. This equation is useful in the design of extended surfaces. L is one of the independent parameters which is selected in the design process. When L is small the efficiency is high, but the effectiveness in the application as measured by Eqs. (2-37) and (2-38) is low.

If there is a concern as to whether a particular configuration is model I or model II, the determination can be made by evaluating $\tanh(mL)$ for the configuration. If $\tanh(mL)$ is 1 or closely approaching 1, the configuration is model I. If it is less than 1, the configuration is model II.

• *Heat Transfer Improvement.* The heat transfer improvement ratio is a measure of the gain in heat transfer realized by the addition of the extended surface. This parameter is defined as

heat transfer improvement ratio

$$R_{es} = \frac{\text{heat transferred with the extended surface}}{\substack{\text{heat transferred by the direct surface} \\ \text{alone without the extended surface}}} \tag{2-29}$$

In the following analysis of the heat transfer improvement from extended surface application, the following simplifying assumptions have been used:

h (convective film heat transfer coefficient) is constant and is the same with and without extended surface;

T_o (temperature of the direct heat transfer surface) is the same with and without extended surface;

T_c (temperature of convective fluid) is constant and is the same with and without extended surface.

In practical applications, there are usually changes in the value of both h and T_o. For an accurate evaluation the effects of the changes in h and T_o should be considered.

A typical application of rods to a flat direct heat transfer surface is illustrated in Figure 2-30. The rods have been placed on the direct heat transfer surface in a rectangular spacing with a horizontal pitch spacing of

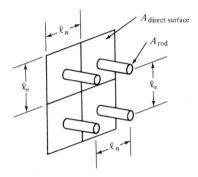

Figure 2-30 Illustrative application of rod extended surface to direct heat transfer surface.

l_n and a vertical pitch spacing of l_v. The direct heat transfer surface area associated with each rod is $l_n \times l_v$. The exposed direct heat transfer surface after the application of the rod is $l_n \times l_v - A$.

The heat transferred with the extended surface application is the sum of

1. heat transferred by the extended surface
2. heat transferred from the exposed (uncovered) direct heat transfer surface

Evaluating the heat transfer improvement ratio on the basis of one rod and the direct heat transfer surface associated with one rod, as illustrated in Figure 2-30, the following amounts of heat transfer are calculated.

The heat transferred by the original direct surface alone without the extended surface is

$$\dot{Q}_{do} = hA_{do}(T_o - T_c) \quad [\text{W}] \tag{2-30}$$

where

$$A_{do} = l_n l_v$$

The heat transferred by the exposed direct heat transfer surface after the application of one rod is

$$\dot{Q}_{de} = hA_{de}(T_o - T_c) \quad [\text{W}] \tag{2-31}$$

where

$$A_{de} = A_{do} - A$$

calling $A/A_{do} = R$. Equation (2-31) can be rewritten as

$$\dot{Q}_{de} = hA_{do}(1 - R)(T_o - T_c) \quad [\text{W}] \tag{2-31a}$$

The heat transferred by the extended surface (one rod) is:
For model I:

$$\dot{Q}_{es} = \sqrt{hPkA}\,(T_o - T_c)\quad [W] \tag{2-32}$$

For model II:

$$\dot{Q}_{es} = \sqrt{hPkA}\,(T_o - T_c)\tanh(mL)\quad [W] \tag{2-33}$$

or

$$\dot{Q}_{es} = hPL(T_o - T_c)\eta_{es}\quad [W] \tag{2-34}$$

The choice between the use of Eq. (2-33) and (2-34) is purely that of convenience. In these equations

h = surface coefficient of convection

A = rod cross-sectional area (normal to heat flow path)

k = thermal conductivity of rod

P = perimeter around rod cross-sectional-area

L = length of rod

$m = \sqrt{hP/kA}$

η_{es} = extended surface efficiency

T_o = temperature of surface

T_c = temperature of fluid

Note that PL in Eq. (2-34) is the "wetted" surface of the rod.

The heat transfer improvement ratio R_{es} *for the model II* configuration
is

$$R_{es} = \frac{\dot{Q}_{es} + \dot{Q}_{de}}{\dot{Q}_{do}}\quad [-] \tag{2-35}$$

Substituting Eqs. (2-33), (2-31a), and (2-30) results in

$$R_{es} = \frac{\left[\sqrt{hPkA}\,(T_o - T_c)\tanh(mL)\right] + \left[hA_{do}(1-R)(T_o - T_c)\right]}{hA_{do}(T_o - T_c)}\quad [-]$$

This simplifies to

$$R_{es} = (1-R) + R\frac{\tanh(mL)}{\sqrt{hA/kP}}\quad [-] \tag{2-36}$$

Using the extended surface efficiency η_{es}

$$R_{es} = (1-R) + R\frac{PL}{A}\eta_{es} \tag{2-37}$$

The heat transfer improvement ratio R_{es} for the model I configuration [as

$\tanh(mL)=1$] becomes

$$R_{es}=(1-R)+\frac{R}{\sqrt{hA/kP}}\quad[-]\qquad\qquad(2\text{-}38)$$

where

$$R=A/A_{do}\quad[-]$$

Example Problem 2-20

A long steel rod ($k=45$ W/m·Δ_1°C) protrudes from a wall. Calculate the heat transfer from the rod to a surrounding fluid under these conditions:

$$\text{rod diameter}=2.5\text{ cm}$$

$$\text{wall temperature }(T_o)=200°C$$

$$\text{convective fluid temperature }(T_c)=40°C$$

$$\text{convective heat transfer film coefficient }(h)=25\text{ W/m}^2\cdot\Delta_1°C$$

SOLUTION

SKETCH

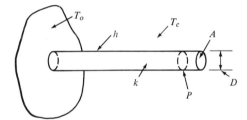

Figure 2-31

EQUATION

$$\dot{Q}=\sqrt{hPkA}\,(T_o-T_c)\quad[\text{W}]\qquad[\text{Eq. (2-32)}]$$

PARAMETERS

$$h=25\text{ W/m}^2\cdot\Delta_1°C\qquad\text{(given)}$$

$$P=2.5\times\pi\times10^{-2}=0.079\quad[\text{m}]$$

$$k=45\text{ W/m}\cdot\Delta_1°C\qquad\text{(given)}$$

$$A=(2.5)^2\frac{\pi}{4}\times1.0^{-4}=0.00049\quad[\text{m}^2]$$

$$T_o=200°C\qquad\text{(given)}$$

$$T_c=40°C\qquad\text{(given)}$$

SUBSTITUTION

$$\dot{Q}=\sqrt{25\times0.079\times45\times0.00049}\ \ (200-40)$$

$$\left([W/m^2\cdot\Delta_1{}^\circ C][m][W/m\cdot\Delta_1{}^\circ C][m^2]\right)^{1/2}[\Delta^\circ C]$$

$$=\left(W^2/\Delta_1{}^\circ C^2\right)^{1/2}[\Delta^\circ C]=[W]$$

$$=\sqrt{0.0435}\ \ 160=0.2086\times160$$

ANSWER

$$\dot{Q}=33.4\ W$$

Example Problem 2-21

If the rod in Example Problem 2-20 were 15 cm long with the end insulated, how much heat would be transferred to the surrounding fluid?

SOLUTION

SKETCH

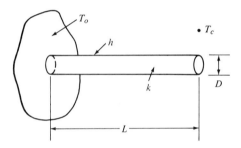

Figure 2-32

EQUATION

$$\dot{Q}=\sqrt{hPkA}\ (T_o-T_c)\tanh(mL)\quad[W/rod]\qquad\left[\text{Eq. (2-33)}\right]$$

PARAMETERS

$h=25\ W/m^2\cdot\Delta_1{}^\circ C$ (from Example Problem 2-24)

$P=0.079\ m$

$k=45\ W/m\cdot\Delta_1{}^\circ C$

$A=0.00049\ m^2$

$T_o=200^\circ C$

$T_c=40^\circ C$

$$m=\sqrt{hP/kA} \quad [1/m]$$

$$=\sqrt{\frac{25\times0.079}{45\times0.00049}} \left(\frac{[W/m^2\cdot\Delta_1{}^\circ C][m]}{[W/m\cdot\Delta_1{}^\circ C][m^2]}\right)^{1/2}=\left(\frac{1}{[m^2]}\right)^{1/2}=\left[\frac{1}{m}\right]$$

$$m=\sqrt{89.6}=9.46 \quad [1/m]$$

$$L=15\times10^{-2}=0.15 \quad [m]$$

$$mL=1.419 \quad [-]$$

$$\tanh(1.419)=0.890= \quad [-]$$

SUBSTITUTION

$$\dot{Q}=\sqrt{25\times0.079\times45\times0.00049}\,(200-40)0.890$$

$$\left([W/m^2\cdot\Delta_1{}^\circ C][m][W/m\cdot\Delta_1{}^\circ C][m^2]\right)^{1/2}[^\circ C-^\circ C][-]$$

$$=0.2087\times160\times0.890 \quad \left([W^2/\Delta_1{}^\circ C^2]\right)^{1/2}[\Delta^\circ C]=[W]$$

ANSWER

$$\dot{Q}=29.7 \text{ W}$$

Example Problem 2-22

If the rod described in Example Problem 2-21 were applied to a flat heat transfer surface to increase the heat transfer, positioned in a square pattern 25 cm between rod centers, what increase in heat transfer would be realized if the h for the flat surface was 20 $W/m^2\cdot\Delta_1{}^\circ C$? Express the increase as the ratio of the heat transfer with the extended surface applied to the heat transfer of the flat plate without the extended surface.

SOLUTION

SKETCH

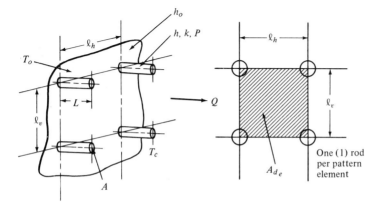

Figure 2-33

EQUATION

$$R_{es} = \frac{\dot{Q}_{es} + \dot{Q}_{de}}{\dot{Q}_{do}} \quad [-] \quad [\text{Eq. (2-35)}]$$

PARAMETERS

$\dot{Q}_{es} = \sqrt{hPkA} \, (T_o - T_c) \tanh(mL) \quad [\text{W/rod}] \quad [\text{Eq. (2-33)}]$

$\dot{Q}_{es} = 29.7 \text{ W/rod}$ (pattern element) (from Example Problem 2-21)

$\dot{Q}_{de} = h_o A_{do} (1 - R)(T_o - T_c) \quad [\text{W}] \quad [\text{Eq. (2-31a)}]$

$\quad h_o = 20 \text{ W/m}^2 \cdot \Delta_1 {}^\circ \text{C} \quad$ (given)

$\quad A_{do} = (l_h)(l_v) \quad [\text{m}^2]$

$\quad\quad l_h = 25 \times 0.01 = 0.25 \text{ m} \quad$ (given)

$\quad\quad l_v = 25 \times 0.01 = 0.25 \text{ m} \quad$ (given)

$\quad A_{do} = (0.25)(0.25) = 0.0625 \text{ m}^2$

$\quad R = A/A_{do} \quad [-]$

$\quad\quad A = 0.00049 \text{ m}^2 \quad$ (from Example Problem 2-20)

$\quad\quad R = 0.00049/0.0625 = 0.00784 \quad [-]$

$\quad T_o = 200 {}^\circ \text{C} \quad$ (from Example Problem 2-20—given)

$\quad T_c = 40 {}^\circ \text{C} \quad$ (from Example Problem 2-20—given)

$\dot{Q}_{de} = (20)(0.0625)(1 - 0.00784)(200 - 40)$

$\quad\quad [\text{W/m}^2 \cdot \Delta {}^\circ \text{C}][\text{m}^2][-][\Delta {}^\circ \text{C}]$

$\dot{Q}_{de} = 198 \text{ W/pattern element}$

$\dot{Q}_{do} = h_o A_{do} (T_o - T_c) \quad [\text{W}] \quad [\text{Eq. (2-30)}]$

$\quad (20)(0.0625)(200 - 40) \quad [\text{W/m}^2 \cdot \Delta_1 {}^\circ \text{C}][\text{m}^2][\Delta {}^\circ \text{C}] = [\text{W}]$

$\dot{Q}_{do} = 200 \text{ W/pattern element}$

SUBSTITUTION

$$R_{es} = \frac{(29.7 + 198)}{(200)} \quad \frac{[\text{W}] + [\text{W}]}{[\text{W}]} = [-]$$

ANSWER

$$R_{es} = 1.139 \quad [-]$$

2-3.2 Fins

The treatment of fins is basically similar to that of rods. Fins are usually separated into the two following configuration types, namely (1) linear, and (2) circumferential. The circumferential fins are in a plane normal to the axis of a cylinder or pipe, and the linear fins are on a flat plate or are parallel to the axis of a cylinder or pipe as illustrated in Figure 2-34.

To calculate the amount of heat transferred by the fin, the basic equation is the same as for rods:

$$\dot{Q}=\eta_{es}S_{w}h(T_{o}-T_{c}) \quad [W] \tag{2-39}$$

where

S_{w} = the surface area of the fin exposed to the convection heat transfer

The equations for evaluating S_{w} are given in Table 2-25.

The configuration models of fins are the same as for rods. Harper and Brown[19] have published fin efficiency curves for the model II configura-

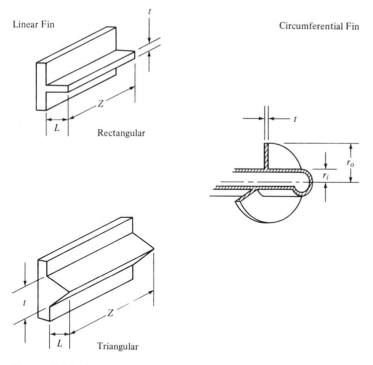

Figure 2-34 Fin configurations.

[19] Harper, W. B. and Brown, D. R., "Mathematical Equations for Heat Conduction in the Fins of Air-Cooled Engines," *NACA Report* 158, 1922.

Table 2-25 EQUATIONS FOR FIN SURFACE AREA EXPOSED TO CONVECTIVE HEAT TRANSFER

TYPE OF FIN	MODEL II	MODEL III
Linear—rectangular	$2Lz$	$(2L+t)z$
Linear—triangular	$2\sqrt{L^2+(t/2)^2}\,z$	$2\sqrt{L^2+(t/2)^2}\,z$
Approximate formula when t is small with respect to L	$2Lz$	$2Lz$
Circumferential	$2\pi(r_o^2-r_i^2)$	$2\pi[(r_o^2-r_i^2)+tr_o]$

tion. These fin efficiency curves are presented in Figures 2-35 and 2-36. When these curves are used for the model III configuration, the length L of a linear fin is increased by $\frac{1}{2}$ of the thickness of the fin as indicated in Eqs. (2-40) and (2-41).

$$L_c=L+t/2 \tag{2-40}$$

$$r_{o_c}=r_o+t/2 \tag{2-41}$$

The error which results from this approximation is a function of $ht/2k$ and $\leqslant 8\%$ when the value of $(ht/2k)^{1/2}\leqslant\frac{1}{2}$.

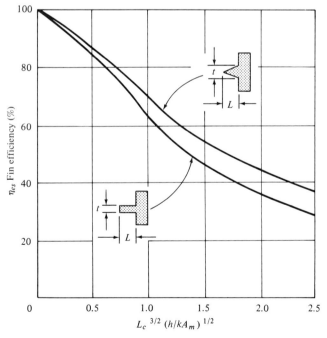

Figure 2-35   Efficiencies of linear rectangular and triangular fins.

The fin efficiency curves for rectangular and triangular fins are based on an assumption that the "z" dimension is large with respect to the "t" dimension illustrated in Figure 2-34. This results in

$$\left(\frac{2h}{kA_m}\right)^{1/2} L^{3/2} = mL$$

This was substituted in the extended surface efficiency Eq. (2-29) for mL to obtain the values given in Figure 2-35.

For model II configuration:
For rectangular fins:

$$L_c = L + t/2$$
$$A_m = t \cdot L_c$$

For triangular fins:

$$L_c = L$$
$$A_m = (t/2)L$$

A similar set of fin efficiency curves for circumferential fins is given in Figure 2-36.

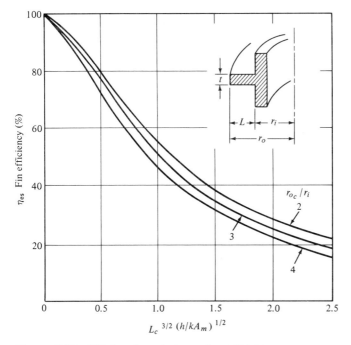

Figure 2-36   Efficiencies of circumferential fins.

For model III configuration:

$$L_c = L + t/2$$
$$r_{o_c} = r_o + t/2$$
$$A_m = t \cdot L_c$$

To use Figure 2-36 to obtain the fin efficiency, the two parameters (1) $L_c^{3/2}(h/kA_m)^{1/2}$ and (2) r_{o_c}/r_i must be evaluated for the problem conditions. The first parameter gives the abscissa position. The second parameter identifies the curve or the interpolation between curves that should be used. The intersection between the two gives the value for the fin surface efficiency.

Example Problem 2-23

A linear triangular fin has the following dimensions:

$$L = 7.5 \text{ cm}$$
$$t = 0.6 \text{ cm}$$
$$z = 30 \text{ cm}$$

Determine the fin efficiency under the following conditions:

$$h = 60 \text{ W/m}^2 \cdot \Delta_1 {}^\circ C$$
$$k = 175 \text{ W/m} \cdot \Delta_1 {}^\circ C$$

SOLUTION

SKETCH

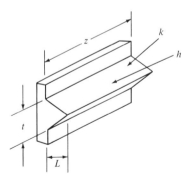

Figure 2-37

PLAN
Use Figure 2-35. We need to evaluate

$$L_c^{3/2} \left(\frac{h}{kA_m} \right)^{1/2}$$

PARAMETERS

$$L_c^{3/2}\left(\frac{h}{kA_m}\right)^{1/2}$$

$L_c = L = 0.075$ m   (given)

$h = 60$ W/m$^2 \cdot \Delta_1°$C   (given)

$k = 175$ W/m$\cdot \Delta_1°$C   (given)

$A_m = (t/2)(L)$ [m^2]   (model III configuration, triangular fins)

$t = 0.6$ cm $= 0.006$ m   (given)

$L = 7.5$ cm $= 0.075$ m   (given)

$$A_m = \frac{(0.006)}{2}(0.075)\ \frac{[m]}{[-]}[m] = [m^2]$$

$A_m = 2.25 \times 10^{-4}$ [m^2]

SUBSTITUTION

$$\frac{h}{kA_m} = \frac{60}{(175)(2.25 \times 10^{-4})}\ \frac{[W/m^2 \cdot \Delta_1°C]}{[W/m \cdot \Delta_1°C][m^2]} = [m^{-3}]$$

$$= 1524\ \ [m^{-3}]$$

$$\left(\frac{h}{kA_m}\right)^{1/2} = 39.04\ \ [m^{-3/2}]$$

$$L_c^{3/2} = (0.075)^{3/2} = 0.021\ \ [m^{3/2}]$$

$$L_c^{3/2}\left(\frac{h}{kA_m}\right)^{1/2} = (0.021)(39.04)\ \ [m^{3/2}][m^{-3/2}] = [-]$$

$$= 0.82\ \ [-]$$

ANSWER

$\eta_{es} = 75\%$

Example Problem 2-24

Circumferential fins around a cylinder have the following dimensions:

$r_o = 8$ cm

$r_i = 5$ cm

$t = 0.25$ cm

Calculate the heat transfer from one fin under the following conditions:

$$h = 35 \text{ W/m}^2 \cdot \Delta_1 °C$$
$$k = 175 \text{ W/m} \cdot \Delta_1 °C$$
$$T_o = 125°C$$
$$T_c = 40°C$$

SOLUTION

SKETCH

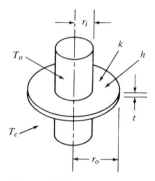

Figure 2-38

EQUATION

$$\dot{Q} = \eta_{es} S_w h (T_o - T_c) \quad [\text{W}] \qquad [\text{Eq. (2-39)}]$$

PARAMETERS
Obtain η_{es} from Figure 2-36

$$r_{o_c} = r_o + t/2 \quad [\text{cm}] \qquad \text{(from Figure 2-40)}$$
$$r_o = 8.0 \quad [\text{cm}] \qquad \text{(given)}$$
$$t = 0.25 \quad [\text{cm}] \qquad \text{(given)}$$
$$r_{o_c} = 8.0 + 0.25/2 \quad [\text{cm}] + [\text{cm}] = [\text{cm}]$$
$$r_{o_c} = 8.125 \quad [\text{cm}]$$
$$r_i = 5.0 \quad [\text{cm}] \qquad \text{(given)}$$
$$r_{o_c}/r_i = 8.125/5.0 \quad [\text{cm}]/[\text{cm}] = [-]$$
$$\cdot r_{o_c}/r_i = 1.625 \quad [-]$$
$$L_c = L + t/2 \quad [\text{cm}] \qquad \text{(from Figure 2-36)}$$
$$L = (r_o - r_i) \quad [\text{cm}]$$
$$r_o = 8.0 \text{ cm} \qquad \text{(given)}$$

$$r_i = 5.0 \text{ cm} \quad \text{(given)}$$

$$L = 8.0 - 5.0 \quad [\text{cm}] - [\text{cm}] = [\text{cm}]$$

$$L = 3.0 \text{ cm}$$

$$t = 0.25 \text{ cm}$$

$$L_c = 3.0 + \frac{0.25}{2} \quad [\text{cm}] + [\text{cm}] = [\text{cm}]$$

$$L_c = 3.125 \quad [\text{cm}] \rightarrow 0.03125 \quad [\text{m}]$$

$$\cdot L_c^{3/2} = (0.03125)^{3/2} = 5.54 \times 10^{-3} \quad [\text{m}^{3/2}]$$

$$\left(\frac{h}{kA_m}\right)^{1/2} = [\text{in consistent units}]$$

$$h = 35.0 \text{ W/m}^2 \cdot \Delta_1{}^\circ\text{C} \quad \text{(given)}$$

$$k = 175 \text{ W/m} \cdot \Delta_1{}^\circ\text{C} \quad \text{(given)}$$

$$A_m = (t)(L_c) \quad [\text{m}^2] \quad \text{(from Figure 2-36)}$$

$$t = 0.25 \text{ cm} = 0.0025 \text{ m} \quad \text{(given)}$$

$$L_c = 0.03125 \text{ m} \quad \text{(calculated above)}$$

$$A_m = (0.0025)(0.03125) \quad [\text{m}][\text{m}] = [\text{m}^2]$$

$$A_m = 7.83 \times 10^{-5} \quad [\text{m}^2]$$

$$\frac{h}{kA_m} = \frac{35}{175(7.83 \times 10^{-5})} \quad \frac{[\text{W/m}^2 \cdot \Delta_1{}^\circ\text{C}]}{[\text{W/m} \cdot \Delta_1{}^\circ\text{C}][\text{m}^2]} = \frac{1}{[\text{m}^3]}$$

$$\frac{h}{kA_m} = 2554.3 \quad \frac{1}{[\text{m}^3]}$$

$$\cdot \left(\frac{h}{kA_m}\right)^{1/2} = 50.54 \quad \frac{1}{[\text{m}^{3/2}]}$$

$$L_c^{3/2}\left(\frac{h}{kA_m}\right)^{1/2} = (5.54 \times 10^{-3})(50.54) \quad [\text{m}^{3/2}]\frac{1}{[\text{m}^{3/2}]} = [-]$$

$$\cdot L_c^{3/2}\left(\frac{h}{kA_m}\right)^{1/2} = 0.280 \quad [-]$$

$$\eta_{es} \text{ (from Figure 2-36)} = 90\%$$

$$S_w = 2\pi\left[(r_o^2 - r_i^2) + (t)(r_o)\right] \times 10^{-4} \quad [\text{m}^2]$$

where

$$r_o = 8\,\text{cm} \quad \text{(given)}$$

$$r_i = 5\,\text{cm} \quad \text{(given)}$$

$$t = 0.25\,\text{cm} \quad \text{(given)}$$

$$S_w = (6.28)\big[(64 - 25) + (0.25)(8)\big] \times 10^{-4}$$

$$[-]\big([\text{cm}^2] - [\text{cm}^2] + [\text{cm}][\text{cm}]\big)[\text{m}^2/\text{cm}^2] = [\text{cm}^2][\text{m}^2/\text{cm}^2] = [\text{m}^2]$$

$$S_w = 0.0258 \quad [\text{m}^2]$$

SUBSTITUTION

$$\dot{Q} = (0.90)(0.0258)(35)(125 - 40)$$

$$[-][\text{m}^2][\text{W}/\text{m}^2 \cdot \Delta_1{}^\circ\text{C}][\Delta^\circ\text{C}] = [\text{W}]$$

ANSWER

$$\dot{Q} = 69.1\ \text{W}$$

Example Problem 2-25

To improve the heat transfer from a 3-cm pipe, circumferential fins are added. The fins are 10.0 cm in diameter, made from 16 gauge (0.16 cm thick) mild steel. The fins are placed on a 1.25-cm pitch. If the value of $h = 11\ \text{W}/\text{m}^2 \cdot \Delta_1{}^\circ\text{C}$, what is the expected increase in the heat transfer that would be realized by adding the fins?

SOLUTION

SKETCH

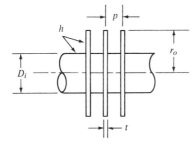

EQUATION

$$R_{es} = \frac{\dot{Q}_{es} + \dot{Q}_{de}}{\dot{Q}_{do}} \qquad [\text{Eq. (2-35)}]$$

PARAMETERS

$$\dot{Q}_{es} = \eta_{es} S_w h \Delta T \quad [\text{W}] \qquad [\text{Eq. (2-39)}]$$

Obtain η_{es} from Figure 2-36.

$$\frac{r_{o_c}}{r_i} = \frac{(r_o + t/2)}{r_i} \quad [-]$$

$$r_o = \frac{\text{o.d.}}{2} = \frac{10.0}{2} = 5.0 \, \text{cm}$$

$$t/2 = \frac{0.16}{2} = 0.08 \, \text{cm}$$

$$r_{o_c} = (5.0 + 0.08) = 5.08 \, \text{cm}$$

$$r_i = \frac{\text{i.d.}}{2} = \frac{3}{2} = 1.50 \, \text{cm}$$

Thus

$$\frac{r_{o_c}}{r_i} = \frac{5.08}{1.50} = 3.39 \quad [-]$$

$$L_c^{3/2} \left(\frac{h}{kA_m} \right)^{1/2} \quad [-]$$

$$L_c = L + t/2 \quad [\text{cm}]$$

$$L = \frac{D_o - D_i}{2} = \frac{(10-3)}{2} = \frac{7}{2} = 3.50 \, \text{cm}$$

$$\frac{t}{2} = \frac{0.16}{2} = 0.08 \, \text{cm}$$

Therefore

$$L_c = (3.50 + 0.08) = 3.58 \, \text{cm} = 0.0358 \quad [\text{m}]$$

$$L_c^{3/2} = (0.0358)^{3/2} = 6.77 \times 10^{-3} \quad [\text{m}^{3/2}]$$

$$\frac{h}{kA_m} = [\text{m}^{-3}]$$

$$h = 11 \quad [\text{W/m}^2 \cdot \Delta_1 {}^\circ\text{C}]$$

$$k = 54 \quad [\text{W/m} \cdot \Delta_1 {}^\circ\text{C}] \text{ at } 20^\circ\text{C} \quad [\text{Table B-5, Appendix B}]$$

$$A_m = (t)(L_c) \quad [\text{m}^2]$$

$$= (0.16)(3.58) = 0.573 \, \text{cm}^2 = 5.73 \times 10^{-5} \quad [\text{m}^2]$$

$$\frac{h}{kA_m} = \frac{11}{(54)(5.73 \times 10^{-5})} = 3555$$

$$\frac{[\text{W/m}^2 \cdot \Delta_1 °\text{C}]}{[\text{W/m} \cdot \Delta_1 °\text{C}][\text{m}^2]} = [\text{m}^{-3}]$$

$$\left(\frac{h}{kA_m}\right)^{1/2} = (3555)^{1/2} = 59.62 \quad [\text{m}^{-3/2}]$$

$$L_c^{3/2}\left(\frac{h}{kA_m}\right)^{1/2} = (6.77 \times 10^{-3})(59.62) \quad [\text{m}^{3/2}][\text{m}^{-3/2}] = [-]$$

$$L_c^{3/2} = 0.404 \quad [-]$$

$$\eta_{es} = 83\% \quad \text{(from curve in Figure 2-36)}$$

$$S_w = 2\pi\left[(r_o^2 - r_i^2) + (t)(r_o)\right] \quad [\text{m}^2] \quad \text{(Table 2-25)}$$

$$= 2\pi\left[(5)^2 - (1.5)^2 + (0.16)(5)\right]$$

$$= [-]\left([\text{cm}^2] - [\text{cm}^2] + [\text{cm}][\text{cm}]\right) = [\text{cm}^2]$$

$$= 2\pi\left[(25 - 2.25) + 0.80\right]$$

Therefore

$$S_w = (6.28)(23.55) = 148 \, [\text{cm}^2] \rightarrow = 0.0148 \quad [\text{m}^2]$$

$$\dot{Q}_{es} = (0.83)(0.0148)(11)(\Delta T) \quad [\text{W}]$$

Therefore

$$\dot{Q}_{es} = 0.135 \, \Delta T \quad [\text{W}]$$

$$\dot{Q}_{de} = hA \, \Delta T \quad [\text{W}]$$

$$A = \pi d_i l \quad [\text{m}^2]$$

$$d_i = 3 \, \text{cm} \rightarrow 0.03 \quad [\text{m}] \quad \text{(given)}$$

$$l = (1.25 - 0.16) = 1.09 \, \text{cm} \rightarrow 0.0109 \quad [\text{m}]$$

$$A = \pi(0.03)(0.0109) = [-][\text{m}][\text{m}] = [\text{m}^2]$$

$$A = 1.03 \times 10^{-3} \quad [\text{m}^2]$$

$$\dot{Q}_{de} = (11)(1.03 \times 10^{-3})(\Delta T) \quad [\text{W/m}^2 \cdot \Delta_1 °\text{C}][\text{m}^2][\Delta °\text{C}] = [\text{W}]$$

Therefore

$$\dot{Q}_{de}=0.0113\,\Delta T \quad [\text{W}]$$

$$\dot{Q}_{do}=hA\,\Delta T \quad [\text{W}]$$

$$A=\pi d_i l \quad [\text{m}^2]$$

$$d_i=3\,\text{cm}\rightarrow0.03\,\text{m} \quad (\text{given})$$

$$l=1.25\,\text{cm}\rightarrow0.0125\,\text{m} \quad (\text{given})$$

$$A=\pi(0.03)(0.0125)=[-][\text{m}][\text{m}]=[\text{m}^2]$$

$$A=1.18\times10^{-3} \quad [\text{m}^2]$$

$$\dot{Q}_{do}=(11)(1.18\times10^{-3})(\Delta T) \quad [\text{W}/\text{m}^2\cdot\Delta_1{}^\circ\text{C}][\text{m}^2][\Delta^\circ\text{C}]=[\text{W}]$$

Therefore

$$\dot{Q}_{do}=0.013\,\Delta T \quad [\text{W}]$$

SUBSTITUTION

$$\mathcal{R}_{es}=\left[\frac{\dot{Q}_{es}+\dot{Q}_{de}}{\dot{Q}_{do}}\right] \quad [-]$$

$$=\left[\frac{0.135\,\Delta T+0.0113\,\Delta T}{0.013\,\Delta T}\right] \quad \frac{[\text{W}]+[\text{W}]}{[\text{W}]}=[-]$$

$$=\left(\frac{0.1463}{0.013}\right)$$

FINAL ANSWER

$$\mathcal{R}_{es}=11.25 \quad [-]$$

2-4 CONDUCTION IN BODIES WITH HEAT SOURCES

Bodies with heat sources are encountered in many branches of engineering. Typical examples are electric coils, electrical resistance heaters, nuclear reactors, and burning fuel beds. The control of the temperatures in such bodies is an important design consideration. In this text the cases of a flat plate and a circular cylinder with internal heat generation will be considered.

2-4.1 Flat Plate with Uniformly Distributed Heat Sources

The simple flat plate with uniformly distributed heat sources is illustrated in Figure 2-40. The sketch in this figure illustrates a section across a long bar or plate of width W. The plate is covered on both sides with an "ideal"

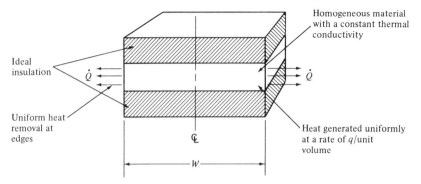

Figure 2-40 Section across a long rectangular bar with a uniformly distributed heat source.

insulation (which transfers no heat). Only the two edges of the plate are exposed for heat transfer. All of the heat generated throughout the volume of the plate flows to the two exposed edges. The heat is removed from the outside edges at a uniform rate. The bar is of a homogeneous material with a constant thermal conductivity. Heat is generated uniformly throughout the volume of the bar.

The temperature contours in the bar are shown in Figure 2-41. As the heat is generated uniformly in the bar, the heat flow is parallel to the insulated sides, and flows from the centerline toward the two edges where the heat is removed. Consequently, the temperature contours are straight lines perpendicular to the insulated sides of the bar. As the amount of heat flowing in the bar toward the two edges where the heat is removed increases from zero to the centerline of the bar to a maximum at the edge, the temperature contours, spaced for the same increment of temperature drop, become closer and closer together toward the edge where heat is removed.

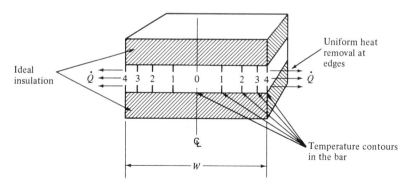

Figure 2-41 Temperature contours in a long rectangular bar with a uniformly distributed heat source.

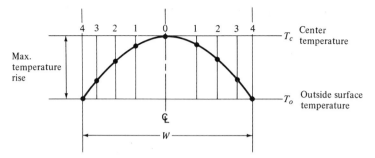

Figure 2-42   The profile of the temperature contours across a section of a long rectangular bar with a uniformly distributed heat source.

The profile of the temperature contours across the section of the bar is shown in Figure 2-42. The peak temperature is at the center. The temperature profile is parabolic in shape.

The maximum temperature rise (from the outside edge to the center) is given by Eq. (2-42).

$$\Delta T_{max} = (T_c - T_o) + \frac{\dot{Q}_v W^2}{8k} \quad [\Delta °C] \tag{2-42}$$

where

T_c = temperature in the center (peak temperature) $[°C]$

T_o = temperature of the outside surface $[°C]$

W = total width of the bar along the heat flow path $[m]$

k = effective thermal conductivity of the bar $[W/m \cdot \Delta_1 °C]$

$\dot{Q}_v$ = volumetric heat generation $[W/m^3]$

The total amount of heat generated is

$$\dot{Q} = \dot{Q}_v \mathcal{V} \quad [W] \tag{2-43}$$

where

$\mathcal{V}$ = volume of the material within which heat is being generated

For a detailed development of Eq. (2-42), the student is referred to authors such as Kreith.[20] Briefly, Eq. (2-42) is obtained from the fundamental equation for the flat plate with uniform heat generation, as given below.

$$k = \frac{d^2 T}{dx^2} + \dot{Q}_v = 0 \tag{2-44}$$

[20] Kreith, Frank, *Principles of Heat Transfer*, pp. 38–42, International Textbook Company, Scranton, Pa., 1958.

The first integration gives the temperature gradient dT/dx distribution

$$\frac{dT}{dx} = \frac{-\dot{Q}_v x}{k} + C_1 \quad (2\text{-}45)$$

The second integration gives the temperature distribution

$$T = \frac{-\dot{Q}_v x^2}{2k} + C_1 x + C_2 \; [\degree C] \quad (2\text{-}46)$$

The integration constants for Eq. (2-46) are evaluated as follows:

- As the temperature is symmetrical about the centerline, T must have the same value at $x = t_{xi}$. Consequently, $C_1 = 0$.
- As the temperature at the center of the bar (where $x = 0$) is T_c, $C_2 = T_c$.

Substituting these integration constants into Eq. (2-46) gives

$$T = \frac{-\dot{Q}_v x^2}{2k} + T_c \quad [\degree C]$$

This equation can be rearranged into

$$T_c - T = \frac{\dot{Q}_v x^2}{2k} \quad [\degree C] \quad (2\text{-}47)$$

Because the maximum temperature difference is between the center and the outside edge of the bar, $W/2$ is substituted for x to determine the maximum temperature difference. This substitution gives Eq. (2-48).

The peak temperature is on the centerline, and is

$$T_p = T_o + \Delta T_{max} \quad [\degree C]$$

$$T_p = T_o + \frac{\dot{Q}_v W^2}{8k} \quad [\degree C] \quad (2\text{-}48)$$

Example Problem 2-26

Air is heated by a long chromium steel bar 2 cm thick by 20 cm wide. Heat is generated in the steel at a rate of 2,000,000 W/m³. If the surface temperature is 500°C, what is the peak temperature within the steel bar? Disregard the heat dissipated from the edges. The thermal conductivity of the steel is constant in the operating temperature range at 24 W/m·Δ_1°C.

SOLUTION

SKETCH

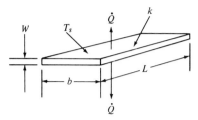

Figure 2-43

EQUATION

$$T_p = \frac{\dot{Q}_v W^2}{8k} + T_o \quad [°C] \qquad \left[\text{Eq. (2-48)}\right]$$

PARAMETERS

$\dot{Q}_v = 2,000,000 \text{ W/m}^3$ (given)

$W = 2 \times 10^{-2} = 0.02 \text{ m}$ (given)

$k = 24 \text{ W/m} \cdot \Delta_1°C$ (given)

$T_o = 500°C$ (given)

SUBSTITUTION

$$T_p = \left[\frac{2,000,000 \times (0.02)^2}{8 \times 24} \right] + 500 \quad \frac{\left[\text{W/m}^3\right]\left[\text{m}^2\right]}{\text{W/m} \cdot \Delta_1°C} + [°C]$$

$$= 4.2 + 500 \quad [\Delta°C] + [°C] = [°C]$$

ANSWER

$T_p = 504.2°C$

2-4.2 Long Cylinder with Uniformly Distributed Heat Sources

The simple long cylinder with uniformly distributed heat sources is illustrated in Figure 2-44. The solid rod is of a homogeneous material with a constant thermal conductivity. Heat is generated uniformly throughout the volume of the rod.

The temperature contours in the bar are shown in Figure 2-45. As the heat is generated uniformly in the bar, the heat flow is from the center of the bar radially outward to the cylindrical outside surface. Consequently,

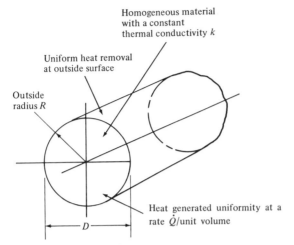

Figure 2-44   Section across a solid rod with uniformly distributed heat source.

the temperature contours are circles concentric about the centerline of the solid rod. As the amount of heat flowing per unit of the conduction heat transfer path increases from zero at the centerline of the rod to a maximum at the outside surface, the concentric temperature contours, spaced for the same increment of temperature drop, become closer and closer together toward the outside surface where the heat is removed.

The profile of the temperature contours across the section of the rod is shown in Figure 2-46. The peak temperature is at the center. The temperature profile is parabolic in shape.

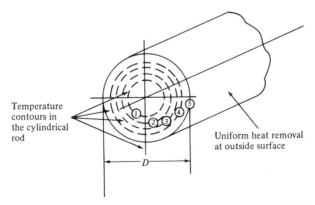

Figure 2-45   Temperature contours in a long cylindrical rod with a uniformly distributed heat source.

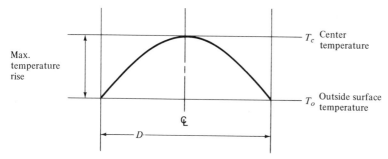

Figure 2-46 Temperature profile along the diameter of a solid rod.

The peak temperature rise (from the outside surface to the centerline of the rod) is given by Eq. (2-49).

$$\Delta T_{max} = (T_c - T_o) = \frac{\dot{Q}_v R^2}{4k} \quad [\Delta °C] \tag{2-49}$$

The maximum temperature rise in a cylindrical rod with a diameter equal to the width W of a flat bar is $\frac{1}{2}$ that of the flat bar.

For a detailed development of Eq. (2-49), the student is referred to authors, such as Kreith.[20] Briefly, Eq. (2-49) is obtained from the fundamental equation for the solid cylindrical rod with uniform heat generation, as given below.

$$k \frac{d}{dr} \left(r \frac{dT}{dr} \right) + \dot{Q}_v r = 0 \tag{2-50}$$

The first integration gives the temperature gradient dT/dr distribution

$$\frac{dT}{dr} = -\frac{\dot{Q}_v r}{2k} + C_1 \tag{2-51}$$

The second integration gives the temperature distribution

$$T = \frac{-\dot{Q}_v r^2}{4k} + C_2 \quad [°C] \tag{2-52}$$

The integration constants for Eqs. (2-51) and (2-52) are evaluated as follows:

- As the temperature gradient is 0 when $r = 0, C_1 = 0$.
- As $T = T_c$ when $r = 0, C_2 = T_c$.

Substituting the integration constant in Eq. (2-52) gives

$$T = \frac{-\dot{Q}_v r^2}{4k} + T_c \quad [°C] \tag{2-53}$$

This equation can be rearranged into

$$T_c - T = \frac{\dot{Q}_v r^2}{4k} \quad [°C]$$

As the maximum temperature difference is between the center and the outside surface of the rod, R is substituted for r to determine the maximum temperature difference. This substitution gives Eq. (2-49).

The peak temperature is on the centerline of the rod, and is

$$T_p = T_o + \Delta T_{max}$$

$$T_p = T_o + \frac{\dot{Q}_v R^2}{4k} \quad [°C] \tag{2-54}$$

Example Problem 2-27

Heat is generated in a long uranium rod at the rate of 75,000 kW/m³. The rod is 5 cm in diameter; the thermal conductivity is 30 W/m·Δ_1°C; and the outside water cooled surface is 130°C. What is the peak temperature in the uranium rod?

SOLUTION

SKETCH

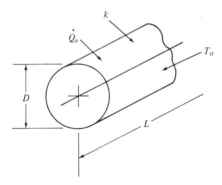

Figure 2-47

EQUATION

$$T_p = T_o + \frac{\dot{Q}_v R^2}{4k} \quad [°C] \qquad \left[\text{Eq. (2-54)}\right]$$

PARAMETERS

$T_o = 130°C$ (given)

$\dot{Q}_v = 75,000 \times 1,000 = 75,000,000 \text{ W/m}^3$ (given)

$R = 5/2 \times 10^{-2} = 0.025 \text{ m}$ (given)

$k = 30 \text{ W/m} \cdot \Delta_1 °C$ (given)

SUBSTITUTION

$$T_p = 130 + \frac{75,000,000 \times (0.025)^2}{4 \times 30} \, [°C] + \frac{[\text{W/m}^3][\text{m}^2]}{[\text{W/m} \cdot \Delta_1 °C]}$$

$$= [°C] + [\Delta°C] = [°C]$$

$$= 130 + 391$$

ANSWER

$T_p = 521°C$

2-4.3 Equations for Bodies with Uniformly Distributed Heat Sources

Arrangements of the equations for bodies with heat sources are given in Table 2-26.

Table 2-26 ARRANGEMENTS OF EQUATIONS FOR BODIES WITH HEAT SOURCES

PARAMETER	FLAT PLATE	CYLINDER
Temperature rise from outside surface $(T_c - T_o)$	$\dfrac{\dot{Q}W^2}{8k}$	$\dfrac{\dot{Q}_v R^2}{4k}$
Peak temperature T_c	$T_o + \dfrac{\dot{Q}_v W^2}{8k}$	$T_o + \dfrac{\dot{Q}_v R^2}{4k}$
T_c	$T_o + \dfrac{\dot{Q}W^2}{8k\mathcal{V}}$	$T_o + \dfrac{\dot{Q}}{4k\pi L}$
Volumetric heat generation rate $\dot{Q}_v$	$\dfrac{8k(T_c - T_o)}{W^2}$	$\dfrac{4k(T_c - T_o)}{R^2}$
$\dot{Q}_v$	$\dfrac{\dot{Q}}{\mathcal{V}}$	$\dfrac{\dot{Q}}{\pi R^2 L}$
Total heat generation rate $\dot{Q}$	$\dot{Q}_v \mathcal{V}$	$\dot{Q}_v \pi R^2 L$
$\dot{Q}$	$\dfrac{8k\mathcal{V}(T_c - T_o)}{W^2}$	$4k\pi L(T_c - T_o)$

2-4.4 Maximum Temperature in an Electric Coil

In the design of an electric coil the peak temperature is important, because the electrical insulation material has temperature limits above which it breaks down. As the current flow through a coil is increased, the rate of heat generation is also increased by the following formula:

$$E = I^2 \mathcal{R} \quad [\text{W}]$$

where

E = generated heat

I = current flow

$\mathcal{R}$ = electrical resistance

Relatively small increases in the coil current can cause relatively large increases in the peak temperature in the coil.

Electric coils can be approximated for thermal analysis by considering them as a homogeneous material with a constant thermal conductivity and with a uniform heat generation throughout their volume. The thermal conduction path passes through alternate layers of wire and electrical insulation. The effective thermal conductivity through the coil is considerably less than that of the wire alone.

The actual peak temperature in the electric coil tends to be a little higher than that calculated using constant thermal conductivity and heat generation. This is because, in the regions of higher temperature in the coil, the temperature profile gradient increases as the thermal conductivity becomes less and the rate of heat generation becomes more.

The peak temperature in an electric coil can be determined by measurements of its electrical resistance at a known isothermal temperature ($\mathcal{R}_s$) and during steady state operation ($\mathcal{R}_m$). Vidmar[21] showed that the peak temperature rise above the outside surface temperature of a coil was $3/2$ times the rise of the average coil temperature above the outside surface temperature.

$$\left(T_p - T_s \right) = \tfrac{3}{2} \left(T_m - T_s \right) \quad [\Delta^\circ \text{C}] \tag{2-55}$$

The details of the development of this equation are given by authors such as Jakob.[22]

The average coil temperature rise above the outside surface temperature $(T_m - T_s)$ can be determined from two measurements of the electrical resistance of the coil.

---

[21]Vidmar, M., "Suggestion of an Addition to the Test Codes on Temperature Rise," *Elektrotehn. u. Maschinenbau*, 26, 49, 65 (1918).

[22]Jakob, M., and Hawkins, G. A., *Elements of Heat Transfer*, pp. 106–108, John Wiley & Sons, Inc., New York, 1957.

1. $\mathcal{R}_m$—under steady state operating conditions.
2. $\mathcal{R}_s$—with the entire coil uniformly at the temperature of the outside surface of the coil under operating conditions.

The basic assumption is that the electrical resistance of the wire in the coil increases linearly with temperature. This is true for the usual temperature range of electric coil operation.

The calculation is then made in the following manner.

$$\mathcal{R}_m = \mathcal{R}_s[1 + \varepsilon(T_m - T_s)] \quad [\Omega]$$

   $\varepsilon =$ the temperature coefficient of electrical resistance of the coil wire.

Solving for $(T_m - T_s)$:

$$(T_m - T_s) = \frac{1}{\varepsilon}\left(\frac{\mathcal{R}_m - \mathcal{R}_s}{\mathcal{R}_s}\right) \quad [\Delta^\circ C] \tag{2-56}$$

Substituting this in Eq. (2-55) results in the following equation for the peak temperature rise above the temperature (T_s) at which the isothermal coil resistance was measured.

$$(T_p - T_s) = \frac{3}{2\varepsilon}\left(\frac{\mathcal{R}_m - \mathcal{R}_s}{\mathcal{R}_s}\right) \quad [\Delta^\circ C] \tag{2-57}$$

The peak temperature in the coil is then

$$T_p = \frac{3}{2\varepsilon}\left[\frac{\mathcal{R}_m - \mathcal{R}_s}{\mathcal{R}_s}\right] + T_s \quad [\Delta^\circ C] + [^\circ C] = [^\circ C] \tag{2-58}$$

If the coil resistance at a uniform temperature were measured at room temperature $(\mathcal{R}_r)$, and this temperature was different from the outside surface temperature of the coil under operating conditions, $\mathcal{R}_s$ for use in Eqs. (2-56), (2-57), and (2-58) must be calculated. The following equation is used:

$$\mathcal{R}_s = \mathcal{R}_r[1 + \varepsilon(T_s - T_r)] \quad [\Omega] \tag{2-59}$$

Example Problem 2-32

An electric coil with a 25-cm i.d. and 30-cm o.d. is immersed in an oil bath. When the coil was uniformly at the oil temperature of 24°C, a current of 0.01 A flowed through the coil with an applied voltage of 0.2 V. When the coil was operating under rated load, the measured resistance of the coil was 22.0 Ω and the surface temperature remained 24°C. Neglecting any end effects of the coil, what was the peak temperature in the coil? The temperature coefficient of the electrical resistance of the coil wire is $0.0043/\Delta_1 ^\circ C$.

SOLUTION

SKETCH

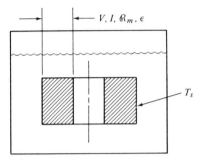

Figure 2-48

EQUATION

$$T_p = \frac{3}{2\varepsilon}\left(\frac{\mathscr{R}_m - \mathscr{R}_s}{\mathscr{R}_s}\right) + T_s \quad [°C] \qquad \left[\text{Eq. (2-58)}\right]$$

PARAMETERS

$\varepsilon = 0.0043 \quad [1/\Delta_1°C] \qquad$ (given)

$I = 0.01 \text{ A} \qquad$ (given)

$V = 0.2 \text{ V} \qquad$ (given)

$\mathscr{R}_m = 22.0 \text{ } \Omega \qquad$ (given)

$\mathscr{R}_s = V/I = 0.2/0.01 = 20.0 \text{ } \Omega$

$T_s = 24°C \qquad$ (given)

SUBSTITUTION

$$T_p = \frac{3}{2 \times 0.0043}\left(\frac{22.0 - 20.0}{20.0}\right) + 24\frac{1}{[1/\Delta_1°C]}\frac{[\Omega - \Omega]}{[\Omega]} + [°C]$$

$$= [\Delta°C] + [°C] = [°C]$$

$$= 34.9 + 24$$

ANSWER

$T_p = 58.9°C$

Example Problem 2-33

An electric coil with a temperature coefficient of $0.005/\Delta_1°C$ is operated in air. When the coil was operating at room temperature (20°C) a current of 0.01 A flowed through the coil with an applied voltage of 0.25 V. When the coil was operated under rated load, the measured resistance was 28 Ω, and the surface temperature was 25°C. Neglecting any end effects of the coil, determine the peak temperature of the coil.

SOLUTION

SKETCH

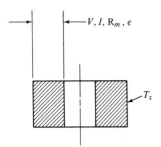

$$-V, I, R_m, \epsilon$$

$$-T_s$$

Figure 2-49

EQUATION

$$T_p = \left(\frac{3}{2\varepsilon}\right)\left[\frac{(\mathcal{R}_m - \mathcal{R}_s)}{\mathcal{R}_s}\right] + T_s \quad [°C] \qquad \left[\text{Eq. (2-58)}\right]$$

where

$$\mathcal{R}_s = \mathcal{R}_r\left[1 + \varepsilon(T_s - T_r)\right] \quad [\Omega] \qquad \left[\text{Eq. (2-59)}\right]$$

PARAMETERS

$$\varepsilon = 0.005 / \Delta_1°C \quad \text{(given)}$$

$$\mathcal{R}_m = 28\ \Omega \quad \text{(given)}$$

$$\mathcal{R}_r = \frac{V}{I} \quad [\Omega]$$

$$V = 0.25\ \text{V} \quad \text{(given)}$$

$$I = 0.01\ \text{A} \quad \text{(given)}$$

$$\mathcal{R}_r = \frac{0.25}{0.01} \quad \frac{[\text{V}]}{[\text{A}]} = [\Omega]$$

$$\mathcal{R}_r = 25.0\ \Omega$$

$$T_s = 25°C \quad \text{(given)}$$

$$T_r = 20°C \quad \text{(given)}$$

EVALUATING $\mathcal{R}_s$

$$\mathcal{R}_s = (25.0)[1 + (0.005)(25 - 20)] \quad [\Omega][[-] + [1/\Delta_1{}^\circ C][\Delta{}^\circ C]]$$

$$= (25.0)(1 + 0.025) \quad [\Omega][[-] + [-]] = [\Omega]$$

$$\mathcal{R}_s = 25.625 \ \Omega$$

SUBSTITUTION

$$T_p = \left[\frac{3}{(2)(0.005)}\right]\left[\frac{(28 - 25.625)}{25.625}\right] + 25$$

$$\left[\left(\frac{1}{1/\Delta_1{}^\circ C}\right)\left(\frac{\Omega - \Omega}{\Omega}\right)\right] + [{}^\circ C]$$

$$= [\Delta{}^\circ C] + [{}^\circ C] = [{}^\circ C]$$

$$= [(300)(0.0927) + 25]$$

$$= (27.8 + 25)$$

ANSWER

$$T_p = 52.8{}^\circ C$$

PROBLEMS

Thermal Conductivity, Conductance, Resistivity, and Resistance

In the following problems, assume no heat loss from the sides of the conduction heat transfer path.

2-1. A bar conducting heat from one end to the other is 30 cm long with a square cross section 2.5 by 2.5 cm. When the temperature difference between the two ends is 10 $\Delta{}^\circ C$ the bar conducts 79 W of heat flow. What is the thermal conductivity (k) of the bar material?

2-2. What is the thermal conductance ($\mathcal{K}$) of the bar of Problem 2-1? What is the thermal conductance ($\overline{\mathcal{K}}$) of the bar?

2-3. If a bar of the geometry of Problem 2-1 with a temperature difference of 15 $\Delta{}^\circ C$ conducted 63.8 W of heat flow, what is the thermal conductivity (k) of the bar material? Of what common metal might the bar be made?

2-4. The thermal conductance ($\overline{\mathcal{K}}$) of a bar with the geometry of Problem 2-1 is 4.25 $W/\Delta{}^\circ C$. It is conducting a heat flow of 42.5 W. What is the thermal conductivity of the bar material?

2-5. What is the thermal conductance ($\mathcal{K}$) of a 25-cm thick concrete wall (1-2-4 mix) that is 3 m high and 10 m long? The heat flow is along the 25-cm dimension.

2-6. What is the thermal conductance ($\mathcal{K}$) of an aluminum bar with a cross-sectional area of 0.0005 m², conduction heat a length of 10 cm?

2-7. What is the thermal resistance ($\mathcal{R}$) of the bar of Problem 2-1?

2-8. What is the thermal resistance ($\mathcal{R}$) of a wooden wall (solid white pine) 1.5-cm thick with an area of 30 m²?

One-Dimensional Linear Conduction Path Element

2-9. The temperature difference between the two ends of a metal bar is 5°C. If the cross-sectional area of the bar is 0.0006 m², the length 30 cm, and the thermal conductivity 250 W/m·°C, determine the heat flow ($\dot{Q}$) in [W].

2-10. The heat flow ($\dot{Q}/A$) through an insulating board 1.5-cm thick (x) is 30 W/m² of the insulating board area. What is the temperature difference between the two surfaces of the insulating board if the conductivity (k) of the board is 0.045 W/m·°C?

2-11. A metal rod with a cross-sectional area of 0.0005 m² is conducting 5 W. One end of the rod is at 260°C. At what distance from this end will the temperature of the rod be 180°C?

2-12. A radiant heating panel is made from a 0.3-cm thick steel plate heated on the back side. The steel has a thermal conductivity of 60 W/m²·°C with a temperature difference of 1°C between the two surfaces. If the panel is radiating 3.0 kW, what is the area of the radiation panel?

2-13. A silver bar 0.3-m long with a 6.5-cm² cross-sectional area conducts 350 W from a 160°C heat source to a 90°C heat sink. If an aluminum bar with one-half the conductivity of silver were to be substituted, what would be the cross-sectional area required to transfer the same amount of heat?

2-14. What is the heat flow ($\dot{Q}$) in [W] through a 0.3-m long brass bar with a cross-sectional area of 7.75 cm² and a thermal conductivity of 111 W/m·°C when the temperature difference between the ends of the bar is 10°C?

Parallel Conduction Path Element Systems

2-15. A house can be characterized by the following geometry:

COMPONENT	AREA [m²]	THERMAL RESISTANCE ($\mathcal{R}$) [m²Δ°C/W]
Solid wall	135	0.70
Ceiling	120	1.00
Floor	120	0.50
Windows	20	0.0035
Doors	10	0.25

What is the conductive heat loss from the house in [W] if the inside surface temperature is 20°C and the outside surface temperature is 5°C?

2-16. In Problem 2-15 what part contributes the major heat loss? Could the total heat loss be reduced 25% by improving the insulation in the walls, floor and ceiling? Why?

2-17. A refrigerator is constructed with the inner and outer walls separated by 10 cm of a foam insulation having a $k=0.02$ W/mΔ_1°C. To prevent moisture from entering the insulation, there is a sealing strip of stainless steel between the inner and outer walls around the four edges of the door and the four edges of the door frame. The sealing strips are made of stainless steel and are 0.5 mm thick. The outside dimensions of the refrigerator are 1 m deep, 1.5 m wide and 2.0 m high. The door frame is flush with the inside walls of the refrigerator. If the temperature of the inside wall is 5°C and the temperature of the outside wall is 25°C, what would be the total heat flow into the refrigerator in [W]? What percentage of the total heat flow is through the insulation?

Series Conduction Path Element Systems

2-18. (a) A composite building wall is composed of face brick 10 cm thick, cement mortar 1 cm thick and a 1-2-4 stone concrete mix 25 cm thick. If the temperature of the exposed surface of the face brick is 0°C and the temperature of the exposed surface of the concrete is 20°, what is the total thermal resistance ($\mathcal{R}_t$) through the wall in [m$^2 \cdot \Delta$°C/W]?
(b) What is the heat flow ($\dot{Q}/A$) in [W/m]?

2-19. (a) Heat is flowing along a 1.45-m long path consisting of the following members in series: (1) a 30-cm length of 3-cm diameter steel rod; (2) a 1.0-m length of standard 2.5-cm diameter steel water pipe; and (3) a 15-cm length of a 5-cm diameter steel rod. What is the total conductive resistance ($\mathcal{R}_t$) in [$\cdot \delta$°C/W]?
(b) When the overall temperature drop across this heat transfer path is 10 Δ°C, what is the heat flow ($\dot{Q}$) in [W]?

2-20. In Problem 2-18(a), what is the temperature at the inside surface of the face brick?

2-21. In Problem 2-19(a), what is the temperature drop between the ends of the 2.5-cm diameter steel pipe?

2-22. In Problem 2-19(a), what is the temperature drop between the ends of the 5-cm diameter steel rod?

2-23. In Problem 2-19(a), what is the temperature drop between the ends of the 3-cm diameter steel rod?

Two-Dimensional Conduction Path

2-24. (a) The temperature of the outside surface of a 20-cm diameter double extra strong bare steel pipe (id = 17.5 cm and od = 22.0 cm) is 295°C, and the inside temperature is 300°C. Using the thick walled pipe formula, calculate the heat loss per meter of pipe.

(b) Using the thin walled pipe formula, 2-24(a) calculate the heat loss per meter of length for the pipe in Problem 2-24(a). What is the percent error in using the thin walled pipe formula for this case without applying the correction factor?

(c) Apply the correction factor from Figure 2-25 to your heat loss per meter calculated using the thin walled pipe formula. How does this answer compare with your answer using the thick walled pipe formula?

2-25. When the pipe in Problem 2-24(a) is insulated with 5 cm of 85% magnesia, the outer temperature of the insulation is 95°C, and the inside of the pipe remains at 300°C. What is the reduction in the heat loss accomplished by insulating the pipe with this method?

(a) Use the thick walled pipe formula.

(b) Use the thin walled pipe formula.

(c) Use the thin walled pipe formula, applying the correction factor(s) from Figure 2-25.

2-26. A 25-cm diameter Schedule 140 steel pipe is carrying steam at 500°C. When it is insulated with a 7.5-cm thick layer of 85% magnesia insulation, the temperature of the outside surface of the insulation is 60°C. An additional 7.5-cm thick layer of asbestos felt (8 laminations/cm) was installed, and its outside surface temperature was 50°C. What was the reduction in heat loss realized by adding the asbestos? What was the temperature at the asbestos/85% magnesia insulation interface?

2-27. A circular electronic component 2 cm in diameter is mounted with good thermal contact in the center of a circular disk 8 cm in diameter. The 30 W of heat generated by the electronic component is conducted to the outside diameter of the circular disk, where the heat is removed by a cooling fluid. If the temperature of the outside diameter of the circular disk is maintained at 35°C and the circular disk is made of steel 3 mm thick, what would be the temperature of the electronic component?

2-28. If the circular disk in Problem 2-27 were of the same thickness, but made of aluminum, what would be the temperature of the electronic component?

2-29. If a copper disk 1 mm thick were added to the steel disk in Problem 2-27, how much would the temperature of the electronic component be reduced?

Extended Surfaces

2-30. A 1.275-cm diameter steel rod is being considered as an extended surface to increase the heat transfer from a flat plate. The rod has a thermal conductivity of 45 W/m·°C.

(a) What is the extended surface efficiency if the rod were 10 cm long and the $h=20$ W/m²·Δ_1°C?
(b) What would be the extended surface efficiency if the length of the rod were increased to 20 cm?
(c) If the flat plate temperature (T_o) was 200°C and the cooling fluid temperature (T_c) was 40°C, what would be the amount of heat transferred by the 10-cm long and 20-cm long rods?
(d) If 10-cm long rods were applied to the flat plate on a 5-cm square pattern, what would be the percent increase in the heat transferred with the application of the rods? Both the heat transfer from the flat plate direct surface and the rod extended surface should be included.

2-31. Circumferential fins are being considered for application to a 2-cm diameter standard water pipe. The fins will be 10 cm in diameter, and will be spaced on a 0.625-cm pitch spacing. One fin would be made from copper with a thickness of 0.0625 cm. The other would be made from steel with a thickness of 0.125 cm.

(a) If the $h=17.5$ W/m²·Δ_1°C, which fin would have the higher efficiency?
(b) What would be the increased heat transfer per meter of length of a pipe equipped with the more efficient fin over the heat transfer of a bare pipe?

2-32. In a tubular heat exchanger, the cooling fluid flows along the outside of the tubes with an $h=20$ W/m²·Δ_1°C. In order to improve the heat transfer from the tubes into the cooling fluid, the use of longitudnal fins is being considered. The red brass tube has an outside diameter (o.d.) of 3 cm. Two forms of longitudnal copper fins are being considered applying 12 fins equally spaced around the tube circumference. Both forms of fins are 2 cm high. One is of constant thickness being 2 mm thick. The other is tapered, with an end thickness of 0.5 mm, and a base thickness at the tube surface of 3.5 mm.

(a) What is the heat transfer rate for the bare tube with a temperature difference between the fluid and the tube surface of 25°C?
(b) What is the heat transfer rate for the same conditions with the fins of constant thickness?
(c) What is the heat transfer rate for the same conditions with the tapered fins?

Bodies with Heat Sources

2-33. An electric furnace is to be heated with electrical resistance elements made of chrome steel ($k = 24$ W/m·Δ_1°C) which will operate with a surface temperature of 600°C. Two cross-sectional geometries are being considered. One is rectangular 0.5 by 5.0 cm. The other is circular with a diameter of 2.0 cm. The heat release is 2000 kW/m^3.

(a) What is the peak internal temperature for the rectangular bar configuration?

(b) What is the peak internal temperature for the circular bar configuration.

2-34. If the electrical current flow were increased, and a peak temperature limit were 800°C for the bars in Problem 2-33 what would be (1) the maximum allowable volumetric heat generation rate and (2) the surface temperature for (a) the rectangular bar? (b) the circular bar?

Electric Coils

2-35. A cylindrical coil 15 cm i.d. by 17.5 cm o.d. by 0.3 m long is immersed in an oil bath. The resistance of the coil was measured by the volt-ammeter method at a room temperature of 20°C. At 6.0 V the current was 0.10 A. The coil was placed in the oil bath and under full load at 110 V, 1.5 A the surface temperature was 38°C. Neglecting the effects of the ends, calculate the following:

(a) The maximum temperature in the coil at full load.
(b) The apparent thermal conductivity of the coil.
The temperature coefficient of the electrical resistance of the coil wire is 0.0045/Δ_1°C.

Chapter 3
Steady State Free
Convection

Convective heat transfer occurs between a solid body and a fluid (liquid or gas). The heat transfer is effected by a combination of molecular conduction within the fluid in combination with energy transport resulting from the motion of fluid particles. The convective process is termed "free convection" when the circulation (or flow) of the fluid is caused by changes in the fluid density resulting from temperature gradients between the surface of the body and the main mass of the fluid. The process is called "forced convection" when the circulation is produced by mechanical means. This chapter is addressed to a discussion of free (or "natural") convection.

Free convection can only occur in gravitational fields because the fluid circulation results from density gradient forces. In space vehicles with a zero gravity flight trajectory (such as orbiting satellites) free convection is nonexistent because there are no density forces.

The convective heat transfer can be expressed in the general form of Fourier's law [Eq. (2-1)] by substituting the convective surface coefficient h for the conductivity k, and the temperature difference between the bulk temperature of the fluid and the surface temperature of the solid body ΔT

for the temperature gradient dT/dx. The resulting equation is

$$\dot{Q}=hA(T_f-T_s)=hA\,\Delta T \quad [\text{W}] \tag{3-1}$$

where

$\dot{Q}=$ heat flow rate $[\text{W}]$

$h=$ convective surface (or "film") coefficient $[\text{W}/\text{m}^2\cdot\Delta_1{}^\circ\text{C}]$

$A=$ surface area $[\text{m}^2]$

$T_f=$ fluid bulk (main mass) temperature $[^\circ\text{C}]$

$T_s=$ solid surface temperature $[^\circ\text{C}]$

In free convection heat transfer application problems, two of the three equation parameters $\dot{Q}$, A, and T are known [(Eq. (3-1)]. In order to solve for the other parameter, the value of h must be determined as outlined in Section 3-3.

Typical ranges in the values of h (the convective surface coefficient) are

air	3 to 7 $[\text{W}/\text{m}^2\cdot\Delta_1{}^\circ\text{C}]$
gases	2 to 20
liquids	30 to 300

The value of h is a function of

- the boundary layer configuration
- the fluid properties
- the temperature difference

The equations for determining h are given in Section 3-2.

3-1 DEVELOPMENT OF THE EQUATION FOR h

In the equations for h, use is made of the following properties of the fluid:

- thermal conductivity (k)
- viscosity (absolute) (μ)
- specific heat at constant pressure (c_p)
- density (ρ)
- coefficient of thermal expansion (β)

These properties are discussed in the following paragraphs.

Thermal conductivity (k) is a measurement of the capacity of a material (fluid) to conduct heat. The units of thermal conductivity are $[\text{W}/\text{m}\cdot\Delta_1{}^\circ\text{C}]$. A detailed discussion of thermal conductivity is given in Section 2.1.

Absolute viscosity (μ), more commonly known as viscosity, is a measurement of the forces that arise within the fluid to oppose any flow of the fluid. The units of absolute viscosity are $[\text{Pa}\cdot\text{s}]$ which can be expressed as [kg/m·s]. The mechanism of viscosity is different between gases and liquids. This results in the viscosity of gases generally increasing with an increase in temperature, while the viscosity of liquids generally decreases with an increase in temperature. However, the measurement units are the same. The higher the value of the viscosity, the greater is the force required to produce the same velocity movement of the fluid. There are numerous methods of experimentally measuring values of viscosity. For further details the student is referred to appropriate discussions of viscosity in fluid mechanics texts and technical papers.

Specific heat at constant pressure (c_p) is usually defined as the amount of heat required to change the temperature of a unit mass of material one degree. As shown by thermodynamics, this amount of heat is different when the fluid is heated (or cooled) at constant pressure (c_p) or at constant volume (c_v). In most heat transfer applications the heat is transferred at constant pressure. Consequently the specific heat at constant pressure (c_p) is usually used in correlations and dimensional analysis. The units of (c_p) are [J/kg·Δ_1°C].

Density (ρ) is a measurement of the amount of mass in a unit volume of material. The units of density are [kg/m^3].

Coefficient of thermal expansion (β) is a measurement of the change in volume per unit change in temperature. When the volume increases with an increase in temperature, the sign of (β) is positive ($+$). When the volume decreases with an increase in temperature the sign of (β) is negative ($-$). The units of (β) are [1/Δ_1°C].

For a perfect gas (which most gases approximate) β is 1/K, where K is the gas temperature on the absolute Kelvin scale (°C+273).

For liquids, the average value of β between two temperature levels, 1 and 2, can be evaluated as follows.

By definition

$$\beta = \left(\frac{\text{change in specific volume per unit temperature difference}}{\text{average specific volume}} \right)$$

where

$$\left(\frac{1}{\rho_2} - \frac{1}{\rho_1} \right) = \text{change in specific volume} \quad [\text{m}^3/\text{kg}]$$

$$(T_2 - T_1) = \text{temperature difference} \quad [\Delta°C]$$

$$\frac{1}{2} \left(\frac{1}{\rho_2} + \frac{1}{\rho_1} \right) = \text{average specific volume} \quad [\text{m}^3/\text{kg}]$$

Substituting in the definition equation,

$$\beta = \frac{(1/\rho_2 - 1/\rho_1)/(T_2 - T_1)}{\frac{1}{2} \left(\frac{1}{\rho_2} + \frac{1}{\rho_1} \right)} \quad \left[\frac{1}{\Delta_1°C} \right]$$

Then combining $(1/\rho_2 - 1/\rho_1)$ and $(1/\rho_2 + 1/\rho_1)$ over the common denominator $\rho_1 \rho_2$, the equation becomes

$$\beta = \frac{2(\rho_1 - \rho_2)}{\Delta T_{1-2}(\rho_1 + \rho_2)} \quad \left[\frac{1}{\Delta_1 {}^\circ C} \right]$$

Then, by using this formula the average value of β can be obtained between two temperatures by using the densities of the liquid at each of the two temperatures.

3-1.1 Dimensionless Numbers

For the determination of h for any set of conditions, use should be made of experimental data. To correlate experimental data from tests of different fluids and flow conditions, dimensional analysis and the pi theorem[1] are applied. Dimensionless ratios or "dimensionless numbers" are developed. The dimensionless numbers used in convective heat transfer are listed below. These numbers were named in honor of early heat transfer investigators.

$$\text{Nusselt number, Nu} = \frac{hL}{k} \quad [-] \tag{3-2}$$

$$\text{Prandtl number, Pr} = \frac{\mu c_p}{k} \quad [-] \tag{3-3}$$

$$\text{Grashof number, Gr} = \frac{L^3 \rho^2 \beta g \, \Delta T}{\mu^2} \quad [-] \tag{3-4}$$

$$\text{Reynolds number, Re} = \frac{VL\rho}{\mu} \quad [-] \tag{3-5}$$

where

$h=$ heat transfer film coefficient $\quad \left[W/m^2 \cdot \Delta_1 {}^\circ C \right]$

$L=$ significant length $\quad [m]$

$k=$ fluid thermal conductivity $\quad \left[W/m \cdot \Delta_1 {}^\circ C \right]$

$\mu=$ fluid viscosity (absolute) $\quad [Pa \cdot s]$

$c_p=$ fluid specific heat at constant pressure $\quad \left[J/kg \cdot \Delta_1 {}^\circ C \right]$

$\rho=$ fluid density $\quad \left[kg/m^3 \right]$

$\beta=$ coefficient of thermal expansion of the fluid $\quad \left[1/\Delta_1 {}^\circ C \right]$

$g=$ acceleration of gravity $\quad \left[m/s^2 \right]$

$\Delta T=$ temperature difference $\quad \left[\Delta {}^\circ C \right]$ or $\left[\Delta K \right]$

$V=$ fluid flow velocity $\quad [m/s]$

[1] Kreith, Frank, *Principles of Heat Transfer*, pp. 245–247, International Textbook Company, Scranton, Pa., 1958.

The *Nusselt number* can be interpreted physically as the ratio of the temperature gradient in the fluid immediately in contact with the surface to a reference temperature gradient. For a given value of the Nusselt number, the convective surface coefficient (h) is directly proportional to the thermal conductivity of the fluid and inversely proportional to the significant length L.

The *Prandtl number* is a ratio of physical properties of the fluid.

The *Grashof number* can be interpreted physically as the ratio of buoyant to viscous forces in the convective system.

The *Reynolds number* can be interpreted physically as the ratio of the fluid flow inertia forces to viscous forces.

3-1.2 Development of the Basic Equation Form

In the usual free convection circumstances, the fluid flow velocities are sufficiently small so that inertia forces in the flow are negligible in comparison with the forces of friction and buoyancy. For the usual free convection circumstances the following three dimensionless numbers apply:

Nusselt
Prandtl
Grashof

The basic equation developed from dimensional analysis[2] for use in determining the value of h is

$$\mathrm{Nu} = C(\mathrm{Gr})^a (\mathrm{Pr})^b \tag{3-6}$$

To evaluate the exponents a and b in Eq. (3-6), the results of tests of free convection with various fluids (both liquids and gases) flowing over single horizontal cylinders and vertical flat plates were used. The exponents a and b were found to be the same, which permits Eq. (3-6) to be written as

$$\mathrm{Nu} = C[\mathrm{Gr} \times \mathrm{Pr}]^d \tag{3-7}$$

By the use of these dimensionless numbers, excellent correlation between a wide variety of fluids and conditions can be obtained as illustrated in the plot of Nu versus (Gr×Pr) presented in Figure 3-1.

To evaluate the exponent d in Eq. (3-7), test data has been plotted on logarithmic coordinates in the form of the Nusselt number versus the product of the Grashof number times the Prandtl number. The resulting plot from Saunders[3] data is shown in Figure 3-2. The range of Gr×Pr most commonly encountered in practical applications is between 10^3 and 10^9.

---

[2] Brown, Aubrey, I., and Marco, Salvatore M., *Introduction to Heat Transfer*, p. 130, McGraw-Hill Book Company, New York, 1958.
[3] Saunders, O. A., *Proc. Roy. Soc.* (London), p. 157, 278–291 (1936).

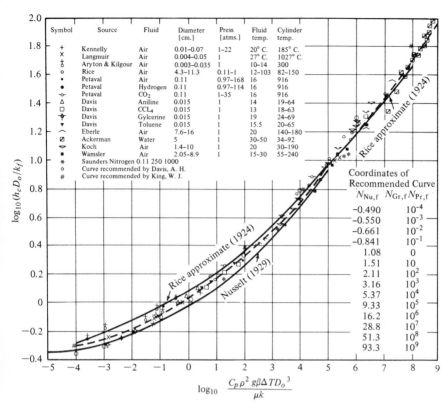

Figure 3-1   Correlation of data for free convection heat transfer from horizontal cylinders in gases and liquids.

The slope of the almost straight line of the test data through this region indicated that the value of the exponent d in Eq. (3-3) is $\frac{1}{4}$ through this region. The slope of the line of the test data changes above $\mathrm{Gr} \times \mathrm{Pr} = 10^9$, indicating a value of $\frac{1}{3}$ for the exponent d. This results in the two following basic equations for determining the value of h for free convection heat transfer:

$$h = C_1 \frac{k}{L} (\mathrm{Gr} \times \mathrm{Pr})^{0.25} \quad \left[\mathrm{W/m^2 \Delta_1 \, ^\circ C}\right] \quad \text{for} \quad 10^3 < (\mathrm{Gr} \times \mathrm{Pr}) < 10^9$$

(3-8)

$$h = C_2 \frac{k}{L} (\mathrm{Gr} \times \mathrm{Pr})^{0.33} \quad \left[\mathrm{W/m^2 \Delta_1 \, ^\circ C}\right] \quad \text{for} \quad 10^9 < (\mathrm{Gr} \times \mathrm{Pr})$$

(3-9)

The product $\mathrm{Gr} \times \mathrm{Pr}$, which appears in both of the previous equations, can be separated into two groups of parameters as indicated by Eq. (3-10).

$$\mathrm{Gr} \times \mathrm{Pr} = \frac{g \beta \rho^2 c_p}{\mu k} (L^3 \Delta T) = \alpha L^3 \Delta T$$

(3-10)

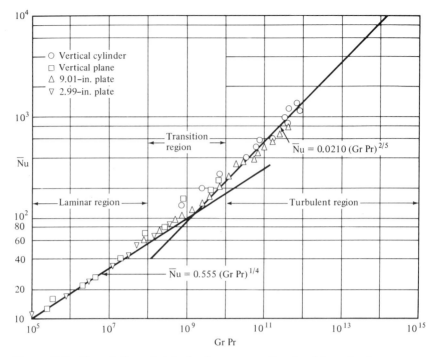

Figure 3-2   Correlation of data for free convection heat transfer from vertical plates and cylinders.

The group $g\beta\rho^2 c_p/\mu k$ contains only fluid properties and has dimensions of $L^{-3}T^{-1}$. It is commonly given the symbol α, and called the "free convection modulus." It is discussed in more detail in Section 3-2.2. Tables of the values of α for various fluids have been published. Values for those fluids used in the free convective heat transfer problems in this text are presented in Appendix B. When values for α cannot be found readily, α can be calculated using the values of the constituent parameters.

The other group of parameters $(L^3 \Delta T)$ contains configuration parameters, the values of which are determined from the application.

3-2 THE EQUATIONS FOR *h*

By substituting $\alpha L^3 \Delta T$ in Eqs. (3-8) and (3-9), and performing algebraic rearrangement, the following two equations are obtained. These are generally used for determining the value of h for free convection.

$$h = C_1 \frac{k}{L^{1/4}} (\alpha \Delta T)^{0.25} \quad \left[\text{W/m}^2\Delta_1 {}^\circ\text{C}\right] \quad \text{for} \quad 10^3 < (\text{Gr} \times \text{Pr}) < 10^9$$
(3-11)

$$h = C_2 k (\alpha \Delta T)^{0.33} \quad \left[\text{W/m}^2\Delta_1 {}^\circ\text{C}\right] \quad \text{for} \quad 10^9 < (\text{Gr} \times \text{Pr})$$
(3-12)

for evaluating C_1 or C_2 refer to Table 3-2.

3-2.1 Selection of the Equation

The first step in determining the value of h for a particular application is to establish which of the two equations should be used. This is done by evaluating $\text{Gr} \times \text{Pr}$, which is usually accomplished by the use of Eq. (3-10).

$$\text{Gr} \times \text{Pr} = \alpha L^3 \Delta T$$

The evaluation of $\text{Gr} \times \text{Pr}$ involves the free convection modulus (α) of the fluid, the significant vertical length (L) of the free convective surface, and the temperature difference (ΔT) between the undisturbed fluid and the free convective surface.

The free convection modulus (α) is evaluated at the average temperature between the undisturbed fluid and the free convection surface directly from tables or by calculations using the various fluid properties as discussed in Section 3-2.2.

The significant vertical length (L) is based upon the actual vertical height, subject to modifications for normalizing the free convection boundary layer conductance for the effects of the

- horizontal width
- vertical height

of the free convection surface area.

1. To obtain the significant vertical length (L) the actual vertical dimension (L_V) of the free convective surface is modified by the horizontal dimension (L_H) of the free convective surface as indicated by Eq. (3-13).

$$\frac{1}{L} = \frac{1}{L_V} + \frac{1}{L_H} \tag{3-13}$$

where

L = significant length

L_V = measured vertical height

L_H = measured horizontal length

2. As the thermal conductance of the free convective boundary layer becomes constant with respect to length after the initial build-up length, the value of the significant vertical length (L) is limited to 0.6 m maximum for all actual values calculated by Eq. (3-13) which are greater than 0.6 m. Values less than 0.6 m are used as calculated.

The computation of L in Eq. (3-13) can be simplified by using the equation

$$L = \phi L_V \tag{3-14}$$

ϕ is a function of the ratio of L_H / L_V, and the values of ϕ are plotted in Figure 3-3.

The determination of the significant length of a sphere can be used as an example of the use of Figure 3-3. The ratio of L_H and L_V of a sphere is $1:1$. From Figure 3-3, $\phi = 0.5$. $L = 0.5 \times$ sphere diameter.

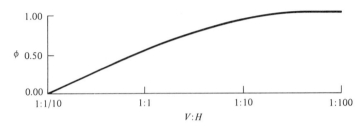

Figure 3-3

The temperature difference (ΔT) is the difference between the undisturbed fluid temperature and the temperature of the free convection surface. Often the temperature difference is not known initially. In such a case a value is assumed. Usually by the nature of the problem the initially assumed value is sufficiently accurate for the evaluation of the $\alpha L^3 \Delta T$ product for the determination of the equation to be used for the calculation of h. When h is calculated, the associated ΔT should be computed. If there is a significant variance with the initially assumed ΔT value, a new value of ΔT should be assumed and new calculations made.

The determination of which equation(s) should be used for the calculation of h is based upon the $\text{Gr} \times \text{Pr}$ product, evaluated by the product of $\alpha L^3 \Delta T$. To calculate h the constant C_1 or C_2 (as the equation requires) are obtained as discussed in Section 3-2.3.

3-2.2 Evaluating the Free Convection Modulus α

The free convection modulus (α) is defined as

$$\alpha = g\beta p^2 c_p / \mu k \quad \left[1/\text{m}^3 \cdot \Delta_1 \text{K}\right] \tag{3-15}$$

Table 3-1 TYPICAL EFFECT OF PRESSURE AND TEMPERATURE ON FLUID PROPERTIES IN THE FREE CONVECTION MODULUS α
$\alpha = g\beta\rho^2 c_p / \mu k \quad [1/\text{m}^3 \cdot \Delta_1 \text{K}]$

	GAS		LIQUID	
PROPERTY	PRESSURE (INCREASING)	TEMP (INCREASING)	PRESSURE (INCREASING)	TEMP (INCREASING)
β	None	None	None	None
ρ	Proportional Increase	Proportional Decrease	None	Very slight Decrease
c_p	None	Very slight	None	Very slight
μ	None	Increase	None	Decrease Usually
k	None	Increase	None	Decrease

Table 3-2 COEFFICIENTS C_1 AND C_2 FOR VARIOUS CONFIGURATIONS

$\alpha L^3 \Delta T$, OR $\dfrac{g\beta\Delta t L^3 \rho^2 c_p}{\mu k}$ (WHERE L IS LIMITED TO 0.6 m)	$10^3 - 10^9$	$> 10^9$
SHAPE AND POSITION	$C_1 [-]$	$C_2 [-]$
Vertical plates	0.55	0.13
Horizontal cylinders (pipes and wires)	0.45	0.11
Long vertical cylinders	0.45–0.55[a]	0.11–0.13[a]
Horizontal plates, warm side facing upward	0.71	0.17
Horizontal plates, warm side facing downward	0.35	0.08
Spheres (L=radius)	0.63	0.15

[a] Turbulence and fluctuating eddies in the flow over long vertical cylinders cause uncertainty in the values of C_1 and C_2.

and has the dimensions of $L^{-3}T^{-1}$. If tables are not available, α can be calculated using the values of the fluid properties involved. The fluid properties should be evaluated at the average temperature between the solid surface temperature and the temperature of the undisturbed fluid. In the use of reference tables for obtaining values of the free convection modulus, the reader should be aware that some authors use the symbol α to denote thermal diffusivity, which is an entirely different parameter and is not involved with convective heat transfer. Confusion can be avoided because the dimensions of thermal diffusivity are $L^2 T^{-1}$ as contrasted with $L^{-3}T^{-1}$ dimensions of the free convection modulus α.

The effect of pressure and temperature upon fluid properties for both gases and liquids is indicated in Table 3-1.

3-2.3 Evaluating C_1 and C_2

The factors C_1 and C_2 evaluate the effect of the solid surface configuration. These factors are dimensionless. Values from Brown and Marco[4] are given in Table 3-2.

Example Problem 3-1

A 16.5-cm diameter horizontal steel pipe supplied with steam at a temperature of 120°C is exposed to still air at 20°C. Determine the surface free convection heat transfer coefficient, h, in [W/m²·Δ_1°C]. (Assume the outer surface of the pipe is at the same temperature as the steam.)

[4] Brown, Aubrey I., and Marco, Salvatore M., *Introduction to Heat Transfer*, p. 166, McGraw-Hill Book Company, New York, 1958.

SOLUTION

SKETCH

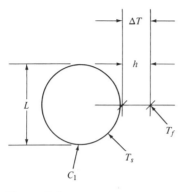

Figure 3-4

EQUATION

$$h = C_1 \frac{k}{L^{1/4}} (\alpha \Delta T)^{0.25} \quad [W/m^2 \cdot \Delta_1 {}^\circ C] \tag{3-11}$$

when $10^3 < (Gr \times Pr) = \alpha L^3 \Delta T < 10^9$. To justify the use of Eq. (3-11), determine the value of $\alpha L^3 \Delta T$ at the average temperature (T_{av}) between the free air and the surface of the pipe.

$$T_{av} = \tfrac{1}{2}(T_s + T_f) \quad [^\circ C]$$

$$T_s = 120 \quad [^\circ C] \quad \text{(given)}$$

$$T_f = 20 \quad [^\circ C] \quad \text{(given)}$$

$$T_{av} = \tfrac{1}{2}(120 + 20) \quad \frac{[^\circ C + {}^\circ C]}{2} = [^\circ C]$$

$$T_{av} = 70^\circ C \rightarrow 343 \text{ K}$$

PARAMETERS

$$\alpha = 51.2 \times 10^6 \quad [1/m^3 \cdot \Delta_1 {}^\circ C] \text{ at } 343 \text{ K} \qquad \text{(Table B-1, Appendix B)}$$

$$L = 16.5 \text{ cm} \rightarrow 0.165 \text{ m} \qquad \text{(given)}$$

$$\Delta T = T_s - T_f \quad [^\circ C]$$

$$\Delta T = 120 - 20 \quad [^\circ C - {}^\circ C] = [\Delta^\circ C]$$

$$\Delta T = 100 \, \Delta^\circ C$$

$$(\alpha L^3 \Delta T) = (51.2 \times 10^6)(0.165)^3 (100) \quad \left[\frac{1}{m^3 \cdot \Delta_1 {}^\circ C}\right] [m^3][^\circ C] = [-]$$

$$= 23.0 \times 10^6, \qquad \text{thus Eq. (3-11) is applicable}$$

PARAMETERS FOR EQ. (3-11)

$C_1 = 0.45$ [—] (Table 3-2)

$k = 0.0295$ W/m·Δ_1°C at 343 K (Table B-1, Appendix B)

$L = 0.165$ m (given)

$\alpha = 51.2 \times 10^6$ [$1/m^3 \cdot \Delta_1$°C]

$\Delta T = 100$°C

SUBSTITUTION

$$h = (0.45) \frac{(0.029)}{(0.165)^{1/4}} \left[(51.2 \times 10^6)(100) \right]^{1/4}$$

$$[-] \frac{[W/m \cdot \Delta_1°C]}{[m]^{1/4}} \; [1/m^3 \cdot \Delta_1°C][°C]^{1/4} = [W/m^2 \cdot \Delta_1°C]$$

$$h = (0.0133)(310.3 \times 10^8)^{1/4}$$

$$= (0.0133)(4.197 \times 10^2)$$

ANSWER

$$h = 5.5 \; W/m^2 \cdot \Delta_1°C$$

Example Problem 3-2

A 9.0-cm diameter horizontal wrought-iron pipe supplied with saturated steam at a temperature of 99°C is immersed in still water at 65°C. Determine the surface free convection heat transfer coefficient, h, in [W/m²·Δ_1°C]. (Assume the outer surface of the pipe is at the same temperature as the steam.)

SOLUTION

SKETCH

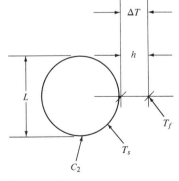

Figure 3-5

EQUATION

$$h = C_2 k (\alpha \Delta T)^{1/3} \quad [\text{W/m}^2 \cdot \Delta_1 {}^\circ\text{C}] \qquad (\text{Eq. 3-12})$$

when $10^9 < (\text{Gr} \times \text{Pr}) = \alpha L^3 \Delta T$. To justify the use of Eq. (3-12), determine the value of $\alpha L^3 \Delta T$ at the average temperature (T_{av}) between the free water and the surface of the pipe.

$$T_{av} = \tfrac{1}{2}(T_s + T_f) \quad [{}^\circ\text{C}]$$

$$T_s = 99{}^\circ\text{C} \qquad (\text{given})$$

$$T_f = 65{}^\circ\text{C} \qquad (\text{given})$$

$$T_{av} = \tfrac{1}{2}(99 + 65) \quad \left[\frac{{}^\circ\text{C} + {}^\circ\text{C}}{2}\right] = [{}^\circ\text{C}]$$

$$T_{av} = 82{}^\circ\text{C} \rightarrow 355 \text{ K}$$

PARAMETERS

$\alpha = 1.09 \times 10^{11} \quad [1/\text{m}^3 \cdot \Delta_1 {}^\circ\text{C}]$ at $82{}^\circ\text{C}$ \qquad (Table B-3, Appendix B)

$L = 9.0 \text{ cm} = 0.090 \text{ m}$ \qquad (given)

$\Delta T = T_s - T_f \quad [\Delta {}^\circ\text{C}]$

$\quad = (99 - 65) \quad [{}^\circ\text{C} - {}^\circ\text{C}] = [\Delta {}^\circ\text{C}]$

$\Delta T = 34 \, \Delta {}^\circ\text{C}$

$(\alpha L^3 \Delta T) = (1.09 \times 10^{11})(0.09)^3 (34) \quad [1/\text{m}^3 \cdot \Delta_1 {}^\circ\text{C}][\text{m}^3][\Delta {}^\circ\text{C}]$

$\qquad = [-] = 2.7 \times 10^9, \qquad$ thus Eq. (3-12) is applicable

PARAMETERS FOR EQ. (3-8)

$C_2 = 0.11 \, [-] \qquad$ (Table 3-2)

$k = 0.673 \text{ W/m} \cdot \Delta_1 {}^\circ\text{C}$ at $82{}^\circ\text{C}$ \qquad (Table B-3, Appendix B)

$\alpha = 1.09 \times 10^{11} \quad [1/\text{m}^3 \Delta_1 {}^\circ\text{C}]$

$\Delta T = 34{}^\circ\text{C}$

SUBSTITUTIONS

$$h = (0.11)(0.673)\left[(1.09 \times 10^{11})(34)\right]^{1/3}$$

$$[-][\text{W/m} \cdot \Delta_1 {}^\circ\text{C}]\left[[1/\text{m}^3 \cdot \Delta_1 {}^\circ\text{C}][\Delta {}^\circ\text{C}]\right]^{1/3} = [\text{W/m}^2 \cdot \Delta_1 {}^\circ\text{C}]$$

$$h = (0.074)[3706 \times 10^9]^{1/3}$$

$$\quad = (0.074)(15.475 \times 10^3)$$

ANSWER

$$h = 1145.2 \text{ W/m}^2 \cdot \Delta_1 {}^\circ\text{C}$$

3-3 THE FREE CONVECTION BOUNDARY LAYER

When a fluid flows across a solid surface, a disturbed flow region extends out from the solid surface/liquid interface. This region is called the boundary layer. The characteristics of the fluid flow in the boundary layer adjacent to the solid surface have a dominant effect upon the convective heat transfer. The major factor in the heat transfer rate is the energy transport resulting from the motion of the fluid particles, as contrasted to the molecular conduction.

The motion of the fluid in the free convective boundary layer is caused by the bouyancy effects resulting from the fluid density variations associated with the boundary layer temperature gradients. Typical temper-

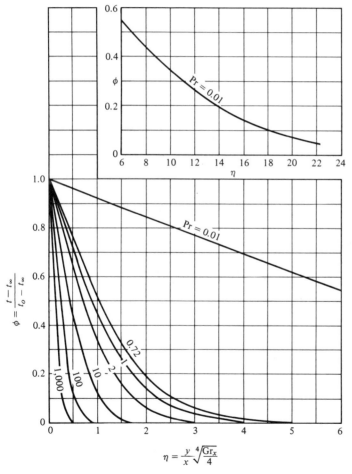

Figure 3-6   Temperature distributions for natural convection on a vertical, isothermal plate for laminar boundary layer flow conditions.

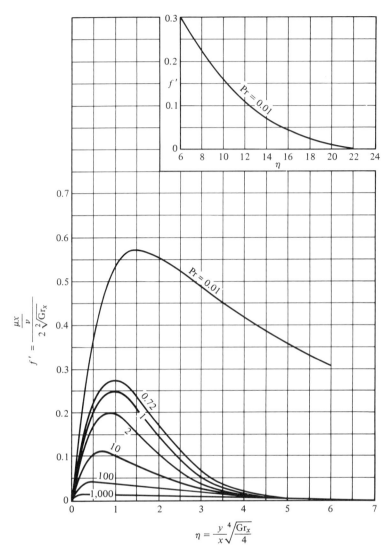

Figure 3-7   Velocity distributions for natural convection on a vertical, isothermal plate for laminar boundary layer flow conditions.

ature and velocity profile distributions for laminar boundary layer flow conditions along a vertical, isothermal plate (from Ostrach[5]), are shown in Figures 3-6 and 3-7. The velocity of the fluid flow increases as the distance from the solid surface becomes greater, until a maximum velocity is reached. At distances beyond this point of maximum velocity, the velocity decreases to that of the undisturbed fluid. The distance at which the flow

[5]Ostrach, S., *NACA Rep.* 1111, 1953.

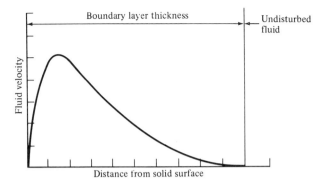

Figure 3-8 **Free convection boundary layer thickness.**

velocity in the boundary layer equals the velocity of the undisturbed fluid is called the "edge" of the boundary layer. The distance between the solid surface and the edge of the boundary layer is called the boundary layer "width" or "thickness," as illustrated by the simplified sketch in Figure 3-8. For a mathematical treatment of the free convection boundary layer, the reader is referred to authors such as Gebhart.[6]

Solid surface configurations have characteristic free convection boundary layer patterns. Figure 3-9 illustrates the typical heating fluid flow patterns for vertical flat plates, the bottom of a horizontal flat plate, the top of a horizontal flat plate, and a horizontal cylinder. In each of these examples the boundary layer flow distribution and the thickness distribution are different. This results in different characteristic thermal conductances of the boundary layers for these configurations. The appropriate factors to apply in the determination of h for various configurations are presented in Section 3-2.3.

The heating and cooling boundary layer patterns are similar for the vertical flat plate and horizontal cylinder, but the flow is in the reverse direction. For heating, the flow is up, and for cooling the flow is down. In the case of the horizontal flat plate, the similarity is between heating on the bottom of a flat plate and cooling on the top of a flat plate; and between heating on the top of a flat plate and cooling on the bottom of a flat plate.

The flow characteristics are well illustrated in the boundary layer of a vertical flat plate, as shown in Figure 3-10. The boundary layer is thin where the fluid has its initial contact with the surface. At this point the boundary layer has its highest thermal conductance, which results in the largest value of h. As the fluid begins to move along the surface, the boundary layer builds up with a laminar flow pattern, and continually increases in thickness, resulting in a reduction in the value of h. Laminar

[6]Gebhart, Benjamin, *Heat Transfer*, Second ed., pp. 316–354, McGraw-Hill Book Company, New York, 1971

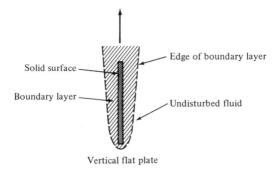

Vertical flat plate

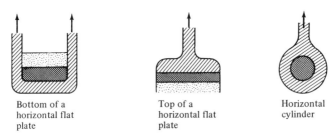

Bottom of a
horizontal flat
plate

Top of a
horizontal flat
plate

Horizontal
cylinder

Figure 3-9 Free convection boundary layer patterns for different solid surface configuration with heating of the fluid.

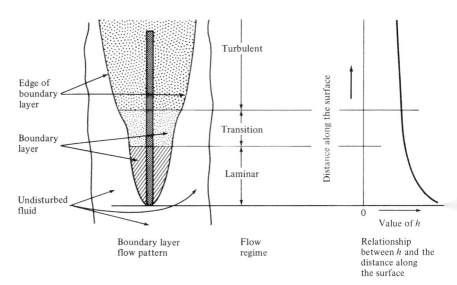

Figure 3-10 Flow characteristics of a boundary layer.

flow is characterized by the particles of the fluid flowing in paths substantially parallel to the solid surface and to each other.

The transitional flow regime begins when cross currents appear in the thickening laminar flow. As these cross currents begin to build up in magnitude the boundary layer thickness increases. The transitional flow boundary layer thickness increases more rapidly than the laminar flow boundary layer. The value of h tends to remain constant as the cross currents in the boundary layer counteract the effect of the increased boundary layer thickness.

The turbulent flow regime begins when eddy currents develop in the boundary layer flow. The turbulent boundary layer maintains a constant thickness and a constant value of h along the solid surface.

This change in the value as the boundary layer builds up through laminar to turbulent flow is considered in the determination of h both in the Grashof number and in the significant length. The Grashof number both selects the equation for h as well as affecting the value of α used in the equation.

3-4 SIMPLIFIED EQUATIONS FOR AIR

As air is often involved in free convection calculations, simplified equations have been developed for air to facilitate the evaluation of h. The simplified equations for vertical flat plates and horizontal cylinders are given in Sections 3-4.1 and 3-4.2 for air at atmospheric (standard atmosphere) pressure. Modification for other pressures is discussed in Section 3-4.3.

3-4.1 Vertical Plane Surfaces

SURFACES LESS THAN 0.6 m IN HEIGHT
Most free convection with air over vertical flat surfaces less than 0.6 m in height fall within the $Gr \times Pr$ range for Eq. (3-11). Equation (3-11) can be rewritten as

$$h = C_1 k \alpha^{1/4} \left(\frac{\Delta T}{L} \right)^{1/4} \quad [W/m^2 \cdot \Delta_1 {}^\circ C] \tag{3-16}$$

Values of $C_1 k \alpha^{1/4}$ for a range of air temperatures at atmospheric pressure are listed in Table 3-3.

$$T_{av} = \text{average temperature} = \frac{T_{surface} + T_{air}}{2}$$

To use the simplified Eq. (3-16), the value of $[C_1 k \alpha^{1/4}]$ at the estimated mean temperature between the wall and the free air can be obtained from Table 3-3 directly.

Table 3-3 VALUES OF $[C_1 k \alpha^{1/4}]$ OVER A TEMPERATURE RANGE FOR ATMOSPHERIC PRESSURE AIR FLOWING OVER A VERTICAL FLAT PLATE

TEMPERATURE T_{av} [°C]	$[C_1 k \alpha^{1/4}][W/(m^{1.75})(\Delta_1 °C^{1.25})]$
−20	1.491
0	1.464
+20	1.437
40	1.410
60	1.383
80	1.356
100	1.329
120	1.302

SURFACES MORE THAN 0.6 m IN HEIGHT

For vertical plane surfaces over 0.6 m in height, Eq. (3-12) is more generally applicable. The actual values of h are often uncertain because of turbulence patterns in the boundary layer. The simplified form of Eq. (3-12) is

$$h = \left[C_2 k \alpha^{1/3} \right] \Delta T^{1/3} \quad \left[W/m^2 \cdot \Delta_1 °C \right] \tag{3-17}$$

Values of $[C_2 k \alpha^{1/3}]$ for a range of temperatures for atmospheric pressure air are listed in Table 3-4.

$$T_{av} = \text{average temperature} = \frac{T_{surface} + T_{air}}{2}$$

To use the simplified Eq. (3-17), the value of $[C_2 k \alpha^{1/3}]$ at the estimated mean temperature between the wall and the free air can be obtained from Table 3-4 directly.

3-4.2 Horizontal Cylinders

FOR DIAMETERS UNDER 0.6 m

The simplified equation for free convection of air over horizontal wires and pipes up to 0.6 m in diameter is based upon Eq. (3-11), as in the case of

Table 3-4 VALUES OF $[C_2 k \alpha^{1/3}]$ OVER A TEMPERATURE RANGE FOR ATMOSPHERIC PRESSURE AIR FLOWING OVER A LARGE VERTICAL FLAT PLATE

TEMPERATURE T_{av} [°C]	$[C_2 k \alpha^{1/3}][W/(m^2)(\Delta_1 °C^{1.33})]$
−20	1.711
0	1.649
+20	1.587
40	1.525
60	1.463
80	1.400
100	1.338
120	1.269

Table 3-5 VALUES OF $[C_1 k\alpha^{1/4}]$ OVER A TEMPERATURE RANGE FOR ATMOSPHERIC PRESSURE AIR FLOWING OVER A HORIZONTAL CYLINDER

TEMPERATURE $T_{av}[°C]$	$[C_1 k\alpha^{1/4}]$ $[W/(m^{1.75})(\Delta_1°C^{1.25})]$
-20	1.211
0	1.190
$+20$	1.168
40	1.146
60	1.124
80	1.102
100	1.080
120	1.058

vertical flat plates up to 0.6 m in height, resulting in Eq. (3-16).

$$h = [C_1 k\alpha^{1/4}] \left(\frac{\Delta T}{D}\right)^{1/4} \quad [W/m^2 \cdot \Delta_1°C] \qquad [\text{Eq. (3-16)}]$$

where

$D =$ cylinder diameter $[m]$

The value of C_1 for horizontal pipes is 0.8125 times the value of C_1 for a vertical flat plate, as determined from the values given in Table 3-2. For the reader's convenience, the values for $[C_1 k\alpha^{1/4}]$ for a horizontal pipe up to 0.6 m in diameter over a range of temperatures are listed in Table 3-5.

$$T_{av} = \text{average temperature} = \frac{T_{surface} + T_{air}}{2}$$

To use the simplified Eq. (3-16), the value of $[C_1 k\alpha^{1/4}]$ at the estimated mean temperature between the wall and the free air can be obtained from Table 3-5 directly.

FOR DIAMETERS OVER 0.6 m
If the horizontal pipe is larger than 0.6 m in diameter, the simplified equation is Eq. (3-17).

$$h = [C_2 k\alpha^{1/3}] \Delta T^{1/3} \quad [W/m^2 \cdot \Delta_1°C] \qquad [\text{Eq. (3-17)}]$$

The value of $[C_2 k\alpha^{1/3}]$ for the horizontal cylinder can be obtained by multiplying the values given in Table 3-4 by 0.8125.

3-4.3 Effect of Pressure

The pressure of the air has an effect upon the convective surface coefficient "h." The values in Tables 3-3, 3-4, and 3-5 are for a standard atmosphere of pressure. The effect of pressure appears in α. Table 3-1 indicates that the effect of pressure enters in the density term of α. The density of a gas at constant temperature is directly proportional to the pressure of the gas. As the density term is squared, α is proportional to the square of the pressure.

In Tables 3-3 and 3-5, α enters to the $\frac{1}{4}$ power. Therefore the value of h obtained using these tables is proportional to the square root of the pressure, in atmospheres. For example, if the actual air pressure is 1.44 atm, the value of h will be 1.20 times that calculated using the values of $[C_1 k \alpha^{1/4}]$ from Tables 3-3 and 3-5.

In Table 3-4 α enters to the $\frac{1}{3}$ power. Therefore, the value of h obtained using this table is proportional to the $\frac{2}{3}$ power of the pressure in atmospheres. For example, if the actual air pressure is 1.44 atm, the value of h will be 1.27 that calculated using the values of $[C_2 k \alpha^{1/3}]$ from Table 3-4.

3-5 EFFECTS OF VARIOUS CONFIGURATIONS

The value of the convective surface coefficient h is affected to a greater or lesser extent by factors influencing the character of the boundary layer. Several typical configurations are discussed in the following paragraphs.

3-5.1 Vertical Cylinders

When the boundary layer thickness is small relative to the diameter of the cylinder, the effect of the transverse curvature of the solid surface becomes small, and the boundary layer flow pattern approaches that of a vertical flat plate. The results of investigations by Sparrow and Gregg[7] indicated that the vertical flat plate solution is within 5% of the actual value of the heat transfer film coefficient h for vertical cylinders of sufficiently large diameter such that

$$\frac{D}{L} \leqslant \frac{35}{\sqrt[4]{\mathrm{Gr}}} \quad [-] \tag{3-18}$$

where Gr is based upon L.

When the diameter of the vertical cylinder is not large enough to meet the criteria of Eq. (3-18), the determination of h by the use of Figure 3-11 is recommended by Gebhart.[7] Figure 3-11 gives a plot of the Nusselt number (Nu) versus the logarithm (to the base 10) of the Rayleigh number (Ra) modified by the D/L ratio. Ra$=$(Gr)(Pr). In this case Gr is based upon the diameter of the vertical cylinder.

[7]Gebhart, Benjamin, *Heat Transfer*, Second Edition, p. 374, McGraw-Hill Book Company, New York, 1971.

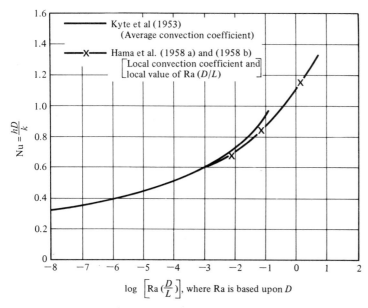

Figure 3-11   Natural convection correlations for vertical cylinders of small diameter.

Example Problem 3-3

A 17.5-cm diameter vertical stainless steel pipe 3 m long has a surface temperature of 125°C in a room which has an air temperature of 15°C. Determine the surface free convection heat transfer coefficient, h, in $[\text{W}/\text{m}^2 \cdot \Delta_1{}^\circ\text{C}]$.

SOLUTION

SKETCH

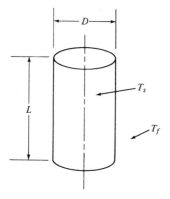

Figure 3-12

EQUATION

$$h=\left[C_2 k\alpha^{1/3}\right](\Delta T)^{1/3} \quad \left[\text{W/m}^2 \cdot \Delta_1 {}^{\circ}\text{C}\right] \quad \left[\text{Eq. (3-17)}\right] \quad\quad (3\text{-}17)$$

when $\dfrac{D}{L} \geqslant \dfrac{35}{(\text{Gr})^{1/4}}$

but $\text{Gr} = \dfrac{(\alpha L^3 \Delta T)}{\text{Pr}}$

PARAMETERS

$\alpha = 49.7 \times 10^6 \left[1/\text{m}^3 \Delta_1 {}^{\circ}\text{C}\right]$ @ 70°C (Table 3, Appendix B-1)

$L = 3.0$ m (given)

$\Delta T = (T_s - T_f) = 110 \Delta {}^{\circ}\text{C}$ (given)

$\text{Pr} = 0.699$ [--] @ 70°C (Table 3, Appendix B-1)

$D = 0.175$ m (given)

$\left[C_2 k\alpha^{1/3}\right] = 1.432$ W/m$^2 \Delta_1 {}^{\circ}\text{C}^{4/3}$ (Table 3-4, p. 140)@ 70°C

SUBSTITUTIONS

$$\text{Gr} = \frac{(49.7\times 10^6)(27)(110)}{(0.699)} \frac{\left[1/\text{m}^3\Delta_1{}^{\circ}\text{C}\right]\left[\text{m}^3\right]\left[\Delta{}^{\circ}\text{C}\right]}{[--]} = [--]$$

$$= 0.211 \times 10^{12} [--]$$

and $\dfrac{35}{(\text{Gr})^{1/4}} = \dfrac{35}{(0.211\times 10^{12})^{1/4}}$

$$= \frac{35}{(0.678\times 10^3)} = 0.052$$

Also

$$\frac{D}{L} = \frac{0.175}{3.0} = 0.58$$

Since

$$\frac{D}{L} = 0.58 > 0.052,$$

Eq. (3-17) is applicable. Therefore,

$$h = [1.432](110)^{1/3} \quad \left[\text{W/m}^2 \cdot \Delta_1{}^{\circ}\text{C}^{4/3}\right]\left[\Delta{}^{\circ}\text{C}\right]^{1/3} = \left[\text{W/m}^2 \cdot \Delta_1{}^{\circ}\text{C}\right]$$

$$= (1.432)(4.79)$$

ANSWER

$$h = 6.86 \text{ W/m}^2 \cdot \Delta_1{}^{\circ}\text{C}$$

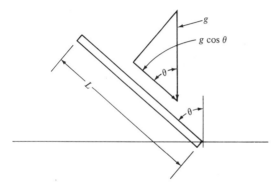

Figure 3-13   Inclined flat plate.

3-5.2 Inclined Flat Plate

The effect of an inclined flat plate, according to Kreith[8] is to reduce the bouyancy effect by the cosine of the angle of inclination from the vertical. The significant dimension is measured along the surface of the inclined plate, as illustrated in Figure 3-13. The effect this has upon Eq. (3-11) is to multiply the calculated h by $(\cos\theta)^{1/4}$ and Eq. (3-12) by $(\cos\theta)^{1/3}$, which is relatively small in either case. This analysis is limited to a "moderately" inclined surface, because it reduces the value of the h of a vertical flat plate progressively as the angle of inclination increases. From Table 3-3 it is seen that the value of C for a horizontal flat plate with the warm side facing upward ($\theta=90°$) is 1.3 times that of the vertical flat plate. So at some angle of inclination the $\cos\theta$ correction ceases to be valid and the value of h increases to that of the horizontal flat plate. Experiments of Vliet[9] suggest that the substitution of $g\cos\theta$ for g in the Grashof number is valid for angles of inclination as large as $60°$.

3-5.3 Enclosed Gas Spaces

Heat transfer by free convection across spaces with parallel plane walls has been treated by the use of an equivalent thermal conductivity term k_c. This term includes the effects of both the convection and conduction in the gas. The heat transfer rate across the gas layer is

$$\dot{Q}=\frac{k_c}{x}A\,\Delta T \quad [\text{W}] \tag{3-19}$$

[8] Kreith, Frank, *Principles of Heat Transfer*, p. 310, International Textbook Company, Scranton, Pa., 1958.
[9] Vliet, G. C., *Journal of Heat Transfer*, Vol. 91, p. 511, 1969.

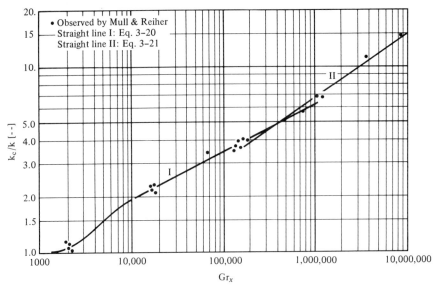

Figure 3-14 Dimensionless representation of heat transfer through horizontal air layers by free convection.

where

k_c = equivalent thermal conductivity $\left[W/m \cdot \Delta_1 °C\right]$

x = width of the gas space $\left[m\right]$

A = heat transfer path cross-sectional area $\left[m^2\right]$

ΔT = temperature difference between the inside surfaces of the space walls $\left[\Delta °C\right]$

Jakob[10] has correlated test data by the use of:

- the ratio k_c/k—where k is the thermal conductivity of the gas in the space at the average temperature
- the Grashof number (Gr_x)—based upon the width of the gas space (x) and the temperature difference between the inside surfaces of the walls

Note: Gr is defined in Section 3-1.1.

HORIZONTALLY DISPOSED AIR SPACE
Test data have been plotted using the parameters discussed above, and equations fitted to the plotted test data, as illustrated in Figure 3-14. Two equations selected to fit the test data are indicated by the line segments I

[10] Jakob, Max, *Heat Transfer*, Vol. I, pp. 534–542, John Wiley & Sons, Inc., New York, 1956.

and II in the plot. The equation for line I through the range of Gr_x from 10,000 to 400,000 is

$$k_c = 0.195k(Gr_x)^{1/4} \quad [\text{W/m}\cdot\Delta_1\text{°C}] \tag{3-20}$$

The equation for line II through the range of Gr_x over 400,000 is

$$k_c = 0.068k\,Gr_x \quad [\text{W/m}\cdot\Delta_1\text{°C}] \tag{3-21}$$

VERTICALLY DISPOSED AIR SPACE

Test data have been plotted using the same parameters as in Figure 3-13. In the case of the vertical space, the ratio of the space height to the space width (H/x) has an effect as indicated in Figure 3-15. The two equations selected to fit the test data are indicated by the groups of line segments I and II. The equation for the group of line segments I through the range of Gr_x from 20,000 to 200,000 is

$$k_c = 0.18k(Gr_x)^{0.25}\left(\frac{H}{x}\right)^{-0.11} \quad [\text{W/m}\cdot\Delta_1\text{°C}] \tag{3-22}$$

The equation for the group of line segments II through the range of Gr_x from 200,000 to 11,000,000 is

$$k_c = 0.065k(Gr_x)^{0.33}\left(\frac{H}{x}\right)^{-0.11} \quad [\text{W/m}\cdot\Delta_1\text{°C}] \tag{3-23}$$

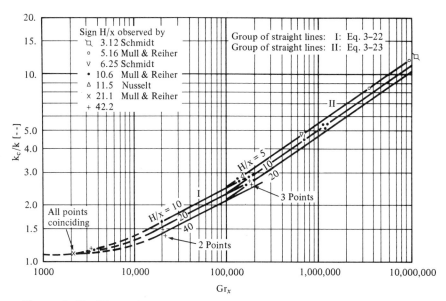

Figure 3-15 Dimensionless representation of heat transfer through vertical air layers by free convection.

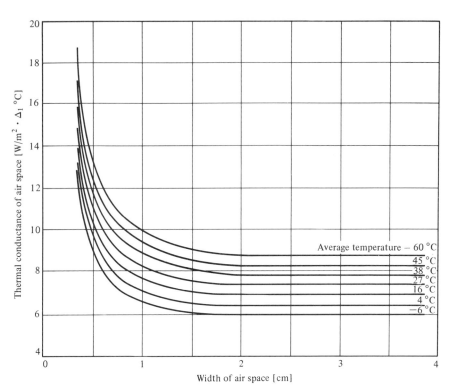

Figure 3-16 Conductance of air spaces for various mean temperatures.

Rowley and Algren[11] investigated vertical air spaces, and obtained data that are presented in Figure 3-16 in the form of a plot of the air space thermal conductance versus the width of the air space.

In vertical air spaces, where two sets of vertical parallel walls form a rectangular cross-sectional duct, the convective surface coefficient h has been found to be independent of the width when the width is 2 cm or greater. Also, the value of h is not subject to the modification of the significant length L as discussed in Section 3-1. The ends of the air space maintain a boundary layer pattern similar to that of a configuration with a long horizontal length.

As the width of the air space becomes smaller than 2 cm, the convective surface coefficient diminishes, and the effect of conduction increases. When the width of the air space has reached 0.3 cm the convective surface coefficient has become essentially 0, and the heat transfer is by pure conduction through the gas.

[11] Rowley, F. B., and Algren, A. B., "Thermal Resistances of Air Spaces," *Trans. ASHVE*, 35, 165–181 (1929).

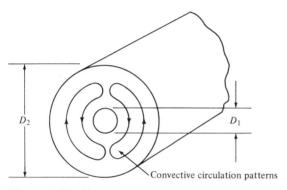

Figure 3-17   Free convection in an annulus with a horizontal axis.

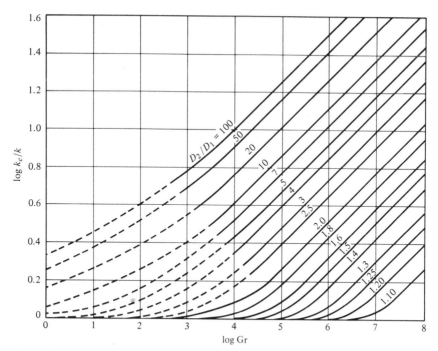

Figure 3-18   Dimensionless representation of heat flow between coaxial horizontal tubes by free convection.

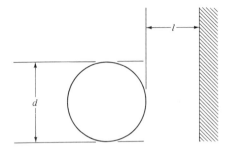

Figure 3-19 Horizontal cylinder close to a wall.

INCLINED SPACE

For inclined plane gas layers, Jakob[12] suggests using a linear interpolation between the formulae for horizontal and vertical layers.

CYLINDRICAL ANNULUS GAS LAYERS

By the use of an equivalent thermal conductivity, as in Section 3-5.3, the heat transfer rate through a cylindrical annulus can be calculated for unit length by the annulus by

$$\frac{\dot{Q}}{L} = \frac{2\pi k_c}{\ln(D_2/D_1)}(T_1 - T_2) \quad [\mathrm{W/m}] \tag{3-24}$$

where D_1 is the inside diameter and D_2 is the outside diameter as illustrated in Figure 3-17. As the width of the annulus $(D_2 - D_1)/2$ becomes smaller the effect of free convection decreases and becomes negligible when the width is reduced to 0.3 cm. Heat is transferred across narrow annuli by the thermal conductivity of the gas.

Plots similar to those presented in Figures 3-14 and 3-15 have been made using a Grashof number Gr_1, based upon the diameter of the inside cylinder. Such a plot for various diameter ratios is given in Figure 3-18.

3-5.4 Cylinder in the Proximity of a Wall

Many applications involve the convective heat transfer from a horizontal cylinder in close proximity to a vertical wall or to a ceiling, or in a horizontal tunnel.

The configuration of a horizontal cylinder close to a vertical wall is illustrated in Figure 3-19.

When $l \geqslant d/4$ there is no effect on h.
When $l < d/4$ there is a reduction in h.

---

[12] Jakob, Max, *Heat Transfer*, Vol. 1, p. 539, John Wiley & Sons, Inc., New York, 1956.

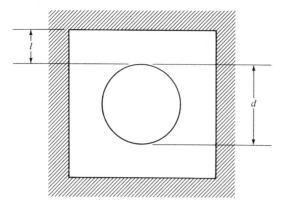

Figure 3-20 Horizontal cylinder in a square horizontal tunnel.

The configuration of a horizontal cylinder in a square horizontal tunnel is illustrated in Figure 3-20.

When $l \geq d$ there is no effect on h.
When $l < d$ there is a reduction in h.
When $l = 1/10\ d$ the value of h has dropped to $0.7h$.

When the horizontal cylinder is under the ceiling, or in a corner formed by the ceiling and the wall, the reduction in h is between the reduction for a horizontal cylinder close to a wall (Figure 3-19) and a horizontal cylinder in a square horizontal tunnel (Figure 3-20).

PROBLEMS

3-1. Describe or define free convection. What forces control the fluid motion?

3-2. Can free convection occur in space vehicles with a zero "g" trajectory? Explain your answer.

3-3. Give a general equation for the rate of heat transfer ($\dot{Q}$) by free convection.

3-4. What factors affect the value of the free convection heat transfer film coefficient (h)?

3-5. What fluid properties are involved in the equations for the free convection heat transfer film coefficient (h)? Explain why these properties are involved and what effect they have.

3-6. What is a dimensionless number? How and why are they used in heat transfer?

3-7. What dimensionless numbers are used in free convection heat transfer? Explain the physical significance of each.

3-8. Give the basic equation for convective heat transfer which was developed from dimensional analysis. How was this equation modified by experimental test results?

3-9. What three regions of laminar boundary layer flow have been identified? What effect do they have on the correlation between Nusselt's number (Nu) and the product of Grashof and Prandtl numbers (Gr×Pr)?

3-10. In the product of Grashof and Prandtl numbers (Gr×Pr), what group of properties is concerned only with the fluid properties and the gravitational field? For what conditions is each of these properties evaluated? What is this group of properties called? What are the dimensional units of this group of properties?

3-11. Give the equation for the free convection heat transfer film coefficient (h) for:
(a) laminar boundary layer conditions
(b) turbulent boundary layer conditions

3-12. What is the parameter C, or C_2 in these equations and how is it evaluated?

3-13. When free convection occurs on a vertical flat plate, is there any interaction between the vertical height and width? If so, explain the reason and indicate how this is considered in the formula for determining the film coefficient (h).

3-14. A window is 1.2 m wide and 20 m tall. If the inside temperature of the glass is 10°C and the air temperature in the room is 20°C, and the pressure is 1 atm, calculate the convective surface coefficient h in [W/m²·Δ_1°C], and the heat transfer rate in [W]. If the window were in a mountain cabin at 2000-m elevation, what would be the heat loss under the same temperature conditions?

3-15. Calculate the heat transferred by natural convection per meter from a horizontal pipe 25 cm in diameter with a surface temperature of 90°C to atmospheric pressure air at 25°C.

3-16. Calculate the convective surface coefficient h between a horizontal 2.5-cm diameter pipe with a surface temperature of 99°C to the following fluids at a temperature of 27°C.
(a) air at 1 atm pressure
(b) hydrogen at 1 atm pressure
(c) carbon dioxide at 1 atm pressure
(d) water
(e) transformer oil

3-17. The bottom of a water tank is heated uniformly by electrical resistance heaters. What is the watt input to the heater if the plate maintains the water temperature at 65°C when the temperature of the plate surface is 75°C?

3-18. Electric windings in the form of a 15-cm cube are immersed in a tank of transformer oil. What is the power loss in watts in the windings if oil temperature is 65°C when the surface of the windings is at 87°C?

3-19. To help reduce heat loss from a house with a floor on joists 6 cm high, plywood was added to the underside of the joists forming a horizontal air space 6 cm high. What would be the equivalent conductivity K_c of this air space?

3-20. The vertical air space in the wall of a house is 9.5 cm wide. If the temperature of one surface is 15°C and the other 0°C, what is the convective heat loss through the air space in $[W/m^2]$?

Chapter 4
Steady State Forced Convection

The convective heat transfer occurring between a solid body and a fluid is termed *forced convection* when the circulation of the fluid (liquid or gas) is caused and controlled by some mechanical means, commonly a pump or a blower. A large portion of heat transfer applications makes use of forced convection, such as the heat exchanger equipment for building heating and cooling systems and for power generation systems, the automobile radiator, cooling internal combustion engines, and many more.

The rate of heat transfer ($\dot{Q}$) for forced convection heat transfer between the fluid stream and a solid wall can be described by a basic equation which has the same form as the free convection Eq. (3-1) presented in Section 3-2. However, h for the forced convection surface or "film" coefficient is evaluated in a significantly different manner. The mechanical forcing of the fluid flow supplants the thermal buoyancy of the free convective system to generate the fluid flow characteristics.

$$\dot{Q} = hA(T_f - T_s) = hA\Delta T \quad [W] \tag{4-1}$$

where

$\dot{Q} =$ heat flow rate $\quad\quad\quad\quad\quad\quad\quad\quad$ [W]
$h =$ convective surface (or film) coefficient $\quad [W/m^2 \cdot \Delta_1 °C]$

A = surface area $\qquad\qquad\qquad\qquad$ [m^2]
T_f = fluid bulk (main mass) temperature $\qquad$ [°C]
T_s = solid surface temperature $\qquad\qquad$ [°C]

In typical forced convective heat transfer application problems, two of the three equation parameters $\dot{Q}$, A, and T in Eq. (4-1) are known. In order to solve for the other parameters, the value of h must be determined.

The usual equations for determining h for a set of conditions are given in Section 4-3. Typical ranges in the values of h are:

$$\text{air and superheated steam 30–300} \qquad [\text{W}/\text{m}^2\,\Delta_1°\text{C}]$$
$$\text{oil 60–3000} \qquad [\text{W}/\text{m}^2\,\Delta_1°\text{C}]$$
$$\text{water 300–10,000} \qquad [\text{W}/\text{m}^2\,\Delta_1°\text{C}]$$

The value of h for a forced convection heat transfer configuration is a function of:

the flow characteristics
the fluid properties
the flow passage geometry
the surface condition

These factors are discussed in the following sections.

Fluid Flow Characteristics

The transfer of heat between a solid boundary and a fluid takes place by a combination of conduction and mass transport. If the solid boundary is at a higher temperature than the fluid, heat flows initially by conduction from the solid boundary to particles of fluid adjacent to the boundary. This transfer of heat to the particles of fluid increases the temperature of those fluid particles. As these heated particles are swept away from the solid boundary by the action of the fluid flow, they carry the heat from the solid boundary with them into regions of the fluid flow, which is at a lower temperature. In these regions the heat is transferred by conduction from hotter fluid particles to cooler fluid particles. The convective mode of heat transfer is closely linked to the transverse fluid motion within the fluid flow.

Fluid Flow Pattern Characteristics

The fluid flow pattern is characterized by being either:

- laminar
- transitional
- fully turbulent

LAMINAR FLOW

In laminar, or streamline flow, the fluid particles flow in paths that are parallel to each other and to the solid surface. A laminar boundary extends from the solid surface out into the fluid flow. The flow velocity on the solid surface is zero, and the flow velocity increases across the boundary layer until it reaches the undisturbed fluid flow velocity. This point is the edge of the boundary layer, and may extend to the center of the pipe or duct in which the fluid is flowing.

Critical flow is the term applied to laminar flow in a pipe or duct when the boundary layer extends across the entire flow duct and the flow velocity is the maximum for the laminar flow mode. Higher flow velocities would introduce turbulence in the laminar flow mode to exist across the full flow duct. Higher flow velocities would cause turbulence to appear in the flow. In the critical flow region, small variations in the flow may cause significant changes in the heat transfer rates. Consequently the critical flow region should be avoided in heat transfer equipment, because the heat transfer performance is subject to significant fluctuations.

TRANSITIONAL FLOW

In transitional flow, a varying amount of turbulence exists. In the transitional flow region, large changes in the fluid flow and heat transfer occur as the degree of turbulence increases from the laminar critical flow pattern to the fully developed turbulent flow pattern.

FULLY TURBULENT FLOW

In fully developed turbulent flow, strong velocity components normal to the nominal direction of flow exist. The path of any individual particle is zig-zag and irregular. On a statistical basis the overall motion of the aggregate of fluid particles is regular and predictable.

Flow Similarities

Osborne Reynolds[1] investigated the problem of the "similar" flow of fluids. In this study he considered the circumstances which would make the flow of different fluids with different velocities similar. He showed that the flow of fluids is similar when a specific ratio of variables is the same. This ratio is dimensionless, and is called the Reynolds number (Re).

$$\text{Re} = \frac{VD\rho}{\mu} = \frac{GD}{\mu} \quad [-] \quad \left[\text{Eq. (3-5)}\right]$$

The Reynolds number was initially discussed in Section 3-1.1. It includes

[1]O. Reynolds, Phil. Transactions Roy. Soc. (London) A, Vol. 174, pt. 111, p. 935 (1883) or "Scientific Papers," Cambridge University Press, Vol. 11, pp. 51–105 (1900–1903).

terms of a length, a velocity, fluid density, and fluid viscosity. These can be arranged in numerous forms, such as substituting the fluid mass flow rate per unit cross-sectional area of the fluid flow for the product of the flow velocity and the density.

The values of the Reynolds numbers for the different flow regimes, that is, laminar, critical, transitional, and full turbulent vary for different geometrical configurations of the fluid flow. For flow inside pipes the ranges of the Reynolds number associated with the four flow regimes are listed below. In this case the Reynolds number is based on the pipe diameter for the length dimension (D), and the average flow velocity for the velocity dimension (V).

FLUID FLOW REGIME AND ASSOCIATED REYNOLDS NUMBER
FOR FLUID FLOW INSIDE PIPES

Laminar flow	<2100	[—]
Critical flow	2100–3000	[—]
Transitional flow	3000–10,000	[—]
Turbulent flow	>10,000	[—]

Fluid Properties

The fluid properties that affect the heat transfer characteristics of a fluid are

k = thermal conductivity
c_p = specific heat at constant pressure
μ = viscosity

Ludwig Prandtl investigated the problem of "similar" heat transfer properties of fluids. In this study he considered the circumstances that would make the properties of the fluids with different thermal conductivities, different specific heats, and different viscosities "similar." He showed that the heat transfer properties are similar when the following dimensionless ratio is the same. This ratio is called the Prandtl number.

$$\mathrm{Pr} = \frac{c_p \mu}{k} \quad [-] \quad \left[\text{Eq. (3-3)}\right]$$

The Prandtl number for gases can be treated as a constant over a wide range of temperatures and pressures. Significant deviations begin as the conditions of the gas begin to approach the critical temperature or pressure. The Prandtl number for several gases is given in Table 4-1. For convenient reference there are also columns of the Prandtl number raised to specific exponential powers usually encountered in formulae for calculating h.

Table 4-1 PRANDTL NUMBERS $c_p\mu/k$ FOR GASES AND VAPORS
AT 1 ATM AND 100°C

	$\dfrac{c_p\mu}{k}$	$\left(\dfrac{c_p\mu}{k}\right)^{0.3}$	$\left(\dfrac{c_p\mu}{k}\right)^{0.4}$	$\left(\dfrac{c_p\mu}{k}\right)^{3/4}$
Air, carbon monoxide, hydrogen, nitrogen,				
oxygen	0.69	0.895	0.862	0.757
Ammonia	0.87	0.959	0.946	0.901
Carbon dioxide, sulphur dioxide	0.74	0.914	0.887	0.798
Ethylene	0.83	0.947	0.928	0.883
Hydrogen sulphide	0.77	0.925	0.900	0.840
Methane	0.79	0.932	0.909	0.855
Steam (low pressure)	0.99	0.997	0.996	0.992

In contrast to gases, the Prandtl number for liquids varies with temperature. Consequently it must be evaluated at the temperatures of the liquid at the conditions of interest. Formulae for calculating h are valid for a specific range in Prandtl numbers. If they are applied outside the specified range, the error in the calculated value of h becomes progressively larger the farther the Prandtl number value is away from the specified range.

The Flow Passage Geometry

Flow passages for fluids in forced convection take many forms. The fluid may be flowing, for example:

inside a pipe
across the outside of a pipe or a wire
across the outside of tubes in a tube bank array
along a flat plate

Each of these will have a different heat transfer characteristic. This is a result, primarily, of the boundary configuration along the solid boundary. The boundary layer characteristics have a dominant effect upon the heat transfer. A general discussion of the forced convection boundary layer for typical flow configurations is given in Section 4-1.

Nusselt investigated the problem of the "similar" film coefficient of fluids and flow geometry. In this study he considered the circumstances that would make the film coefficients of different fluids with different thermal conductivities in ducts of different diameters (or some characteristic dimension) "similar." He showed that the following dimensionless ratio provided this similarity. This ratio is called the Nusselt number.

$$\text{Nu} = \frac{hD}{k} \quad [-] \qquad [\text{Eq. (3-2)}]$$

The Surface Condition

The surface condition has an effect upon the film coefficient, primarily in two manners:

adding an insulating layer
causing increased turbulence

The heat transfer can be reduced by an insulating layer on the solid/fluid interface surface. Such an insulating layer can be in the form of scale or other deposits upon the surface. It also can be in the form of roughness of the surface of such a nature that causes static fluid to be held along the surface, introducing the conductive resistance of the static layer thickness.

The heat transfer can be increased by surface roughness or irregularities of a nature which introduce higher levels of turbulence into the boundary layer.

4-1 THE BOUNDARY LAYER

As fluid flows along the surface of a solid, the flow near the surface is affected by viscous forces, which cause substantial velocity changes. The region of substantial velocity change is called the boundary layer. Its thickness is usually defined as the distance from the solid surface at which the fluid velocity is 90% of the undisturbed free stream velocity.

In the boundary layer, the fluid particles adjacent to the solid surface stick to it and have zero velocity. (This is not strictly true in rarified gases when the mean free path of the molecules is significant compared with the boundary layer thickness.) The velocity profile in the fluid immediately adjacent to the solid surface is dominated by viscous shearing forces, and this extends out from the surface to the edge of the laminar flow zone as illustrated in Figure 4-1. Next to the laminar zone, a turbulent zone forms,

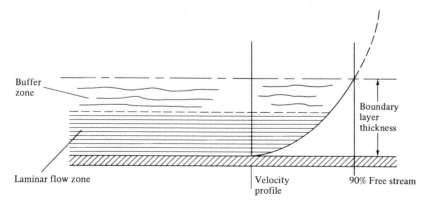

Figure 4-1   Turbulent boundary layer characteristics.

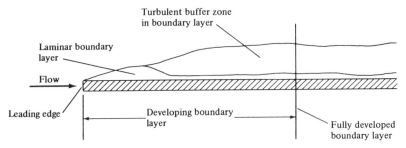

Figure 4-2   Typical boundary layer for flow along a flat plate.

with turbulent shear being superimposed on the viscous shear. This zone is usually called the buffer zone.

The boundary layer is not uniform in character along the entire length of a surface. The boundary layer develops as the fluid flows along the solid surface. At the initial point of contact with the solid surface the boundary layer is thin, building up along the surface until a fully developed boundary layer is reached. This is illustrated in Figure 4-2 for a flat plate—the boundary layer buildup along a flat plate. Because of this change in the boundary layer, short surfaces have higher average heat transfer film coefficients than do long surfaces under the same flow conditions.

4-1.1 The Boundary Layer Along a Flat Plate

When the fluid flows along a flat plate the boundary layer builds up from the leading edge to a fully developed boundary layer that does not change with a further increase of length of the flat plate. This boundary layer is illustrated in Figure 4-2.

4-1.2 The Boundary Layer in Flow Through Pipes

Typical profiles of laminar, transitional, and fully turbulent flow in pipes is illustrated in Figure 4-3. The boundary layer thickness is indicated for the transitional and fully turbulent flow. There will be a difference in the velocity profiles between isothermal flow and flow with heating or cooling. During heating the fluid in the boundary layer is less viscous than isothermal, and during cooling it is more viscous. This affects the thickness of the laminar zone in the boundary layer and the heat transfer rate.

Fluid flow inside a pipe will have a buildup to a fully developed boundary layer along the pipe. Because the developing boundary layer is thinner than the fully developed boundary layer it will have a high film coefficient. Consequently, the average film coefficient will be higher for a short pipe than for a long pipe.

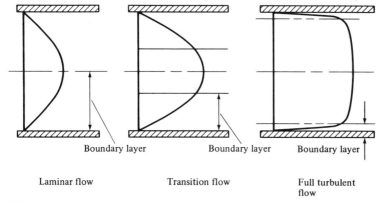

Figure 4-3 Fluid flow velocity profiles in pipes.

Any changes to a flow duct, such as sudden enlargements, sudden contractions, elbows, and turns or coils will have an effect upon the boundary layer. In general these changes increase the flow turbulence and increase the film coefficient.

4-1.3 The Boundary Layer in Flow Across a Cylinder

Typical boundary layers for laminar and for turbulent flow across a cylinder is shown in Figure 4-4. The fluid flow, as it reaches the cylinder, is split and passes around the cylinder. The point on the cylinder surface where the fluid flow splits is called the stagnation point.

The typical characteristics of the boundary layer begin at the stagnation point where the nominal flow direction is normal to the solid surface. At this point the flow velocity is zero along the surface of the cylinder. To one side or the other a flow velocity along the surface of the cylinder builds up a laminar boundary layer. In laminar flow the boundary layer remains laminar over the entire surface of the cylinder. As the flow velocity increases, changing the nominal flow regime from laminar to turbulent, the thickness of the laminar boundary layer increases until a turbulent "buffer" zone appears, resulting in turbulence in the wake downstream from the cylinder. With further flow velocity increases, the boundary layer separates from the surface of the cylinder and large eddies appear in the wake downstream from the cylinder.

In general, the thinner the boundary layer, the higher the value of the local film coefficient (h). The highest local values of h are in the region of the thin laminar boundary layer. The local values of h in the separation zone are usually low because the fluid particles in the separation zone immediately adjacent to the cylinder surface are more or less captive in the

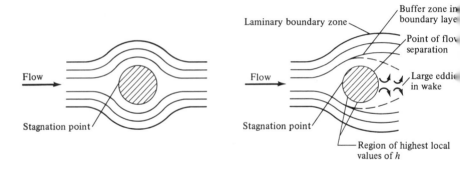

Laminary boundary zone

Buffer zone in boundary layer

Point of flow separation

Flow

Flow

Large eddies in wake

Stagnation point

Stagnation point

Region of highest local values of h

Laminar flow

Turbulent flow

Figure 4-4 Typical boundary layers for flow across a cylinder.

eddy pattern and have no direct contact with the undisturbed fluid stream flow.

When the fluid flows through a bank of pipes, the wake of the preceding pipes has a major effect upon the boundary layer of the following pipes. Both the spacing distances of the bank pattern as well as the arrangement "in line" or "staggered" influence the flow pattern. The flow pattern for various tube bank configurations is illustrated on pages 185 and 186.

4-2 THE EVALUATION OF THE CONVECTIVE HEAT TRANSFER FILM COEFFICIENT h

The convective heat transfer film coefficient (h) can be evaluated by the use of an equation based on theoretical principles or experimental data. Equations based on theoretical principles have been found to be limited to the laminar flow regime and to fluids with a Prandtl number of 1.0 (as discussed on page 164). For other conditions an experimentally evaluated factor applied to the theoretical equation is required. As the majority of heat transfer occurs at conditions other than laminar flow with a fluid having a Prandtl number of 1, experimental data are required to evaluate h even when theoretical formulae are used.

The most effective method of using experimental data for predicting the convective heat transfer film coefficient h for any combination of heat transfer conditions, flow regimes, and fluid properties is through the use of dimensionless ratios. This is discussed on page 125.

Many investigators have found by plotting test data on coordinates of various combinations of dimensionless numbers that equations for determining h can be developed from curves giving the best fit with the experimental data point plot. In this manner the important similarities between fluids and heat transfer conditions can be used to accurately apply an experimental determination of *h* under one set of conditions to all other sets of conditions wherein it is applicable.

Dimensionless numbers were discussed in Section 3-1.1 under the development of the equation for *h* for free convection. In forced convection the fluid flow velocities are sufficiently high so the inertia forces in the flow are important. The buoyancy forces are negligible in comparison with the forces of inertia and friction. For the usual forced convection circumstances the following three dimensionless numbers apply:

$$\text{Nusselt} = hD/k \quad [-]$$
$$\text{Prandtl} = \mu c_p/k \quad [-]$$
$$\text{Reynolds} = DV\rho/\mu \quad [-]$$

The conventional generalized basic equation for use in determining the value of *h* is

$$\text{Nu} = C(\text{Re})^a(\text{Pr})^b \quad [-] \tag{4-2}$$

The numerical values for the constant *C* and the exponents *a* and *b* are determined by obtaining the "best fit" to experimental data. The values of *C*, *a*, and *b* are different for various flow configurations, such as flow inside pipes, flow along a flat plate, flow across the outside of a cylinder, and flow through a bank of tubes passing across the outside of the tubes, as well as being different between gases and liquids for the same configuration. Equations for the evaluation of *h* for the various flow configuration conditions and fluids are given in Section 4-2.1 for gases and in Section 4-2.2 for liquids.

Osborne Reynolds[2] suggested that momentum and heat in a fluid are transferred in the same way so that in geometrically similar systems, fluid friction and heat transfer should be proportional. Reynolds' reasoning was that from a molecular viewpoint fluid friction and heat conduction are controlled by the same law of motion. If two neighboring layers in a gas moving along a plane wall have different velocities, then in the course of normal molecular motion, molecules of the faster layer pass into the slower layer and are retarded, and molecules in the slower layer pass into the faster layer and are accelerated. A change of momentum occurs between the molecules, and this change in momentum acts like viscous friction. If a temperature difference exists between the two layers, the molecules will transfer heat between the two layers.

This theoretical concept is known as the Reynolds analogy, and from it the following equation for the heat transfer film coefficient *h* can be

---

[2]O. Reynolds, Phil. Transactions Roy. Soc. (London) A, Vol. 174, pt. 111, p. 935 (1883) or "Scientific Papers," Cambridge University Press, Vol. 11, pp. 51–105 (1900–1903).

derived.

$$h = \left(\frac{f}{8}\right) V c_p \rho \quad \left[\text{W/m}^2 \cdot \Delta_1 {}^{\circ}\text{C}\right] \tag{4-3}$$

where

    f = friction factor as shown in Figure 4-37
    V = velocity   [m/s]
    c_p = specific heat at constant pressure   [J/kg$\cdot\Delta_1{}^{\circ}$C]
    ρ = density   [kg/m^3]

This equation requires the Prandtl number to be 1.0, and in the derivation of this equation the flow in the fluid layers is treated as streamline flow. To establish correlations between test data and this theoretical model investigators have added a dimensionless function ψ to Eq. (4-3), resulting in

$$h = \frac{f}{8} V c_p \rho \psi \quad \left[\text{W/m}^2 \cdot \Delta_1 {}^{\circ}\text{C}\right] \tag{4-4}$$

where

    ψ = a dimensionless function that accounts for the deviation
        of conditions from those upon which Reynolds' analogy
        is based

    The theoretical model from the Reynolds analogy with the dimensionless correlation function [Eq. (4-4)] is not generally used for the evaluation of the heat transfer film coefficient h for a particular set of conditions. Instead, the generally accepted method, to evaluate the heat transfer film coefficient for a particular set of conditions, is to use Eq. (4-2), evaluating the constant C and the exponents a and b from appropriate plots of applicable test data.
    Forced convection heat transfer encompasses a wide range of conditions, including type of fluid (gas or liquid), flow passage configuration, and flow regime (laminar, transitional, or turbulent). As this text is applications oriented, rather than subject oriented, the following discussion relative to the evaluation of the convective film coefficient h has been organized, first in accordance with the type of fluid (gas or liquid), then by the flow passage configuration, and finally by the flow regime. As transitional flow regimes are usually avoided in heat exchanger design and are relatively complex, the scope of the following discussion in this text is limited to the laminar and turbulent flow regimes.

4-2.1 Equation for h with Gases

It is characteristic of gases that the Prandtl number for each gas remains nearly constant over a wide range of conditions. Consequently Eq. (4-2)

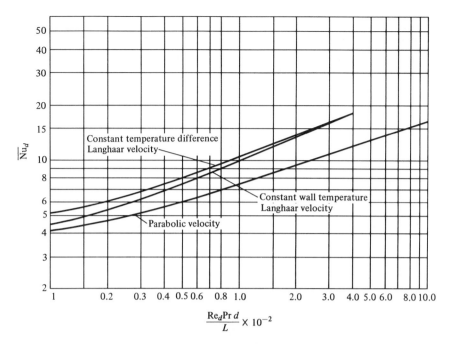

Figure 4-5 Mean Nusselt number with respect to tube length for gases in laminar flow. [SOURCE: W. M. Kays, "Numerical Solutions for Laminar Flow Heat Transfer in Circular Tubes," Trans. A.S.M.E., Vol. 77 (1955)]

can be written as

$$\text{Nu} = C(\text{Re})^a \quad [-] \tag{4-5}$$

GAS FLOW INSIDE PIPES

• *Laminar Flow.* Extensive numerical analysis has been made by Kays[3] for laminar flow of gases in the range of $\text{Re}(\text{Pr})D/L$ from 0 to 300. In this analysis Kays used actual velocity profiles, calculated earlier by Langhaar, because the assumption of a parabolic or a uniform profile is not satisfactory. The results of Kays' analysis are presented in Figure 4-5. The mean Nusselt number (the ordinate of the plot) is defined as

$$\overline{\text{Nu}_D} = \frac{\bar{h}D}{k} \tag{4-6}$$

where k is taken at the average fluid temperature. To determine the value of h for a particular set of conditions, the value of the abscissa for Figure 4-5 is calculated. Then by use of the curve considered most appropriate, a

[3]W. M. Kays, "Numerical Solution for Laminar Heat Flow Transfer in Circular Tubes," Trans. A.S.M.E., Vol. 77, pp. 1265–1274 (1955).

value of $\overline{\mathrm{Nu}_D}$ is obtained. From this, using Eq. (4-6), the value of h can be calculated. More extensive discussions on laminar flow heat transfer have been published by authors such as Kreith[4] and Simonson.[5]

Example Problem 4-1

Calculate the average value for h for atmospheric pressure air at 20°C flowing through a 1-cm diameter pipe 100 cm long at a velocity of 2 m/s with a parabolic velocity.

SOLUTION

SKETCH

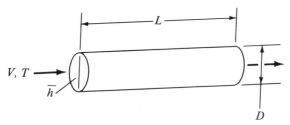

Figure 4-6

EQUATION

$$\bar{h} = \overline{\mathrm{Nu}}\frac{k}{D} \quad \left[\mathrm{W/m^2 \cdot \Delta_1 °C}\right] \quad \left[\mathrm{Eq.\ (3\text{-}2)}\right]$$

PLAN
Evaluate $\overline{\mathrm{Nu}}$ using the intersection of the value of $\mathrm{Re\,Pr}(D/L)$ and the parabolic velocity curve in Figure 4-5.

PARAMETERS
Evaluate $\mathrm{Re\,Pr}(D/L)$ for the problem condition

$$\mathrm{Re} = \frac{VD\rho}{\mu} \quad [-]$$

$V = 2$ m/s   (given)

$D = 1 \times 0.01 = 0.01$ m   (given)

---

[4] Frank Kreith and W. Z. Black, *Basic Heat Transfer*, pp. 221–226. Harper & Row, New York, 1980.
[5] John R. Simonson, *An Introduction to Engineering Heat Transfer*, pp. 87–91. McGraw-Hill, New York, 1967.

ρ at $20°C = 1.210 \text{ kg/m}^3$ (Appendix B, Table B-1)

μ at $20°C = 1.875 \times 10^{-5} \text{ Pa·s}$

$$\text{Re} = \frac{2 \times 0.01 \times 1.210}{1.875 \times 10^{-5}} \quad \frac{[\text{m/s}][\text{m}][\text{kg/m}^3]}{[\text{Pa·s}]} = \frac{[\text{kg·m/s}^2]}{[\text{Pa·m}^2]}$$

$$= \frac{[\text{N}]}{[\text{N/m}^2][\text{m}^2]} = \frac{[\text{N}]}{[\text{N}]} = [-]$$

$\text{Re} = 1290 \quad [-]$

$\text{Pr} = 0.710 \quad [-]$ (Appendix B, Table B-1)

$D = 0.01 \text{ m}$ (given)

$L = 0.10 \text{ m}$ (given)

$$\text{Re Pr}\frac{D}{L} = 1290 \times 0.710 \times \frac{0.01}{0.10} \quad [-][-][\text{m}]/[\text{m}] = [-]$$

$$= 91.6 \quad [-]$$

$$\text{Re Pr}\frac{D}{L} \times 10^{-2} = 0.92 \quad [-]$$

Using Figure 4-5, the intersection of the abscissa of 0.92 with the curve for the parabolic velocity gives $\text{Nu} = 7.3$.

SUBSTITUTION

$$\overline{h} = \overline{\text{Nu}}\frac{k}{D} \quad [\text{W/m}^2 \cdot \Delta_1°\text{C}]$$

$\overline{\text{Nu}} = 7.3 \quad [-]$

$k_{20°C} = 0.0256 \quad [\text{W/m} \cdot \Delta_1°\text{C}]$

           (Appendix B, Table B-1)

$D = 0.01 \text{ m} \qquad$ (given)

$$\overline{h} = \frac{7.3 \times 0.0256}{0.01} = \frac{[-][\text{W/m} \cdot \Delta_1°\text{C}]}{[\text{m}]} = [\text{W/m}^2 \cdot \Delta_1°\text{C}]$$

ANSWER

$$\overline{h} = 18.7 \quad [\text{W/m}^2 \cdot \Delta_1°\text{C}]$$

• *Turbulent Flow.* Test data for gas flow inside straight pipes, such as shown in Figure 4-7, evaluate the exponent a and constant C in Eq. (4-5) as 0.8 and 0.0204, respectively. The resulting fit with the experimental data is indicated by line $A - A$. Substituting these values for the constant C and

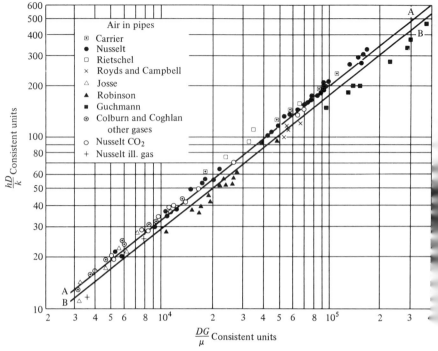

Figure 4-7 Heat transfer data for air inside pipes. (SOURCE: W. H. McAdams, *Heat Transmission*, 3rd ed. McGraw-Hill, New York, 1954)

the exponent a into Eq. (4-5) gives

$$\frac{hD}{k} = 0.0204 \left(\frac{DV\rho}{\mu} \right)^{0.8} \quad [-] \qquad (4-7)$$

Solving Eq. (4-7) for h gives the following equation for air flowing in turbulent flow inside straight pipes:

$$h = 0.0204 \frac{k}{D} \left(\frac{DV\rho}{\mu} \right)^{0.8} \quad [\text{W}/\text{m}^2 \cdot \Delta_1 \,^{\circ}\text{C}] \qquad (4-8)$$

For turbulent flow in coiled pipes the film coefficient h is higher than in straight pipes. Based on experiments Jeschke[6] proposed the equation

$$\frac{hD}{k} = \left(0.039 + \frac{0.138D}{D_H} \right) \left(\frac{DGc_p}{k} \right)^{0.76} \quad [-] \qquad (4-9)$$

where

D_H = diameter of the helix coil

[6]D. Jeschke, Z. Ver deut. Ing., Vol. 69, p. 1526 (1925).

For ordinary use, the h from Eq. (4-8) can be multiplied by $[1+3.5(D/D_H)]$.

To apply Eq. (4-8) to other gases, the numerical constant 0.0204 must be modified. Based upon Eq. (4-4), the modification of the numerical constant consists of multiplying it by the ratio of the Prandtl number of the gas concerned to that of air (0.69) raised to the exponent power b. From experimental data the value of the exponent b has been found to be 0.4. The numerical constant to substitute in Eq. (4-8) for use with a gas other than air is

$$C = 0.0204 \left(\frac{N_{Pr}}{0.69} \right)^{0.4} \quad [-] \qquad (4\text{-}10)$$

The Prandtl numbers of several gases are given in Table 4-1.

For an example of heat transfer in another gas, superheated steam is used. A plot of test data for superheated steam is shown in Figure 4-8. Modifying Eq. (4-8) for superheated steam results in

$$h = 0.0236 \frac{k}{D} \left(\frac{DG}{\mu} \right)^{0.8} \quad [\text{W/m}^2 \cdot \Delta_1 °\text{C}] \qquad (4\text{-}8a)$$

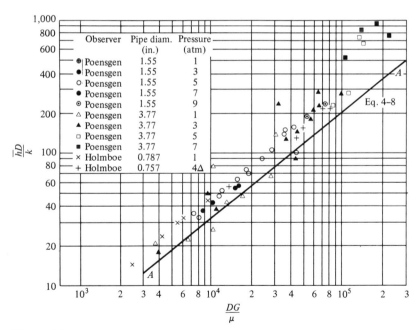

Figure 4-8   Data on cooling of superheated steam (without condensation) not corrected for gas radiation, compared with Eq. (4-8a). (SOURCE: W. H. McAdams, *Heat Transmission*, 3rd ed. McGraw-Hill, New York, 1954)

GAS FLOW ON FLAT SURFACES

In flow along a flat plate without being bounded by the walls of a channel or duct, the nature of the boundary layer determines whether the "fluid flow" is laminar or turbulent. The nature of the boundary layer is affected significantly by the geometry of the leading edge of the flat plate. Blunt leading edges can produce turbulent flow right from the start. Compared to tubes and ducts, relatively little experimental work has been done concerning the flow of a fluid parallel to a heating or cooling surface when the fluid is not bounded by the walls of a channel or duct.

The Reynolds number* used for characterizing the flow regime at any point along the surface of a flat plate is

$$\text{Re}_x = \frac{xV_0\rho}{\mu} \quad [-] \tag{4-11}$$

where

$x =$ the distance along the flow path over the surface from
  the leading edge of the flat plate   [m]
$V_0 =$ the velocity of the undisturbed fluid stream relative
  to the flat plate   [m/s]
$\rho =$ the gas density   [kg/m^3]
$\mu =$ the gas viscosity   [Pa·s]

The values of ρ and μ are determined at the average temperature between the flat plate surface and the undisturbed fluid stream.

In configurations, such as illustrated in Figure 4-9, wherein the flow over the first portion of the plate is laminar, calculations of the heat transfer should consider both flow regimes when they are present. Treating the plate as turbulent flow only, when a significant amount of laminar flow is present, may result in errors as large as 30%. The equations given in the following sections for flat surfaces are for gas flow velocities less than Mach 0.3. Equations for heat transfer in high speed flow should be used if the flow velocities exceed Mach 0.3.

· *Laminar Flow.* Laminar flow exists in the boundary below the critical Reynolds number for flow along an unbounded flat surface. Experimental values of the critical Reynolds number have ranged between 100,000 and 500,000 depending upon the starting conditions, with 300,000 usually being considered as a fair average of the critical Reynolds number value.

---

*It should be noted that for flow along an unbounded flat surface the Reynolds number is a function of the distance along the surface, while for flow bounded by walls of a channel or duct the Reynolds number is constant with respect to the length of the channel or duct.

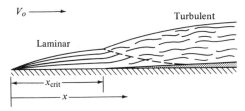

Figure 4-9 Forced flow over a flat plate.

To evaluate the heat transfer film coefficient for laminar flow of air, Jakob[7] suggests

$$\bar{h}=0.592\frac{k}{L}(\mathrm{Re}_L)^{0.5}\quad\left[\mathrm{W/m^2\cdot\Delta_1{}^\circ C}\right]\tag{4-12}$$

where

$\bar{h}$ = average heat transfer film coefficient
k = conductivity of the air at the average temperature
    between plate surface and the undisturbed stream
L = the total length of the flat plate in the direction
    of the air flow

To apply Eq. (4-12) to another gas, the numerical constant 0.590 must be modified by multiplying it by the ratio of the Prandtl number of the gas to that of air (0.69) raised to the exponent $1/3$ (cube root). Equation (4-12) then becomes

$$\bar{h}=0.592\frac{k}{L}(\mathrm{Re}_L)^{1/2}\left(\frac{\mathrm{Pr}}{0.69}\right)^{1/3}\quad\left[\mathrm{W/m^2\cdot\Delta_1{}^\circ C}\right]\tag{4-13}$$

When hydrodynamic starting sections are applied to the flat plates, the heat transfer equations can be modified as discussed by Jakob.[7]

• *Turbulent Flow.* The local convection film coefficient for air, based upon Johnson and Rubesin[8] is

$$h_x=0.0260\frac{k}{x}(\mathrm{Re}_x)^{0.8}\quad\left[\mathrm{W/m^2\cdot\Delta_1{}^\circ C}\right]\tag{4-14}$$

In order to obtain the heat transferred between the flat plate and the air flow, an average value of h should be used. The following formula for the average value of h considers both the laminar and turbulent flow sections along the flat plate, and assumes there is an instantaneous

[7] M. Jakob, *Heat Transfer*, Vol. I, p. 554. Wiley, New York, 1956.
[8] H. A. Johnson and M. W. Rubesin, "Aerodynamic Heating and Convective Heat Transfer," Trans. A.S.M.E., Vol. 68, p. 124 (1946).

transition between laminar and turbulent flow at the critical Reynolds number.

$$\bar{h} = 0.0326 \frac{k}{L} \left[\text{Re}_L^{0.8} - \text{Re}_{\text{crit}}^{0.8} + 18.2(\text{Re}_{\text{crit}})^{0.5} \right] \quad \left[\text{W/m}^2 \cdot \Delta_1 ^\circ\text{C} \right] \quad (4\text{-}15)$$

where Re_{crit} is the critical Reynolds number.

Both Eqs. (4-14) and (4-15) can be modified for use with another gas by multiplying the numerical constant by the ratio of the Prandtl number of the gas to that of air (0.69) raised to the $1/3$ power, as on page 171 for Eq. (4-13).

Example Problem 4-2

Atmospheric pressure air at 50°C is flowing over a 30°C flat plate 3 m long at a velocity of 10 m/s. Using Eq. (4-15) calculate the value of $\bar{h}$ when the critical Reynolds number $\text{Re}_{\text{crit}} = 300{,}000$.

SOLUTION

EQUATION

$$\bar{h} = 0.0326 \frac{k}{L} \left[\text{Re}_L^{0.8} - \text{Re}_{\text{crit}}^{0.8} + 18.2(\text{Re}_{\text{crit}})^{0.5} \right] \quad \left[\text{W/m}^2 \cdot \Delta_1 ^\circ\text{C} \right]$$

$$\left[\text{Eq. } (4\text{-}15) \right]$$

PARAMETERS

$k = $ (obtain from Table B-1, Appendix B)

Interpolate for average temperature.

$$T_{av} = \frac{50 + 30}{2} = 40^\circ\text{C} \rightarrow 313 \text{ K}$$

k	T
0.02624	300
	313
0.03003	350

$$k = k_{300} + \Delta k$$

$$k_{300} = 0.02624 \text{ W/m} \cdot \Delta_1 ^\circ\text{C}$$

$$\Delta k = (0.03003 - 0.02624) \frac{(313 - 300)}{(350 - 300)}$$

$$= 0.00379 \left[\frac{13}{50} \right]$$

$$\Delta k = 0.00099$$

$k = 0.02624 + 0.00099$

$k = 0.02723 \text{ W/m} \cdot \Delta_1 \degree \text{C}$

$L = 3 \text{ m}$ (given)

$\text{Re}_L = \dfrac{LV\rho}{\mu}$ $[-]$ $\big[\text{Eq. (4-11)}\big]$

$V = 10 \text{ m/s}$ (given)

$\rho = $ (obtain from Table B-1, Appendix B)

Interpolate for average temperature $40 \degree \text{C} \rightarrow 313$ K.

ρ	T
1.1774	300
	313
0.9980	350

$\rho = \rho_{300} + \Delta\rho \quad [\text{kg/m}^3]$

$\rho_{300} = 1.1774 \text{ kg/m}^3$ (Table B-1)

$\Delta\rho = (1.1774 - 0.9980)\dfrac{(313 - 300)}{(350 - 300)}$

$\quad = 0.1794 \left[\dfrac{13}{50}\right]$

$\Delta\rho = 0.0466$

$\rho = 1.1774 - 0.0466$

$\rho = 1.1308 \text{ kg/m}^3$

$\mu = $ (obtain from Table B-1, Appendix B)

Interpolate for average temperature $40 \degree \text{C} \rightarrow 313$ K.

μ	T
1.983	300
	313
2.075	350

$\mu = \mu_{300} + \Delta\mu \quad [\text{Pa} \cdot \text{s}]$

$\mu_{300} = 1.983 \times 10^{-5} \text{ Pa} \cdot \text{s}$

$\Delta\mu = (2.075 - 1.983) \times 10^{-5}\dfrac{(313 - 300)}{(350 - 300)}$

$\quad = (0.092)\dfrac{13}{50} \times 10^{-5}$

$\Delta\mu = 0.024 \times 10^{-5} \text{ Pa} \cdot \text{s}$

$$\mu = 1.983 \times 10^{-5} + 0.024 \times 10^{-5} \ \text{Pa} \cdot \text{s}$$

$$\mu = 2.007 \times 10^{-5} \ \text{Pa} \cdot \text{s}$$

$$\text{Re}_L = \frac{(3)(10)(1.1308)}{2.007 \times 10^{-5}} \quad \frac{[\text{m}][\text{m/s}][\text{kg/m}^3]}{\text{Pa} \cdot \text{s}}$$

$$\text{Re}_L = 1.69 \times 10^6 \quad [-] \frac{[\text{kg/ms}]}{[\text{kg/ms}^2][\text{s}]} = \frac{[\text{kg/ms}]}{[\text{kg/ms}]} = [-]$$

$$\text{Re}_{\text{crit}} = 300,000 \quad [-] \quad \text{(given)}$$

SUBSTITUTION

$$\bar{h} = 0.0326 \frac{0.02723}{3} \left[1.69 \times 10^{6^{0.8}} - 300,000^{0.8} + 18.2(300,000)^{0.5} \right]$$

$$[-] \frac{\text{W/m} \cdot \Delta_1 {}^{\circ}\text{C}}{\text{m}} = [-] = \left[\text{W/m}^2 \cdot \Delta_1 {}^{\circ}\text{C} \right]$$

$$= 0.00029(96,008 - 24,082 + 9,969)$$

$$= 0.00029(81,895)$$

ANSWER

$$\bar{h} = 23.7 \ \text{W/m}^2 \cdot \Delta_1 {}^{\circ}\text{C}$$

GAS FLOW ACROSS A CYLINDER

The two characteristics of primary interest in heat transfer between a gas and the surface of a cylinder in crossflow are

- the average value of h
- the local value of h at any point around the surface of the cylinder

The average value of h, based upon the total surface area of the cylinder, is used to calculate the total amount of heat that is transferred for the set of conditions of interest. For the majority of problems, this is the only value of h that need be considered.

The local value of h becomes important if the local temperatures in the boundary layers or in the wall of the cylinder have temperature limits that should not be exceeded. In these cases the local value of h may become the controlling factor in the design for the heat transfer between the gas and the cylinder.

The gas flow pattern across the cylinder is characterized by the use of a Reynolds number that is based upon the

- diameter of the cylinder
- the velocity of the undisturbed gas stream, relative to the cylinder
- the gas properties at the average temperature between the gas free stream and the surface of the cylinder

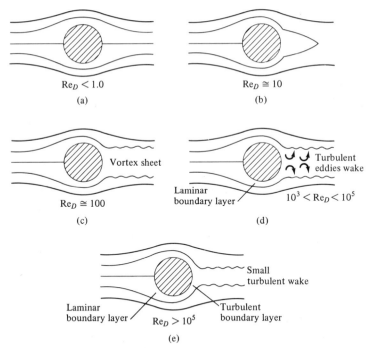

$Re_D < 1.0$

(a)

$Re_D \cong 10$

(b)

Vortex sheet

$Re_D \cong 100$

(c)

Laminar
boundary layer

Turbulent
eddies wake

$10^3 < Re_D < 10^5$

(d)

Small
turbulent wake

Laminar
boundary layer

Turbulent
boundary layer

$Re_D > 10^5$

(e)

Figure 4-10 Sketches illustrating flow pattern for crossflow over a circular cylinder at various Reynolds numbers. (SOURCE: F. Kreith, *Principles of Heat Transfer*, 3rd ed., p. 462. Harper & Row, New York, 1973)

The characteristic gas flow patterns as related to the Reynolds number are illustrated in Figure 4-10. Both the boundary layer characteristics and associated values of Reynolds number are markedly different from those for flow inside a circular pipe, as discussed on page 160. The key points in the gas flow across the cylinder are discussed in the following paragraphs.

At Reynolds numbers of the order of unity or less, the fluid flow adheres to the surface of the cylinder and the streamlines of the flow follow potential flow predictions. Heat is transferred by conduction alone.

At Reynolds numbers of the order of 10, two weak eddies appear in the rear of the cylinder. Heat is transferred mainly by conduction.

At Reynolds numbers of the order of 100, vortices separate alternately from both sides of the cylinder, forming a wake pattern that extends a considerable distance downstream from the cylinder. This vortex pattern is called the von Karman vortex-streets, because Theodore von Karman made extensive studies of the shedding of vortices from bluff bodies. An increased portion of the heat is transferred by convection.

At Reynolds numbers in the range of 10^3 to 10^5, the laminar boundary layer usually extends from the stagnation point around the cylinder for 80° to 85°. At this point boundary layer separation occurs with the formation of turbulent eddies in the wake. Almost all of the heat transfer is by convection.

At Reynolds numbers greater than 10^5, a turbulent boundary layer extends over the surface of the cylinder beyond the laminar boundary layer. The separation point between the turbulent boundary layer and the cylinder moves further aft, reducing the size of the wake.

The characteristics of the average value of h as well as the local value of h are both a function of the gas flow boundary layer pattern around the cylinder. These values of h are related to the Reynolds number of the gas flow, as discussed in the following sections.

• *The Average Value of h for Circular Cylinders.* The equations presented in this section for the average value of h are based upon many measurements taken by various investigators. These results for the flow of air normal to a cylinder are shown in Figure 4-11 as a plot of Nusselts number versus Reynolds number. The curve drawn through the test data points has a continually increasing slope with an increase of Reynolds number. There is no discontinuity in a transition zone between laminar and turbulent flow, as was characteristic of flow inside circular pipes.

For convenience, this curve is described mathematically by dividing it into a number of segments, each of which is described by an equation giving a straight line in the log-log plot of Figure 4-11. The equation used has the following form:

$$\bar{h} = \frac{k}{D} B (\mathrm{Re})^n \quad \left[\mathrm{W/m^2 \cdot \Delta_1 {}^\circ C} \right] \tag{4-16}$$

where

$$k = \text{thermal conductivity of the fluid}$$
$$D = \text{diameter of the cylinder}$$
$$\mathrm{Re} = \text{Reynolds number based upon the diameter of the cylinder}$$
$$\text{and the undisturbed free stream velocity of the fluid}$$
$$B \text{ and } n = \text{dimensionless numbers}$$

The values of k and Re are at the film temperature, which is taken as the average between the surface of the cylinder and the fluid free stream. The values of B and n for applicable ranges of Reynolds number, as established by Hilpert,[9] are given in Table 4-2.

For a gas having a Prandtl number different from air, Eq. (4-16) should be modified as indicated in Eq. (4-17).

$$\bar{h} = \frac{k}{D} B (\mathrm{Re})^n \left(\frac{\mathrm{Pr}}{0.69} \right)^{1/3} = 1.13 \frac{k}{D} B (\mathrm{Re})^n (\mathrm{Pr})^{1/3} \quad \left[\mathrm{W/m^2 \cdot \Delta_1 {}^\circ C} \right]$$
$$\tag{4-17}$$

[9] R. Hilpert, Forschung a.d. Geb. d. Ingenieurwes, Vol. 4, p. 215 (1933).

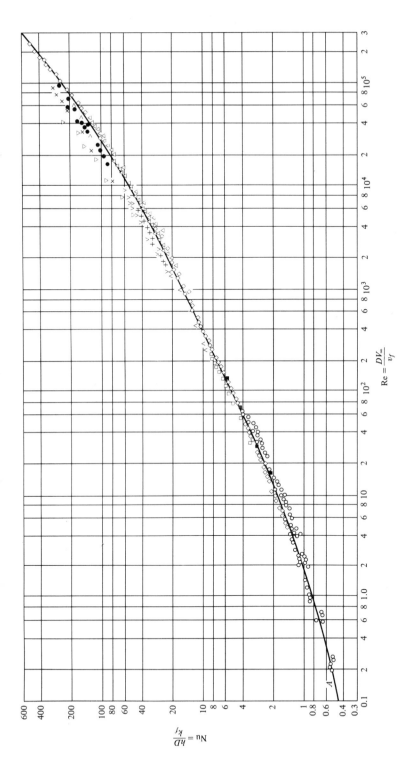

Figure 4-11   Average Nusselt number versus Reynolds number for a circular cylinder in air, placed normal to the flow. (SOURCE: W. H. McAdams, *Heat Transmission*, 3rd ed. McGraw-Hill, New York, 1954) D = diameter of cylinder, V_∞ = fluid full stream velocity, ν_f = fluid kinematic viscosity, h = heat transfer film coefficient, k_f = fluid thermal conductivity, Re = Reynolds number, Nu = Nusselt number

Table 4-2   CONSTANTS FOR EQ. (4-16)

RANGE OF Re	n	B
1–4	0.330	0.891
4–40	0.385	0.821
40–4000	0.466	0.615
4000–40,000	0.618	0.174
40,000–250,000	0.805	0.0239

SOURCE: R. Hilpert.[9]

Example Problem 4-3

A 5-cm o.d. pipe is placed normal to a stream of air at 20°C and 1 atm pressure moving at a velocity of 5 m/s. The outside surface of the tube is maintained at 60°C by electric heaters inside the pipe. Determine the surface coefficient and the heat loss per unit length.

SOLUTION

SKETCH

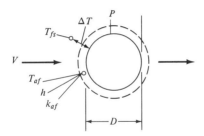

Figure 4-12

EQUATIONS

$$(A)\ \bar{h} = \frac{k_{af}}{D} B\mathrm{Re}^n \quad [W/m^2 \cdot \Delta_1{}^\circ C] \quad [Eq.\ (4\text{-}16)]$$

$$(B)\ \dot{Q}/L = hP\Delta T \quad [W/m] \quad [Eq.\ (4\text{-}1)]$$

PARAMETERS FOR (A)

$T_{af} = 40°C \rightarrow 313\ K$ (given)

$k_{af} = 0.02723\ W/m \cdot \Delta_1{}^\circ C$ air at 40°C (Table B-1, Appendix B)

$B = 0.174 \quad [-]$ (From Table 4-2)

$$\mathrm{Re} = \frac{VD\rho}{\mu} \quad [-]$$

$V = 5.0 \quad [\text{m/s}] \qquad \text{(given)}$

$D = 5.0 \quad [\text{cm}] \rightarrow 0.05 \quad [\text{m}] \qquad \text{(given)}$

$\rho = 1.1308 \text{ kg/m}^3 \text{ air at 1 atm } 40°\text{C}$

$\mu = 2.007 \text{ Pa} \cdot \text{s} \qquad \text{(Table B-1, Appendix B)}$

$$\text{Re} = \frac{(5.0)(0.05)(1.1308)}{(2.007 \times 10^{-5})} \quad \frac{[\text{m/s}][\text{m}][\text{kg/m}^3]}{[\text{Pa} \cdot \text{s}]}$$

$$= \frac{[\text{kg} \cdot \text{m/s}^3]}{[\text{Pa} \cdot \text{m}^2]} = \frac{[\text{N}]}{[\text{N}]} = [-]$$

$\text{Re} = (1.41 \times 10^4) \quad [-]$

$n = 0.618 \quad [-] \qquad \text{(From Table 4-2)}$

$D = 5.0 \text{ cm} \rightarrow 0.05 \text{ m} \qquad \text{(given)}$

SUBSTITUTION FOR (A)

$$\bar{h} = \frac{(0.02723)}{(0.05)} (0.174)(1.41 \times 10^4)^{0.618}$$

$$\frac{[\text{W/m} \cdot \Delta_1°\text{C}]}{[\text{m}]} [-] = [\text{W/m}^2 \cdot \Delta_1°\text{C}]$$

ANSWER FOR (A)

$\bar{h} = 34.74 \quad [\text{W/m}^2 \cdot \Delta_1°\text{C}]$

PARAMETERS FOR (B)

$P = \pi D \quad [\text{m}]$

$D = 0.05 \quad [\text{m}] \qquad \text{(given)}$

$P = \pi(0.05)$

$P = 0.1571 \quad [\text{m}]$

$\Delta T = T_s - T_{f_s} \quad [\Delta°\text{C}]$

$T_s = 60 \quad [°\text{C}] \qquad \text{(given)}$

$T_{f_s} = 20 \quad [°\text{C}] \qquad \text{(given)}$

$\Delta T = (60) - (20) \quad [°\text{C}] - [°\text{C}] = [\Delta°\text{C}]$

$\Delta T = (40) \quad (\Delta°\text{C})$

SUBSTITUTION FOR (B)

$$\dot{Q}/L=(34.74)(0.1571)(40) \quad [W/m^2 \cdot \Delta_1 {}^\circ C][m][\Delta {}^\circ C]=[W/m]$$

ANSWER FOR (B)

$$\dot{Q}/L=218.3 \quad [W/m]$$

• *The Average Value of h for Noncircular Cylinders.* The average value of h based upon the wetted surface of a noncircular cylinder can be calculated using Eq. (4-16) or (4-17) with the constants n and B as given in Table 4-3. This table presents experimental data from Hilpert and Reiher as correlated by Jakob for use with these equations. The Reynolds number listed in Table 4-3 is based upon the dimension of the noncircular cylinder normal to the direction of the gas flow in lieu of the diameter of the circular cylinder.

• *The Local Value of h.* The unit surface conductance is not uniform around the surface of the cylinder because of variations in the boundary layer and the fluid flow pattern around the surface of the cylinder. The local film coefficient characteristically has two minima around the cylinder in the region of Reynolds numbers wherein the boundary layer has a transition from laminar to turbulent flow without separation in the boundary layer.

The first minimum occurs at the transition point where the laminar has reached its maximum thickness. As the transition from laminar to

Table 4-3 CONSTANTS FOR NONCIRCULAR CYLINDER CONFIGURATIONS FOR USE IN EQS. (4-16) AND (4-17)

FLOW DIRECTION AND PROFILE		N_{Re} FROM	N_{Re} TO	n	B	OBSERVER
→	◇	5,000	100,000	0.588	0.222	Hilpert
→	◯	2,500	15,000	0.612	0.224	Reiher
→	◇	2,500	7,500	0.624	0.261	Reiher
→	◯	5,000	100,000	0.638	0.138	Hilpert
→	◯	5,000	19,500	0.638	0.144	Hilpert
→	□	5,000	100,000	0.675	0.092	Hilpert
→	□	2,500	8,000	0.699	0.160	Reiher
→	I	4,000	15,000	0.731	0.205	Reiher
→	◯	19,500	100,000	0.782	0.035	Hilpert
→	◯	3,000	15,000	0.804	0.085	Reiher

SOURCE: Jakob[10], Hilpert[9], and Reiher.[11]

[10]M. Jakob, *Heat Transfer*, Vol. 1, p. 562. Wiley, New York, 1956.
[11]O. Knoblauch and H. Reiher, "Waermeuebertragung," in W. Wien and F. Harms, Handbuch d. Experimentalphysik, Vol. 9, pt. 1, p. 189, Leipzig (1925).

turbulent flow in the boundary layer progresses, the local film coefficient increases, reaching a maximum approximately at the point where the boundary layer becomes fully turbulent.

The second minima is at the separation point where the turbulent boundary layer separates from the surface of the cylinder. This point is about 140° from the stagnation point (where the fluid flow vector impinges upon the cylinder). The local film coefficient then increases to another maximum at the rear stagnation point. A plot of Nusselts number versus the angular position from the stagnation point for various Reynolds numbers from 70,800 to 219,000 is shown in Figure 4-13 presenting data from Giedt[12].

• *The Effect of Free Stream Turbulence.* Turbulence in the free stream of a gas flowing across a cylinder has a significant effect upon the heat transfer. In general, the average value of h increases with increased turbulence levels as discussed in the following paragraphs. The equations for h given in the previous sections [Eqs. (4-16) and (4-17)] are for conditions of no free stream turbulence.

Turbulence in the free stream of the fluid flow is characterized primarily by

- intensity
- scale

The intensity of the turbulence is represented by the energy of the fluid flow velocity vectors normal to the primary flow velocity vector. It is usually quantized by a percent of the total dynamic energy of the fluid flow.

The scale is a measurement of the geometric size of the turbulence eddies. It is usually referred to in qualitative terms, such as "small" scale or "large" scale. The reference dimension for comparison is the diameter of the cylinder.

The free stream turbulence increases the local value of h both at the stagnation point and in the laminar boundary layer. When the free stream turbulence causes an earlier transition from the laminar to a turbulent boundary layer, significant increases in the local value of h can result. Little, if any, effect has been noted in the region aft of the boundary layer separation point.

In general the relative size of the "scale" of the turbulence has little, if any, effect. However, the intensity level does, and the effect is a function of the Reynolds number of the flow. At Reynolds numbers up to 2000, only small increases are noted. The effect increases with increasing Reynolds

[12] W. H. Giedt, "Investigation of Variation of Point Unit—Heat-Transfer Coefficient Around a Cylinder Normal to an Air Stream," Trans. A.S.M.E., Vol. 71 (1949).

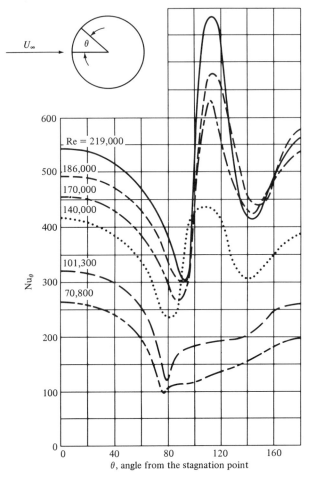

Figure 4-13 Variation of the local Nusselt number on the surface of a long cylinder normal to the flow. [SOURCE: W. H. Giedt, "Investigation of Variation of Point Unit Heat Transfer Coefficient Around a Cylinder Normal to an Air Stream," Trans. A.S.M.E., Vol. 71 (1949)]

number, and large increases in h can be experienced at Reynolds numbers over 140,000 with turbulence of low intensity. References concerned with the effect of turbulence are indicated in Table 4-4 as a guide to the reader. A quantative treatment of the effects of turbulence is outside the scope of this introductory text.

GAS FLOW THROUGH TUBE BANKS
Many form of heat exchangers have a gas flow through a tube bundle passing across the outside surface of the tubes. The outside surface of the

Table 4-4 SOME INVESTIGATIONS OF THE EFFECT
OF TURBULENCE UPON THE HEAT TRANSFER IN
GAS FLOW ACROSS CYLINDERS

REYNOLDS NUMBER RANGE	SCOPE OF INVESTIGATION	INVESTIGATOR
400–20,000	Various intensities	Comings et al.[13]
70,000–220,000	Two intensities (1 and 4%)	Giedt[14]
125,000–310,000	Intensities up to 2.7%	Kestin and Maeder[15]

[13] E. W. Comings, J. T. Clapp, and J. F. Taylor, Ind. Eng. Chem., Vol. 40, p. 1076 (1948).
[14] W. H. Giedt, J. Aeron. Sci., Vol. 18, p. 725 (1951).
[15] J. Kestin and P. F. Maeder, NACA Tech. Note 4018 (1957).

tubes may be plain, or it may have fins to provide extended surface for the heat transfer into the gas stream. Both the plain tubes and finned tubes will be discussed in the following sections.

• *Plain Tube Bundles.* The heat transfer in gas flow through tube bundles across the outside of the tubes depends largely upon the Reynolds number of the gas flow and the tube bank geometry (arrangement and tube spacing).

The Reynolds number is based upon the

• mass flow rate through the minimum free area between tubes
• diameter (outside) of the tube

The tube bank geometry is described by the tube arrangement and the tube spacing. The arrangement of the tubes is categorized in being either "in line" or "staggered." These two arrangements are illustrated in Figures 4-14 and 4-15, which also give the tube spacing designations.

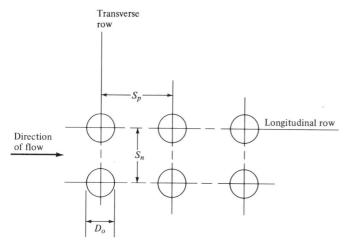

Figure 4-14 Sketch illustrating nomenclature for in-line tube arrangements.

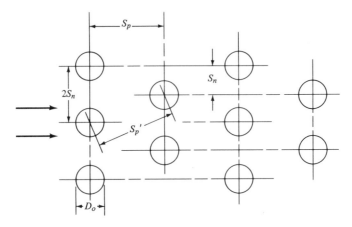

Figure 4-15   Sketch illustrating nomenclature for staggered tube arrangements.

The "in-line" tube bank geometry and spacing nomenclature is shown in Figure 4-14. The tubes are in a rectangular pattern. The tube spacing (center to center) parallel to the fluid flow direction is designated S_p. The tube spacing (center to center) normal to the fluid flow direction is designated S_n. The tube diameter is designated D_o. In describing the tube bank geometry, the term "pitch" is often used. The pitch of the tube pattern is the ratio of the tube spacing (center-to-center) to the tube diameter, as indicated below:

parallel to the fluid flow $= S_p/D_o$
normal to the fluid flow $= S_n/D_o$

The minimum flow cross-sectional area for determining the air mass flow rate (G) or the flow velocity (V) for evaluating the gas flow Reynolds number is at the point where the distance between the tube surfaces is $(S_n - D_o)$.

The "staggered" tube bank geometry and spacing nomenclature is shown in Figure 4-15. The tubes are arranged in an isosceles triangular pattern, with the tube outside diameter designated as D_o. The length of the side of the triangular pattern is designated as S_p', and the height of the triangular pattern, which is parallel to the fluid flow direction, is designated as S_p. The base of the isosceles triangle is designated as $2S_n$ because it is twice the tube spacing normal to the fluid flow. The minimum free flow area between the tubes, for use in the calculation of the fluid flow Reynolds number, is based upon the smaller of the two flow cross-sectional areas, where the distances between the tube surfaces are

1. $2S_n - D_o$
2. $2[S_p' - D_o]$

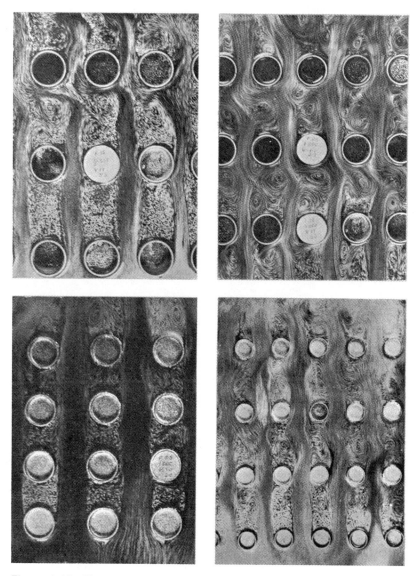

Figure 4-16 Flow patterns for in-line tube bundles. [SOURCE: R. D. Wallis, "Photographic Study of Fluid Flow Between Banks of Tubes," Engineering, Vol. 148 (1933)]

The general characteristics of the gas flow pattern through an in-line tube bank is shown in Figure 4-16 and a staggered tube bank in Figure 4-17. As the Reynolds number of the gas flow increases the in-line tube bank shows changes in the heat transfer characteristics analogous to the changes in flow through the inside of tubes as the flow regime changes from laminar through transitional flow to turbulent flow. In contrast, the

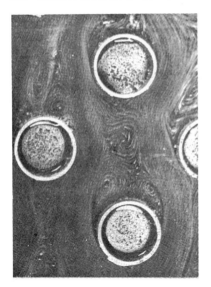 

Figure 4-17 Flow patterns for staggered tube bundles. [SOURCE: R. D. Wallis, "Photographic Study of Fluid Flow Between Banks of Tubes," Engineering, Vol. 148 (1933)]

staggered tube bank shows changes in the heat transfer characteristics analogous to the changes in flow across the outside of a single cylinder.

Investigators have generally correlated the experimental data into two groups of correlations for

- tube banks with 10 or more rows
- tube banks with less than 10 rows

Table 4-5 GRIMISON'S CORRELATION FOR HEAT TRANSFER FOR TUBE BANKS OF 10 ROWS OR MORE

		S_n/D							
		1.25		1.5		2.0		3.0	
ARRANGEMENT	S_p/D	C	n	C	n	C	n	C	n
In-line	1.25	0.348	0.592	0.275	0.608	0.100	0.704	0.0633	0.752
	1.5	0.367	0.586	0.250	0.620	0.101	0.702	0.0678	0.744
	2.0	0.418	0.570	0.299	0.602	0.229	0.632	0.198	0.648
	3.0	0.290	0.601	0.357	0.584	0.374	0.581	0.286	0.608
Staggered	0.6	—	—	—	—	—	—	0.213	0.636
	0.9	—	—	—	—	0.446	0.571	0.401	0.581
	1.0	—	—	0.497	0.558				
	1.125	—	—	—	—	0.478	0.565	0.518	0.560
	1.25	0.518	0.556	0.505	0.554	0.519	0.556	0.522	0.562
	1.5	0.451	0.568	0.460	0.562	0.452	0.568	0.488	0.568
	2.0	0.404	0.572	0.416	0.568	0.482	0.556	0.449	0.570
	3.0	0.310	0.592	0.356	0.580	0.440	0.562	0.421	0.574

SOURCE: From Grimison.[17]

• *Tube Banks with 10 or More Rows.* To calculate the average value of h for air flowing through tube banks Grimison[17] correlated test results available for the Reynolds number range of 2000 to 40,000 and for banks of 10 rows or more, using Eq. (4-18).

$$\bar{h} = \frac{k}{D_o} C(\mathrm{Re})^n \quad \left[\mathrm{W/m^2 \cdot \Delta_1 \,^\circ C} \right] \tag{4-18}$$

The air properties are evaluated at the average temperature between the tube wall and the air stream. The values of C and n given for in-line and staggered arrangements with various tube spacings are presented in Table 4-5.

Example Problem 4-4

Atmospheric pressure air at 10°C is flowing through a staggered tube bank with 12 rows of 1.5-cm diameter tubes arranged in a pattern (see Figure 4-15) of $S_n/D_o = 1.5$ and $S_p/D_o = 1.25$. The temperature of the tube surface is 100°C. The air approach velocity to the tube bank is 5 m/s. Calculate the average heat transfer film coefficient $\bar{h}$ using the Grimison correlation.

---

[16] R. D. Wallis, "Photographic Study of Fluid Flow Between Banks of Tubes," *Engineering*, Vol. 148 (1933).
[17] E. D. Grimison, *Trans. ASME*, Vol. 59, p. 583 (1937).

SOLUTION

SKETCH

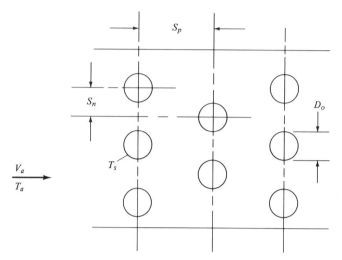

Figure 4-18

EQUATION

$$\bar{h} = \frac{k}{D_o} C(\mathrm{Re})^n \quad [\mathrm{W/m^2 \cdot \Delta_1 ^\circ C}] \qquad [\text{Eq. (4-18)}]$$

PARAMETERS

$k = (\text{interpolate from Table B-1 in Appendix B for } T_{\mathrm{av}})$

$$T_{\mathrm{av}} = \frac{T_a + T_s}{2}$$

$T_a = 10^\circ C \qquad (\text{given})$

$T_s = 100^\circ C \qquad (\text{given})$

$$T_{\mathrm{av}} = \frac{10 + 100}{2} = 55^\circ C \rightarrow 328 \text{ K}$$

$k = k_{300} + \Delta k \quad [\mathrm{W/m \cdot \Delta_1 ^\circ C}]$

k	T
0.02624	300
	328
0.03003	350

$k_{300} = 0.02624 \text{ W/m} \cdot \Delta_1 {}^\circ\text{C}$

$\Delta k = (0.03003 - 0.02624) \dfrac{(328 - 300)}{(350 - 300)}$

$\qquad = 0.00379 \left[\dfrac{28}{50} \right]$

$\Delta k = 0.00212 \text{ W/m} \cdot \Delta_1 {}^\circ\text{C}$

$k = 0.02624 + 0.00212$

$k = 0.02836 \text{ W/m} \cdot \Delta_1 {}^\circ\text{C}$

$D_o = 1.5 \text{ cm} \rightarrow 0.015 \text{ m}$ (given)

$C = 0.505 \quad [-] \qquad$ Table 4-5 at $S_n/D = 1.5$

$\qquad\qquad\qquad\qquad$ staggered $S_p/D = 1.25$

Calculate Re for maximum velocity.

$\text{Re} = \dfrac{V_m D_o \rho}{\mu} \quad [-]$

$V_m =$ maximum velocity in tube bundle

$\qquad = V_a \dfrac{2 S_n}{W_m} \quad [\text{m/s}]$

$V_a = 5 \text{ m/s} \qquad$ (given)

$S_n = 1.5 D_o \qquad$ (given)

$W_m =$ minimum flow passage width

For the location and value of W_m, compare (Figure 4-15)

$2 S_n - D_o = W'_m$

$2(S'_p - D_o) = W''_m$

$W'_m = 2 S_n - D_o$

$S_n = 1.5 \, D_o \qquad$ (given)

$W'_m = 2(1.5 D_o) - D_o$

$W'_m = 2 D_o$

$W''_m = 2(S'_p - D_o)$

$\qquad\qquad S'_p = \left[S_p^2 + S_n^2 \right]^{1/2}$

$\qquad\qquad S_p = 1.25 D_o \qquad$ (given)

$\qquad\qquad S_n = 1.5 D_o \qquad$ (given)

$$S'_p = \left[(1.25D_o)^2 + (1.5D_o)^2\right]^{1/2}$$
$$= \left[1.56D_o^2 + 2.25D_o^2\right]^{1/2}$$
$$= \left[3.81D_o^2\right]^{1/2}$$
$$S'_p = 1.95D_o$$
$$W'''_m = 2(S'_p - D_o) = 2(1.95D_o - D_o) = 1.90D_o$$

Therefore, smaller flow cross section W_m is at

$$W'''_m = 2(S'_p - D_o) \quad \text{and} \quad 1.90D_o$$

$$V_m = V_a \frac{2S_n}{W_m} \quad [\text{m/s}]$$

$$V_a = 5 \text{ m/s} \quad \text{(given)}$$
$$S_n = 1.5D_o \quad \text{(given)}$$
$$W_m = 1.9D_o \quad \text{(calculated above)}$$
$$V_m = \frac{(5)(2)(1.5D_o)}{1.90D_o} \quad \frac{[\text{m/s}][-][\text{cm}]}{[-][\text{cm}]} = [\text{m/s}]$$

$$V_m = 7.9 \text{ m/s}$$
$$D_o = 1.5 \text{ cm} \rightarrow 0.015 \text{ m} \quad \text{(given)}$$

$\rho = \left[\text{interpolate from Table B-1, Appendix B for average temperature } (T_{av})\right]$

$$T_{av} = \frac{T_a + T_s}{2} = \frac{10 + 100}{2} = 55°\text{C} \rightarrow 328 \text{ K}$$

$$\rho = \rho_{300} - \Delta\rho \quad [\text{kg/m}^3]$$

ρ	T
1.1774	300
	328
0.9980	350

$$\rho_{300} = 1.1774 \text{ kg/m}^3$$

$$\Delta\rho = (1.1774 - 0.9980)\frac{(328 - 300)}{(350 - 300)}$$
$$= 0.1794\left[\frac{28}{50}\right]$$

$$\Delta\rho = 0.1005 \text{ kg/m}^3$$
$$\rho = 1.1774 - 0.1005 \quad [\text{kg/m}^3] - [\text{kg/m}^3]$$

$\rho = 1.0769 \quad [\text{kg/m}^3]$

$\mu = (\text{interpolate from Table B-1, Appendix B for } T_{av} = 328 \text{ K})$

$\mu = \mu_{300} + \Delta\mu \quad [\text{Pa·s}]$

μ	T
1.983×10^{-5}	300
	328
2.075×10^{-5}	350

$\mu_{300} = 1.983 \times 10^{-5} \text{Pa·s}$

$\Delta\mu = (2.075 - 1.983) \times 10^{-5} \dfrac{(328 - 300)}{(350 - 300)}$

$= 0.092 \times 10^{-5} \left[\dfrac{28}{50} \right] \text{Pa·s}$

$\Delta\mu = 0.0515 \times 10^{-5} \text{ Pa·s}$

$\mu = 2.035 \times 10^{-5} \; [\text{Pa·s}]$

$\text{Re} = \dfrac{(7.9)(0.015)(1.0769)}{2.035 \times 10^{-5}} \quad \dfrac{[\text{m/s}][\text{m}][\text{kg/m}^3]}{[\text{Pa·s}]}$

$= \dfrac{[\text{kg/m·s}]}{[\text{kg/ms}^2][\text{s}]} = [-]$

$\text{Re} = 6271 \quad [-]$

$n = 0.554 \quad [-] \qquad \text{(Table 4-5)} \qquad S_n / D = 1.5$

$\qquad\qquad\qquad\qquad \text{(staggered)} \qquad S_{p_1} / D = 1.25$

SUBSTITUTION

$\bar{h} = -\dfrac{(0.02836)}{(0.015)} (0.505)(6271)^{0.554} \quad \dfrac{[\text{W/m·}\Delta_1\text{°C}]}{[\text{m}]} [-][-]$

$= (0.9548)(127.0)$

ANSWER

$\bar{h} = 121.3 \quad [\text{W/m}^2 \cdot \Delta_1 \text{°C}]$

TUBE BANKS WITH LESS THAN 10 ROWS

The heat transfer coefficients for tube banks having fewer than 10 rows are greater. For more information in this area, the reader is referred to Jones and Monroe[18] and Gram, Mackey, and Monroe.[19]

• *For Gases Other Than Air.* The average heat transfer coefficient for gases other than air passing through a tube bank can be obtained by the use of a modification of Eq. (4-18) as indicated in Eq. (4-19).

$$\bar{h} = \frac{k}{D_o} C(\mathrm{Re})^n \left(\frac{\mathrm{Pr}}{\mathrm{Pr}_{air}} \right)^{1/3} = \frac{k}{D_o} C(\mathrm{Re})^n \left(\frac{\mathrm{Pr}}{0.69} \right)^{1/3} \qquad (4\text{-}19)$$

$$\bar{h} = 1.13 \frac{k}{D_o} C(\mathrm{Re})^n (\mathrm{Pr})^{1/3} \quad \left[\mathrm{W/m^2 \cdot \Delta_1 {}^\circ C} \right]$$

Values for C and n are obtained from Table 4-5.

Example Problem 4-5

If atmospheric hydrogen at 10°C were flowing through the staggered tube bank of Example Problem 4-4 with the same tube surface temperature and approach velocity, what would be the average heat transfer film coefficient $\bar{h}$ using the Grimison correlation?

SOLUTION

SKETCH
Refer to the sketch in Example Problem 4-4.

EQUATION

$$\bar{h} = 1.12 \frac{k}{D_0} (\mathrm{Re})^n \mathrm{Pr}^{1/3} \quad \left[\mathrm{W/m^2 \cdot \Delta_1 {}^\circ C} \right] \qquad [\text{Eq. (4-19)}]$$

PARAMETERS

$k = $ (interpolate from Table B-2, Appendix B for T_{av})

$T_{av} = 328$ K   (Example Problem 4-4)

$k = k_{300} + \Delta k \quad \left[\mathrm{W/m \cdot \Delta_1 {}^\circ C} \right]$

k	T
0.182	300
	328
0.206	350

$$k_{300} = 0.182 \quad [W/m \cdot \Delta_1 °C]$$

$$\Delta k = (0.206 - 0.182) \frac{(328 - 300)}{(350 - 300)}$$

$$= 0.024 \left[\frac{28}{50} \right]$$

$$\Delta k = 0.013 \quad [W/m \cdot \Delta_1 °C]$$

$$k = 0.182 + 0.013$$

$$k = 0.195 \quad [W/m \cdot \Delta_1 °C]$$

$$D_o = 1.5 \text{ cm} \rightarrow 0.015 \text{ m} \qquad \text{(given)}$$

$$C = 0.505 [-] \qquad \text{(Table 4-5)} \qquad S_n/D = 1.5$$

$$\text{(staggered)} \qquad S_p/D = 1.25$$

Calculate Re for maximum velocity.

$$Re = \frac{V_m D_o \rho}{\mu} \quad [-]$$

V_m is where flow cross section W_m is $1.90 D_o$ (from Example Problem 4-4).

$$V_m = V_a \frac{2S_n}{W_m} \quad [m/s]$$

$$V_a = 5 \text{ m/s} \qquad \text{(given)}$$

$$S_n = 1.5 D_o \qquad \text{(given)}$$

$$W_m = 1.9 D_o \qquad \text{(from Example Problem 4-4)}$$

$$V_m = \frac{(5)(2)(1.5 D_o)}{1.9 D_o} \quad \frac{[m/s][-][cm]}{[-][cm]} = [m/s]$$

$$V_m = 7.9 \text{ m/s}$$

$$D_o = 1.5 \text{ cm} \rightarrow 0.015 \text{ m} \qquad \text{(given)}$$

$$\rho = \text{(interpolate from Table B-2 for } T_{av} = 328 \text{ K)}$$

$$\rho = \rho_{300} - \Delta\rho \quad [kg/m^3]$$

ρ	T
0.08185	300
	328
0.07016	350

$$\rho_{300} = 0.08185 \quad [kg/m^3]$$

$$\Delta\rho = (0.08185 - 0.07016) \frac{(328 - 300)}{(350 - 300)}$$

$$= 0.01169 \left[\frac{28}{50} \right]$$

$\Delta\rho = 0.00655 \quad [\text{kg/m}^3]$

$\rho = 0.08185 - 0.00655$

$\rho = 0.07530 \quad [\text{kg/m}^3]$

$\mu = $ (interpolate from Table B-2, Appendix B for $T_{av} = 328$ K)

$\mu = \mu_{300} + \Delta\mu \quad [\text{Pa} \cdot \text{s}]$

μ	T
8.963×10^{-6}	300
	328
9.954×10^{-6}	350

$\mu_{300} = 8.963 \times 10^{-6} \quad [\text{Pa} \cdot \text{s}]$

$\Delta\mu = (9.954 - 8.963) \times 10^{-6} \dfrac{(328 - 300)}{(350 - 300)}$

$\quad = (0.991 \times 10^{-6}) \left[\dfrac{28}{50} \right]$

$\Delta\mu = 0.555 \times 10^{-6} \quad [\text{Pa} \cdot \text{s}]$

$\text{Re} = \dfrac{(7.9)(0.015)(0.07530)}{0.555 \times 10^{-6}} \quad \dfrac{[\text{m/s}][\text{m}][\text{kg/m}^3]}{[\text{Pa} \cdot \text{s}]}$

$\quad = \dfrac{[\text{kg/m} \cdot \text{s}]}{[\text{kg/m} \cdot \text{s}^2][\text{s}]} = [-]$

$\text{Re} = 16{,}077 \quad [-]$

$n = 0.554 \quad [-] \qquad \text{(Table 4-5)} \qquad S_n/D = 1.5$

$\qquad\qquad\qquad\qquad \text{(staggered)} \qquad S_p/D = 1.25$

$\text{Pr} = $ (interpolate from Table B-2, Appendix B for $T_{av} = 328$)

$\text{Pr} = \text{Pr}_{300} - \Delta\text{Pr} \quad [-]$

Pr	T
0.706	300
	328
0.697	350

$\text{Pr}_{300} = 0.706 \quad [-]$

$\Delta\text{Pr} = (0.706 - 0.697) \dfrac{(328 - 300)}{(350 - 300)}$

$\quad = 0.009 \left[\dfrac{28}{50} \right]$

$$\Delta Pr = 0.005 \quad [-]$$
$$Pr = 0.706 - 0.005$$
$$Pr = 0.701 \quad [-]$$

SUBSTITUTION

$$\bar{h} = (1.13)\frac{0.195}{0.015}(0.505)(16{,}077)^{0.554}(0.701)^{1/3}$$

$$[-]\frac{[W/m\cdot\Delta_1{}^\circ C]}{[m]}[-][-][-]=[W/m^2\cdot\Delta_1{}^\circ C]$$

ANSWER

$$\bar{h} = 1410 \quad [W/m^2\cdot\Delta_1{}^\circ C]$$

• *Tubes with Extended Surfaces.* Tubes with extended surfaces are used extensively in building heating and air conditioning. Standard tube bundles are manufactured, and the heat transfer performance for each is readily available from the manufacturer. Some basic performance of tube bundles with extended surface is given by Kays and London.[20]

HEAT TRANSFER IN HIGH SPEED FLOW

High speed aircraft, vehicles that pass through the earth's atmosphere during flight, and rocket motor exhaust nozzles are typical examples of equipment designs involving heat transfer in high speed flow. A common design requirement is to control and limit the temperature of the solid body structure as dictated by the characteristics of the material from which it is made. The scope of the following discussion of heat transfer in high speed flow is limited to gas flow.

The heat transfer in high speed flow of gases is complex, deviating from the forced convection heat transfer discussed in Section 4-2.1 when compressibility effects become significant. The compressibility effects are a function of the gas flow velocity with respect to the velocity of sound in the gas. The compressibility effects usually become significant at gas flow velocities greater than 0.5 times the velocity of sound.

The primary compressibility effect is that of the adiabatic compression and the associated temperature increase of the gas as its flow velocity (with respect to the solid surface) is reduced from the free stream velocity to the low velocities in the boundary layer adjacent to the solid surface (approaching zero at the solid surface). The temperature of the gas at the

---

[20] W. M. Kays and A. L. London, *Compact Heat Exchangers*, National Press, Palo Alto, CA, 1955.

point of zero velocity with respect to the solid body is called the stagnation temperature. The stagnation temperature is the sum of the free stream gas temperature and the adiabatic compression temperature rise as indicated in Eq. (4-20).

$$T_{o\alpha} = T_\alpha \left(1 + \frac{\gamma - 1}{2} M_\alpha^2\right) \quad [\text{K}] \qquad (4\text{-}20)$$

where

$T_{o\alpha}$ = adiabatic stagnation temperature   [K]
T_α = free stream gas temperature   [K]
γ = ratio of gas specific heat at constant
     pressure to constant volume   [—]
M_α = free stream Mach number (velocity/speed
     of sound)   [—]

Although the processes in the boundary layer in high speed flow are not adiabatic and vary with the configuration, the general practice has been to relate the heat transfer processes to an adiabatic boundary layer for simplicity.

Plots of the stagnation temperature as a function of flight velocity and altitude in the earth's atmosphere are presented in Figure 4-19. For example, the stagnation temperature for an SST aircraft flying at 800 m/s at an altitude of 25,000 m would be 550 K, and for a missile with a flight velocity of 4200 m/s at 45,000-m altitude, the stagnation temperature would be 4300 K.

• *Flow Parameters and Flow Regimes.* In high speed flow at least two parameters in addition to the Reynolds number must be considered. These are

- the flow velocity relative to the local speed of sound (Mach number)
- the mean free path of the molecule relative to the size of the solid surface (Knudsen number)

As long as the gas can be treated as a continuum, the parameters governing the convective heat transfer are the Reynolds, Prandtl, and Mach numbers. When the gas pressure becomes sufficiently low, such as at high altitudes in the earth's atmosphere, the gas molecular mean free path becomes long relative to the dimensions of the boundary layer and the effect of the relative coarseness of the molecular structure upon the heat transfer should be considered. The Knudsen number is a measure of the degree of the relative coarseness of the molecular structure, and is defined below.

$$\text{Kn} = \lambda / L \quad [—]$$

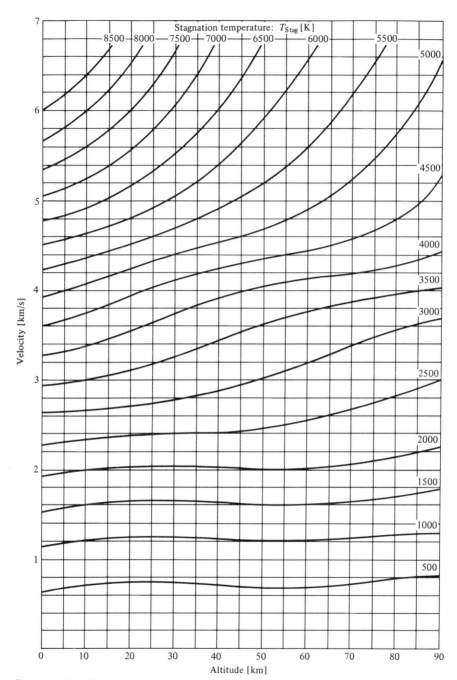

Figure 4-19 The relationship between stagnation temperature, flight velocity, and altitude in the earth's atmosphere.

where

λ = the molecular mean free path length   [m]
L = a significant length of the solid body   [m]

Tsien[21] related the Knudsen number to the Reynolds number and the Mach number by

$$\mathrm{Kn} = 1.26\sqrt{\gamma M / \mathrm{Re}} \quad [-] \tag{4-21}$$

where Kn and Re are both based on the same characteristic length.

Kn = Knudsen number $[-]$
Re = Reynolds number $[-]$
γ = ratio of specific heats $[-]$

The gas flow, with respect to the molecular mean free path length, has been divided into four regions, namely

- continuum
- slip flow
- transition
- free molecular flow

In the continuum region the mean free path of the gas molecule is small with respect to the boundary layer. It is limited to

$\mathrm{Re} \gg 1.0$ and $M / \mathrm{Re} < 0.1$

In the slip flow region the mean free path is of the order of 1 to 10% of the boundary layer thickness. The molecules immediately adjacent to the solid surface do not stick to it, as in the continuum region, but slide along it with a definite tangential velocity.

In the transition region, the collisions between (a) molecules and (b) molecules and the solid surface are of equal importance.

In the free molecular flow region no boundary layer exists. Molecule to molecule collisions are negligible in comparison with the molecule to solid surface collisions. This region is defined as

$M / \mathrm{Re} > 3$

The gas flow, with respect to its velocity, has been divided into three regions, namely

- subsonic
- supersonic
- hypersonic

[21] Tsien, H. S., "Supraerodynamics, Mechanics of Rarefied Gases," J. Aero. Sci., Vol. 13, pp. 653–664 (1946).

In subsonic flow laminar boundary layers build up along the solid surface becoming turbulent boundary layers as discussed in Section 4.1. As the flow velocity becomes greater than mach 0.5 the adiabatic free stream dynamic temperature rise becomes significant, and the stagnation temperature becomes significantly higher than the free stream temperature. No shock waves are present in the flow.

In supersonic flow (free stream velocities greater than mach 1.0) shock waves are present in the gas flow around the solid body. One of the main shock waves is the detached bow shock wave. The gas velocity in the free stream is supersonic in front of the bow shock wave, but is subsonic between the bow shock wave and the solid surface. The shock waves have a significant effect upon the gas flow adjacent to the solid surface and the heat transfer to the solid surface. The shock wave patterns in the gas flow around solid bodies such as a sphere, cone, cylinder, small flat plate, and so forth, are a function of the configuration geometry and the free stream Mach number.

Hypersonic flow involves high Mach numbers with small perturbations (slender solid bodies). Hypersonic flow is defined as the flow region in which the factor

$$M_\alpha(t/L) \geqslant 1$$

where

M_α = the free stream Mach number
t/L = the thickness to length ratio of the body in the flow field

These regions in the earth's atmosphere are indicated by Kreith[22] by the plot on coordinates of Reynolds number versus Mach number shown in Figure 4-20.[23-29] The solid lines indicate boundaries between the different flow regimes. The dotted lines indicate the altitude for a body having a characteristic dimension L (for the Reynolds number) of unity. For bodies

[22] F. Krieth, *Principles of Heat Transfer*, p. 472. Harper & Row, New York, 1962.

[23] R. M. Drake and G. H. Backer, "Heat Transfer from Spheres to a Rarefied Gas in Supersonic Flow," Trans. A.S.M.E., Vol. 74 (1952).

[24] F. M. Saver, "Convective Heat Transfer from Spheres in Free Molecule Flow," J. Aero. Sci., Vol. 18, pp. 353–354 (1951).

[25] J. R. Stalder and M. O. Creager, "Heat Transfer to Bodies in a High Speed Rarefied Gas in Supersonic Flow," Univ. of Calif., Inst. Eng. Res. Rep. M. E., pp. 150–191 (1952).

[26] L. L. Karanau, "Heat Transfer from Sphere to a Rarefied in Subsonic Flow," Trans. A.S.M.E., Vol. 77 (1955).

[27] S. A. Schaaf and P. L. Chambre, "Flow of Rarefied Gases," *High Speed Aerodynamics and Jet Propulsion Series*, Vol. IV, pt. G. Princeton University Press, Princeton, N.J., 1956.

[28] A. K. Oppenheim, "Generalized Theory of Convective Heat Transfer in a Free Molecule Flow," J. Aero. Sci., Vol. 20, pp. 49–57 (1953).

[29] J. R. Stalder and D. Jukoff, "Heat Transfer to Bodies Travelling at High Speeds in the Upper Atmosphere," N.A.C.A., Rep. No. 944 (1944).

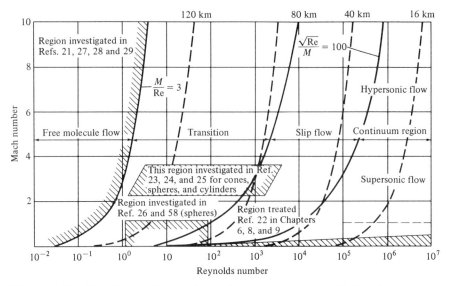

Figure 4-20 The flow regimes of gas dynamics. (SOURCE: F. Kreith, *Principles of Heat Transfer*, 1st ed., p. 472. International Textbook Co., Scranton, Pa., 1958)

with a larger or smaller characteristic dimension, the corresponding Re number would be larger or smaller in a direct ratio.

• *Common Solid Body Configurations and the Associated Flow Patterns.* The major interests in high speed gas flow heat transfer to solid bodies are for the following applications:

- low drag vehicles going through the earth's atmosphere (satellite launch vehicles and missiles)
- high drag vehicles going through the earth's atmosphere (re-entry vehicles)
- supersonic aircraft flying through the earth's atmosphere
- rocket exhaust nozzles (De Laval)

The configuration for the lowest drag solid body at high speeds is that of a sharp pointed cone with a cone angle optimized for the flight Mach number. However, the tip of such a cone is subject to extremely high heat transfer from the gas, and the tip material is difficult to cool. To limit the temperature of the tip, a spherical segment nose is substituted for the sharp point. The larger the spherical radius the lower the heat transfer rate to the nose, but the greater the aerodynamic drag upon the solid body.

The configuration for a high drag solid body is that of a flat plate normal to the free stream gas velocity vector or a large radius spherical segment. The choice is often dictated by the aerodynamic flight require-ments of the relative position of the center of pressure and the center of gravity for flight stability.

The configuration elements of a supersonic aircraft include

- cones (nose)
- cylinders (leading edges of the wings)
- essentially flat surfaces parallel to the air flow

Consequently the common configurations investigated for high speed flight vehicles are

- spheres
- cones
- cylinders
- flat plates

In rocket exhaust nozzles the heat transfer flux varies along the nozzle length, being a maximum in the region of the nozzle throat. Often the heat transfer flux is reduced to an extent by placing an annulus of cooler gas adjacent to the nozzle wall surface.

4-2.2 Equation for *h* with Liquids

The basic equation for *h* with liquids is similar to that for gases. However, because of the markedly different comparative values of key fluid proper-ties, such as thermal conductivity and viscosity, the equations giving the best fit with experimental data for evaluating the heat transfer film coefficient for liquids are not necessarily the same as for gases.

Because of the significant differences in heat transfer characteristics and physical properties between different types of liquids, liquids are usually divided into the three categories listed below for heat transfer purposes. The range in Prandtl number ($\Pr = \mu c_p / k$) is indicated for each category.

- water and similar to water, $1.0 < \Pr < 20$
- significantly more viscous than water, $\Pr > 20$
- liquid metals, $\Pr < 0.1$

The equations for *h* of each of these three categories of liquids are discussed in the following sections, which are organized according to the heat transfer configuration, in a manner similar to that used for gases.

EQUATIONS FOR *h* FOR WATER AND FLUIDS SIMILAR TO
WATER, $1.0 < \Pr < 20$

• *Water Flow Inside Pipes—Laminar Flow (Re < 2100).* Heat transfer coef-ficients for laminar flow are considerably smaller than for turbulent flow.

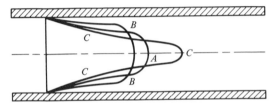

Figure 4-21   Effect of heat transfer on velocity profiles in fully developed laminar flow. Curve A, isothermal flow; curve B, heating of liquid or cooling of gas; curve C, cooling of liquid or heating of gas.

Laminar flow forced convection heat transfer is seldom used with water. Heat transfer in laminar (streamline) flow has been investigated less extensively than turbulent flow.

The heat flow mechanism in pure laminar flow is conduction. In isothermal flow the velocity profile of the liquid flow becomes parabolic. As soon as the conditions are nonisothermal, the viscosity of the fluid changes with the temperature gradient, resulting in modifications in the flow velocity profile as illustrated in Figure 4-21.

In addition to the boundary layer changes, free convective forces appear. These result in differences between horizontal pipes and vertical pipes, as well as between upward flow and downward flow in vertical pipes.

When buoyancy forces have no significant effect upon the liquid flow pattern, the Nusselt number for a fully developed laminar flow of a constant property fluid with a constant heat flux, as shown by Bayley,[30] is a constant.

$$\text{Nu} = 4.364 \quad [-] \tag{4-22}$$

or

$$h = 4.364 \frac{k}{D} \quad [\text{W/m}^2 \cdot \Delta_1 {}^\circ\text{C}] \tag{4-23}$$

and for a constant wall surface temperature is

$$\text{Nu} = 3.66 \quad [-] \tag{4-24}$$

or

$$h = 3.66 \frac{k}{D} \quad [\text{W/m}^2 \cdot \Delta_1 {}^\circ\text{C}] \tag{4-25}$$

In applications where the properties of the fluid vary and buoyancy forces become significant, such factors should be considered by the formula. For example, for a given set of conditions the effect of buoyancy

[30] F. J. Baylay, J. M. Owen, and A. B. Turner, *Heat Transfer*, pp. 189–192. Barnes & Noble Books, New York, 1972.

forces would be different between flow in vertical and horizontal pipes. The Martinelli/Boelter[31] equation for a vertical pipe with a constant surface temperature is one example of laminar flow equations.

$$\text{Nu}=1.75F_1\left[\frac{\pi}{4}RP\left(\frac{d}{l}\right)\pm0.0722\left(\frac{d}{l}GP\right)^nF_2\right]^{1/3} \tag{4-26}$$

The authors give values of the exponent n and the factors F_1 and F_2 for the fluids investigated. The buoyancy term (following the $\pm$ sign) is $+$ for cooled fluids flowing downward or heated fluids flowing upward, and $-$ for the opposite combinations.

Tests on the heating of water, glycerol, and several oils in streamline flow were correlated by King[32] by the following formula within the range of values of Gc_p/kL from 8 to 3500.

$$h=2.53\frac{k}{D}\left(\frac{Gc_p}{kL}\right)^{1/3}\quad\left[\text{W/m}^2\cdot\Delta_1{}^\circ\text{C}\right] \tag{4-27}$$

where

$h=$ film coefficient based upon the arithmetical mean
temperature difference between the fluid and the
tube wall $\qquad\qquad$ $[\text{W/m}^2\cdot\Delta_1{}^\circ\text{C}]$
$G=$ mass flow rate $\qquad\qquad$ $[\text{kg/m}^2\cdot\text{s}]$
$D=$ diameter of pipe (inside) $\qquad\qquad$ $[\text{m}]$
$L=$ length of pipe $\qquad\qquad$ $[\text{m}]$
$c_p=$ specific heat $\qquad\qquad$ $[\text{J/kg}\cdot\Delta_1\ {}^\circ\text{C}]$
$k=$ thermal conductivity of liquid $\qquad\qquad$ $[\text{W/m}\cdot\Delta_1\ {}^\circ\text{C}]$

For the flow in a circular pipe, Eq. (4-27) can be rewritten as

$$h=2.34\left(\frac{V\rho c_p k^2}{DL}\right)^{1/3}\quad\left[\text{W/m}^2\cdot\Delta_1{}^\circ\text{C}\right] \tag{4-28}$$

where

$V=$ average velocity $\qquad\dfrac{4G}{\pi D^2\rho}=[\text{m/s}]$

$\rho=$ density $\left[\text{kg/m}^3\right]$

The effect of tube length on the film coefficient for laminar flow is discussed in footnote 33.

[31] R. C. Martinelli and L. M. K. Boelter, University of California Publications in Engineering, Vol. 5, p. 23 (1942).
[32] W. S. King, Mech. Eng., Vol. 54, p. 412 (June 1932).
[33] A. P. Colburn and O. A. Hougen, Ind. Eng. Chem., Vol. 22, p. 522 (1930).

Example Problem 4-6

Water flowing upward in a 2 m long vertical pipe of 5-cm diameter with an average velocity of 90 m/h is being heated. The temperature of the surface of the pipe is 50°C and the water temperature is 30°C. Obtain a value of h using the King correlation.

SOLUTION
For the King correlation use Eq. (4-28) arrangement.

EQUATION

$$h = 2.34 \left(\frac{V \rho c_p k^2}{DL} \right)^{1/3} \quad [W/m^2 \cdot \Delta_1 °C] \quad [\text{Eq. (4-28)}]$$

PARAMETERS

$$V = 90 \text{ m/h} \rightarrow \frac{90}{3600} = 0.025 \quad [\text{m/s}]$$

Evaluate fluid properties at the average temperature (wall/fluid).

$$T_{av} = \frac{T_w + T_f}{2} \quad [°C]$$

$$T_w = 50°C \quad (\text{given})$$

$$T_f = 30°C \quad (\text{given})$$

$$T_{av} = \frac{50 + 30}{2} = \frac{80}{2} = \frac{[°C] + [°C]}{2} = [°C]$$

$$T_{av} = 40 \, [°C]$$

$\rho = 993.0 \text{ kg/m}^3$ (Table B-3, Appendix B at 40°C)

$c_p = 4.177 \text{ kJ/kg} \cdot \Delta_1 °C$ (Table B-3, Appendix B)

$k = 0.633 \text{ W/m} \cdot \Delta_1 °C$ (Table B-3, Appendix B)

$D = 5.0 \text{ cm} \rightarrow 0.050 \text{ m}$ (given)

$L = 2 \text{ m}$ (given)

SUBSTITUTION

$$h_c = 2.34 \left[\frac{(0.025)(993.0)(4.177)(0.633)^2}{(0.050)(2)} \right]^{1/3}$$

$$[-] \left[\frac{[\text{m/s}][\text{kg/m}^3][\text{kJ/kg} \cdot \Delta_1 °C][\text{W/m} \cdot \Delta_1 °C]^2}{[\text{m}][\text{m}]} \right]^{1/3}$$

$$= [-] \left[\frac{[kJ/s][W^2]}{[m]^6 [\Delta_1{}^\circ C]^3} \right]^{1/3} = \left[[W/m^2 \cdot \Delta_1{}^\circ C]^3 \right]^{1/3}$$

$$= [W/m^2 \cdot \Delta_1{}^\circ C]$$

$$= 2.34 [411.3]^{1/3}$$

ANSWER

$$h_c = 17.4 \quad [W/m^2 \cdot \Delta_1{}^\circ C]$$

• *Water Flow Inside Pipes—Turbulent Flow.* In the following discussion, basic equations will be presented for two conditions of the magnitude between the wall and the liquid bulk, namely

- small temperature differences
- large temperature differences

In the determination of h, it should be understood that the equations given represent nominal conditions and configurations.

To apply the equations for h to noncircular ducts, an equivalent diameter for the noncircular duct is used in evaluating Reynolds number and Nusselts number. The equivalent diameter is calculated by the following formula

$$D_e = 4 \frac{\text{duct flow cross-sectional area}}{\text{duct wetted perimeter}} \quad [m] \tag{4-29}$$

For a circular annulus,

$$D_e = 4 \left(\frac{A}{P} \right) \quad [m]$$

$$A = \pi/4 (D_o^2 - D_i^2) \quad [m^2]$$

$$P = \pi (D_o + D_i) \quad [m]$$

$$D_e = 4 \frac{(\pi/4)(D_o^2 - D_i^2)}{\pi (D_o + D_i)} \quad [-] \frac{[m^2]}{[m]} = [m]$$

$$D_e = \frac{(D_o^2 - D_i^2)}{(D_o + D_i)} \quad [m] = D_o - D_i \quad [m] \tag{4-30}$$

Example Problem 4-7

Calculate the equivalent diameter (D_e) for an annulus passage with the outside diameter (D_o) 15 cm and the inside diameter (D_i) 9 cm.

SOLUTION

SKETCH

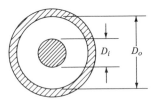

Figure 4-22

EQUATION

$$D_e = D_o - D_i \quad [\text{cm}] \qquad [\text{Eq. (4-30)}] \text{ for a circular annulus}$$

PARAMETERS

$$D_o = 15 \text{ cm} \qquad (\text{given})$$

$$D_i = 9 \text{ cm} \qquad (\text{given})$$

SUBSTITUTION

$$D_e = 15 - 9 \quad [\text{cm}] - [\text{cm}]$$

$$= 6 \quad [\text{cm}]$$

ANSWER

$$D_e = 6 \quad [\text{cm}]$$

The value of h is affected by

- heat transfer surface condition: roughness, cleanliness
- length of the pipe
- turbulence in the flow entering the pipe
- the pipe entrance configuration
- elbows or turbulence generators in the pipe
- coiled pipe

The amount of heat flow from the liquid into the wall is computed using Eq. (4-1)

$$\dot{Q} = hA\Delta T \quad [\text{W}]$$

The *temperature difference* (ΔT) in Eq. (4-1) is the difference between the temperature of the inside surface of the pipewall and the "bulk" temperature of the liquid.

The *heat transfer area* (A) is the surface area of the liquid/wall interface.

The *convective surface coefficient* (h) is evaluated as discussed in the following sections.

The expected error from the use of these equations for h is less than 20%. If more accurate evaluation of h is required, test data closely matching the configuration of interest should be sought, or tests made. By fitting curves to such test data, limited extrapolations of high accuracy can be made.

• *The Equation for h with Small Temperature Differences.* The basic equation developed from dimensional analysis in the form of

$$\mathrm{Nu} = c(\mathrm{Re})^a (\mathrm{Pr})^b \tag{4-31}$$

can give satisfactory correlations. Test data for turbulent water flow inside pipes were correlated by McAdams[34] with the following equation:

$$h = 0.023 \frac{k}{D} \left(\frac{DV_p}{\mu} \right)^{0.8} \left(\frac{c_p \mu}{k} \right)^{0.4} \quad \left[\mathrm{W/m^2 \cdot \Delta_1 {}^\circ C} \right] \tag{4-32}$$

Kays[35] suggests

$$h = 0.0155 \frac{k}{D} \left(\frac{DV_p}{\mu} \right)^{0.83} \left(\frac{c_p \mu}{k} \right)^{0.5} \quad \left[\mathrm{W/m^2 \cdot \Delta_1 {}^\circ C} \right] \tag{4-33}$$

Equation (4-32) is generally known as the "McAdams" equation. It applies to the following conditions:

10,000 < Re < 150,000
0.7 < Pr < 120
pipe length $> 60L/D$
temperature difference across film not large (less than 10°)

The properties of the fluid should be evaluated at the "bulk" temperature. The bulk temperature is defined as the arithmetic mean of the "cup" temperatures at the beginning and at the end of the heat transfer section. The "cup" temperature is the equilibrium temperature of a sample taken uniformly across the pipe section when the hotter and colder portions of the liquid flow have been thoroughly mixed.

The McAdams equation is a good general equation which gives a value of h usually within 20% of the actual. When this equation is applied to the water heating test data shown in Figure 4-23, the equation is

[34] W. H. McAdams, *Heat Transmission*, 3rd ed. McGraw-Hill, New York, 1954.
[35] W. M. Kays, *Convective Heat and Mass Transfer*, p. 173. McGraw-Hill, New York, 1966.

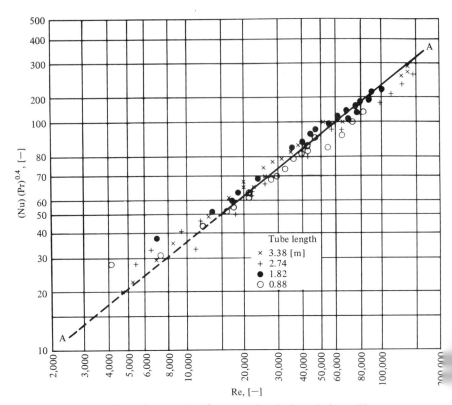

Figure 4-23 Heating of water in tubes ranging in length from 59 to 224 diameters. (SOURCE: W. H. McAdams, *Heat Transmission*, 3rd ed. McGraw-Hill, New York, 1954)

centered on the test data in the Reynolds number range of 20,000 to 40,000. Above 40,000 the curve is progressively high, and below 20,000 it is progressively low. By changing the exponents and the constant, a better fit for this particular set of data could be obtained for selected ranges of Reynolds numbers.

The test data plots in Figures 4-23 and 4-24 are typical. They show a spread of h at a given Reynolds number. The more varied the conditions under which the test data were taken, the greater the spread of the test data, in general. Some of the spread in the test data is probably due to slight differences in the condition of the heat transfer surface, some due to the accuracy in the measurement of the heat transfer rate as well as the temperatures involved. Any curve going through the center of the data would then be subject to a basic accuracy of $\frac{1}{2}$ the data point spread. For example, if the test data spread were 30% total, then the accuracy of the calculated value of h would be within $\pm 15\%$.

From the example presented in Figure 4-23 it is seen that Eq. (4-32) fits the test data well through a range of Reynolds numbers from 10,000 to

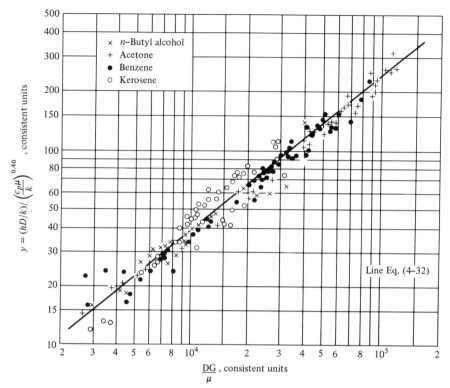

Figure 4-24   Heating of four liquids from data of Sherwood and Petrie. (SOURCE: W. H. McAdams, *Heat Transmission*, 3rd ed. McGraw-Hill, New York, 1954)

50,000. Beyond this range the equation becomes increasingly in error. There is no sudden relative change of slope between the test data and the equation. Consequently, an equation can be used outside of the recommended range of applicability, with an increase in the magnitude of the expected error in the calculated value of h.

On the other hand, if appropriate test data are available, modifications can be made to the constant (0.023) or the exponent 0.4 in Eq. (4-32) to obtain a better fit with the test data. Extrapolations to the conditions of interest can then be made using the modified equation.

Various investigators have endeavored to develop equations with an increased range of application and accuracy. For example, the equations developed by Prandtl and von Karman are given below. Comparisons of these equation and the McAdams equations with the experimental data obtained by Eagle and Ferguson[36] is shown by the plots in Figure 4-25.

[36] A. Eagle and R. M. Ferguson, Proc. Roy. Soc. (London) A, Vol. 127, pp. 540–566 (1930); Engineering, Vol. 130, pp. 691, 788, 821 (1930); Proc. Inst. Mech. Engrs. (London), Vol. 2, p. 985 (1930).

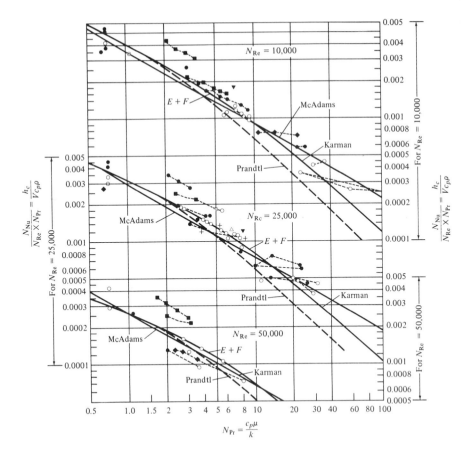

Figure 4-25 Heat transfer numbers for pipes by various equations and by experimental data of Eagle and Ferguson.[36] (SOURCE: A. I. Brown and S. M. Marco, *Introduction to Heat Transfer*, 3rd ed., p. 134. McGraw-Hill, New York, 1958)

The McAdams equation:

$$N_{Nu} = 0.023 N_{Re}^{0.8} N_{Pr}^{0.4}$$

The Prandtl equation:

$$N_{Nu} = \frac{0.04 N_{Re}^{3/4} N_{Pr}}{1 + 1.74 N_{Re}^{-1/3}(N_{Pr} - 1)}$$

The von Kármán equation:

$$N_{\text{Nu}} = \frac{0.04 N_{\text{Re}}^{3/4} N_{\text{Pr}}}{1 + N_{\text{Re}}^{-1/8}\left\{N_{\text{Pr}} - 1 + \ln\left[1 + \frac{5}{6}(N_{\text{Pr}} - 1)\right]\right\}}$$

· *The Equation for h with Large Temperature Differences.* The change in the boundary layer, due to the fluid viscosity change with the temperature gradient, affects the film conductance for large temperature differences. In many practical problems the wall temperature and the bulk temperature are not directly available. In such a case a trial-and-error iterative solution becomes necessary. For this type of a problem, Kays and London[37] presented a method to determine h by using the Station number (St) for fluids having a Prandtl number (Pr) in the range between 0.5 and 100. The Stanton number definition and relationships are given in Eq. (4-34).

$$St = \frac{\text{Nu}_f}{\text{Re}_f \times \text{Pr}_f} = \frac{h}{Gc_p} = 0.023 \text{Re}_f^{-0.2} \text{Pr}_f^{-2/3} \quad [-] \tag{4-34}$$

The equation for h is

$$h = Gc_p \, St \quad [W/m^2 \cdot \Delta_1 {}^\circ C] \tag{4-35}$$

where

$$G = \dot{M}/A$$

and

$\dot{M}$ = fluid mass flow rate through the duct $[kg/s]$

A = the duct cross-sectional area $[m^2]$

To evaluate the film coefficient (h) by Eq. (4-35) using the properties of the fluid film, the Stanton number (St) can be calculated by the relationship from Eq. (4-34)

$$St = 0.023 \, \text{Re}_f^{-0.2} \text{Pr}_f^{-2/3} \quad [-]$$

The values of Re and Pr are calculated using the fluid properties at the film temperature (which is considered to be the arithmetic average of the bulk and surface temperatures).

When the specific heat (c_p) value has a significant change with temperature, the c_p should be evaluated at the film temperature. For initial calculations the evaluation of the c_p at the bulk temperature is satisfactory.

To evaluate the film coefficient (h) by Eq. (4-35) using the fluid bulk temperature properties, Eq. (4-35) must be modified to take into consideration the effects of the film differing from the bulk temperature. This was

[37]W. M. Kays and A. L. London, "Compact Heat Exchangers—A Summary of Basic Heat Transfer and Flow Friction Design Data," Tech. Rep. 23, Stanford University, 1954.

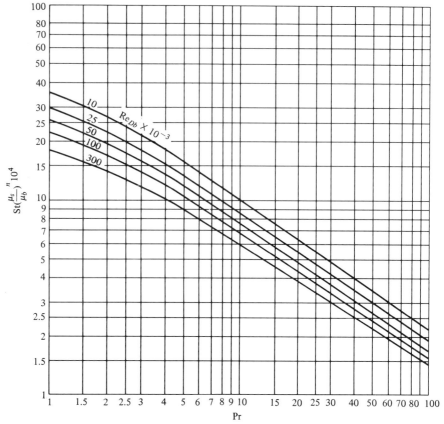

Figure 4-26 Variation of the Stanton number with Prandtl number for various values of the bulk Reynolds number. Note: $n=0.36$ for heating; $n=0.20$ for cooling. (SOURCE: W. M. Kays and A. L. London, *Compact Heat Exchangers*. National Press, Palo Alto, CA, 1955)

done by Kays and London by introducing the ratio of the viscosity of the fluid at the bulk temperature to the viscosity at the surface temperature. The relationships between the Re and Pr at bulk temperatures and the parameter $St(\mu_s/\mu_b)$ is given in Fig. 4-26. Equation (4-35) is modified as shown below

$$h = \left[St(\mu_s/\mu_b)^n \right] Gc_p(\mu_b/\mu_s)^n \quad \left[W/m^2 \cdot \Delta_1{}^\circ C \right] \quad (Eq.\ 4\text{-}35a)$$

The value of the exponent n for the viscosity ratio for use in Eq. (4-35a) is

liquid heating $n = 0.36$
liquid cooling $n = 0.20$

By using Fig. 4-26 and estimating the viscosity ratio (if the temperatures are not known) an initial value of h can readily be obtained. By then using this value of h, the bulk and wall temperatures can be calculated, and the estimated bulk temperatures and viscosity ratio verified or new values be assumed for the next iteration of the value of h.

Example Problem 4-8

Calculate the value of h for water flowing at a velocity of 5 m/s in a tube 1 cm in diameter with an average (bulk) water temperature of 30°C and a tube wall temperature of 90°C.

SOLUTION

SKETCH

Figure 4-27

PLAN
Calculate by:

(A) Using the Stanton number

$$h_c = Gc_p \text{ St} \quad [\text{W/m}^2 \cdot \Delta_1 °\text{C}] \qquad [\text{Eq. (4-35)}]$$

based on the average temperature of the film.

(B) Using the $\text{St}(\mu_s/\mu_b)^n$ from Figure 4-26

$$h_c = \left[\text{St} \left(\frac{\mu_s}{\mu_b} \right)^n \right] Gc_p \left(\frac{\mu_b}{\mu_s} \right)^n \quad [\text{W/m}^2 \cdot \Delta_1 °\text{C}] \quad \text{Eq. } [(4\text{-}35a)]$$

EQUATION FOR (A)

$$h = Gc_p \text{ St} \quad [\text{W/m}^2 \cdot \Delta_1 °\text{C}] \qquad [\text{Eq. (4-35)}]$$

PARAMETERS FOR (A)
ρ, c_p, and St to be evaluated at the average temperature (T_{av}) of the film.

$$T_{av} = \frac{T_b + T_s}{2}$$

$T_b = 30°\text{C}$ (given)

$T_s = 90°\text{C}$ (given)

$$T_{av} = \frac{30 + 90}{2} \quad \frac{[°\text{C}] + [°\text{C}]}{2} = [°\text{C}]$$

$T_{av} = 60°\text{C}$

(C-1) $\quad G = \rho V \quad [kg/m^2 \cdot s]$

$\rho = 983.3 \quad [kg/m^3] \qquad$ (Table B-3, Appendix B at 60°C)

$V = 5 \quad [m/s] \qquad$ (given)

$G = (983.3)(5) \quad [kg/m^3][m/s] = [kg/m^2 \cdot s]$

$G = 4917 \quad [kg/m^2 \cdot s]$

$c_p = 4183 \quad [J/kg \cdot \Delta_1 °C] \qquad$ (Table B-3, Appendix B at 60°C)

To determine the Stanton number (St):

$$St = 0.023 \, Re_b^{-0.2} Pr_b^{-2/3} \quad [-]$$

(C-2) $\quad Re = \dfrac{DG}{\mu} \quad [-]$

$D = 1 \, cm \rightarrow 0.01 \, m \qquad$ (given)

$G = 4917 \quad [kg/m^2 \cdot s]$

$\mu = 0.475 \times 10^{-3} \quad [kg/m \cdot s] \qquad$ (Table B-3, Appendix B at 60°C)

$Re = \dfrac{(0.01)(4917)}{(0.475 \times 10^{-3})} \quad \dfrac{[m][kg/m^2 \cdot s]}{[kg/m \cdot s]} = [-]$

$Re = 23{,}356 \quad [-]$

$Re^{-0.2} = 0.1338 \quad [-]$

$Pr = 3.04 \quad [-] \qquad$ (Table B-3, Appendix B at 60°C)

$Pr^{-2/3} = 0.4765 \quad [-]$

(C-3) $\quad St = (0.023)(0.1338)(0.4765)$

$$[-][-][-] = [-]$$

$$St = 0.001466 \quad [-]$$

SUBSTITUTION FOR (A)

(C-4) $\quad h = (4917)(4183)(0.001466)$

$$[kg/m^2 s][J/kg \cdot \Delta_1 °C][-] = [J/s][1/m^2 \cdot \Delta_1 °C]$$

ANSWER FOR (A)

$h = 30{,}152 \quad [W/m^2 \cdot \Delta_1 °C]$

EQUATION FOR (B)

$$h_c = \left[St \left(\frac{\mu_s}{\mu_b} \right)^n \right] G c_p \left(\frac{\mu_b}{\mu_s} \right)^n \quad [W/m^2 \cdot \Delta_1 °C] \quad [Eq.\ (4\text{-}35a)]$$

where $St(\mu_s/\mu_b)^n$ is obtained from Figure 4-26, entering at the Pr value, moving up the Re value, and then left across to read the St $(\mu_s/\mu_b)^n \times 10^4$ value.

PARAMETERS FOR (B) FROM FIGURE 4-26

$\qquad$ Pr$=5.42$ $[-]$ $\qquad$ (Table B-3, Appendix B at 30°C)

(C-5) $\quad$ Re$=\dfrac{DG}{\mu}$ $[-]$

$\qquad$ $D=1$ cm$\rightarrow 0.01$ m $\quad$ (given)

$\qquad$ $G=4917$ $\left[\text{kg/m}^2\cdot\text{s}\right]$ $\quad$ (C-1)

$\qquad$ $\mu=0.803\times 10^{-3}$ $\left[\text{kg/m}\cdot\text{s}\right]$ $\quad$ Table B-3, Appendix B at 30°C

$\qquad$ Re$=\dfrac{(0.01)(4917)}{(0.803\times 10^{-3})}$ $\dfrac{[\text{m}]\left[\text{kg/m}^2\cdot\text{s}\right]}{[\text{kg/m}\cdot\text{s}]}=[-]$

$\qquad$ Re$=61{,}230$ $[-]$

Then entering Figure 4-26 and reading the value for St $(\mu_s/\mu_b)^n$

$\qquad$ $St\left(\dfrac{\mu_s}{\mu_b}\right)^n\times 10^4=11.5$ $[-]$

$\qquad$ $St\left(\dfrac{\mu_s}{\mu_b}\right)^n=1.15\times 10^{-3}$ $[-]$

$\qquad$ $G=4980$ $\left[\text{kg/m}^2\cdot\text{s}\right]$

$\qquad$ $c_p=4177$ $\left[\text{J/kg}\cdot\Delta_1°\text{C}\right]$ $\qquad$ (Table B-3, Appendix B at 30°C)

$\qquad$ $\mu_b=0.803\times 10^{-3}$ $\left[\text{kg/m}\cdot\text{s}\right]$ $\qquad$ (Table B-3, Appendix B at 30°C)

$\qquad$ $\mu_s=0.318\times 10^{-3}$ $\left[\text{kg/m}\cdot\text{s}\right]$ $\qquad$ (Table B-3, Appendix B at 90°C)

$\qquad$ $n=0.36$ $[-]$ $\qquad$ (exponent for heating)

SUBSTITUTION FOR (B)

$\qquad$ $h_c=(1.15\times 10^{-3})(4980)(4177)\left(\dfrac{0.803\times 10^{-3}}{0.318\times 10^{-3}}\right)^{0.36}$

$\qquad\qquad$ $[-]\left[\text{kg/m}^2\cdot\text{s}\right]\left[\text{J/kg}\cdot\Delta_1°\text{C}\right]\dfrac{[\text{kg/m}\cdot\text{s}]}{[\text{kg/m}\cdot\text{s}]}$

$\qquad\qquad$ $=[\text{J/s}]\left[1/\text{m}^2\cdot\Delta_1°\text{C}\right]=\left[\text{W/m}^2\cdot\Delta_1°\text{C}\right]$

$\qquad\qquad$ $=(23{,}920)(2.525)^{0.36}=(23{,}920)(1.396)$

ANSWER

$$h_c = 33,390 \quad \left[\text{W}/\text{m}^2 \cdot \Delta_1 {}^\circ\text{C} \right]$$

When the wall temperature is known, the Sieder-Tate[38] equation for viscous fluids can be used.

$$h = 0.027 \frac{k}{D} (\text{Re})^{0.8} (\text{Pr})^{0.33} \left(\frac{\mu}{\mu_s} \right)^{0.14} \quad \left[\text{W}/\text{m}^2 \cdot \Delta_1 {}^\circ\text{C} \right] \tag{4-36}$$

where all liquid properties are evaluated at the bulk temperature except the viscosity at the wall temperature μ_s.

CONFIGURATION FACTORS AFFECTING THE VALUE OF *h*

• *The Effect of the Heat Transfer Surface Condition.* Any roughing of the heat transfer surface that causes an increase in the boundary layer turbulence will increase the film coefficient. However, the increase in the heat transfer due to the roughness is accompanied by a relatively larger increase in the frictional resistance to the flow. Consequently, with a fixed pressure drop available, an increased roughness of the surface will result in a decrease in the flow rate and the amount of heat transferred. Experimental studies of water flowing in rough electrically heated tubes spanning the Reynolds number range of 10,000 to 500,000 were made by Dipprey and Sabersky.[39] Increases of heat transfer rates over that for smooth surfaces by as much as the factor of 3 were obtained. However, the increase in the friction loss in the fluid flow was always greater, except in the region of transition between the effectively "smooth" and the "completely rough" flow characteristics.

A microscopic roughing of the heat transfer surface which does not cause an increase in the boundary layer turbulence can add a conduction resistance in the form of microscopic stagnant pockets of liquid between the solid wall and the fluid boundary layer. The lower the thermal conductivity of the fluid, the greater will be the effect.

Any fouling of the inside surface of the pipe with rust, scale, grease, gas bubbles, and so forth can materially reduce the film conductance by introducing a conduction resistance between the solid wall and the fluid boundary layer. Water with high mineral content can deposit scale to such an extent that the heat transfer performance becomes seriously degraded.

• *The Effect of the Length of the Pipe.* When a flow entering a pipe has a uniform velocity, a boundary layer will develop as illustrated in Figure 4-28. Progressively from the pipe's entrance, the boundary layer is initially

[38] E. M. Sieder and C. E. Tate, Ind. Eng. Chem., Vol. 28, p. 1429 (1936).
[39] D. F. Dipprey and R. H. Sabersky, Int. J. Heat Mass Transfer, Vol. 6, p. 329 (1963).

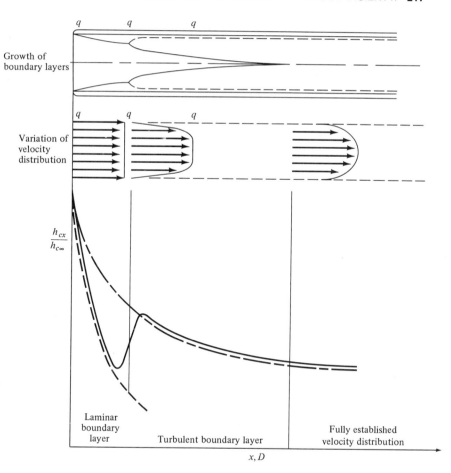

Figure 4-28 Velocity distribution and variation of unit convection conductance near the entrance of a tube for a fluid in turbulent flow. (SOURCE: F. Kreith, *Principles of Heat Transfer*, 1st ed., p. 339. International Textbook Co., Scranton, Pa., 1958)

a thin laminar layer with a high film conductance. The laminar layer increases in thickness, with a reduction in conductance, until it goes through a transition, becoming turbulent. The film conductance increases during this transition to a turbulent boundary layer. As the turbulent boundary layer becomes fully developed, the film conductance drops slightly, reaching a constant value for the fully developed turbulent boundary layer.

The film coefficient equations, such as Eqs. (4-32) and (4-33), are for conditions of the fully developed turbulent boundary layer, being applicable to pipes of lengths of $50L/D$ or longer. When the pipes are shorter, the average h will be higher because of the relative effect of the local

Table 4-6 FILM COEFFICIENT FACTORS FOR SHORT PIPES

Length of tube in terms of diameters	1	5	10	15	20	30	40	60	100	200
Shortness factor	2.0	1.32	1.2	1.15	1.1.3	1.10	1.08	1.057	1.040	1.025

increased values of h near the pipe entrance. As an approximation McAdams[22] suggests the factor of

$$1+\left(\frac{D}{L}\right)^{0.7} \quad [-] \tag{4-37}$$

for pipes shorter than L/D of 60 with the following fluid flow parameters:

- sharp edge (square) fluid flow entrance
- flow Reynolds number greater than 10,000
- flow Prandtl number between 0.7 and 120
- no turbulence in the entering fluid stream

For determining the average film coefficient of short pipes, the calculated values for long pipes should be multiplied by the shortness factor given by Eq. (4-37). The values of the shortness factor [Eq. (4-37)] for various L/D's given in Table 4-6 are for the above conditions. Turbulence in the entering fluid stream as well as deviations in the entrance geometry from a sharp edge can significantly affect the values of the shortness factor.

• *The Effect of Turbulence in the Flow Entering the Pipe.* Turbulence in the flow entering the pipe generally tends to reduce the length of the developing laminar boundary layer, resulting in a turbulent boundary layer formation in a shorter length. The effect of the turbulence in the flow entering the pipe does not extend beyond the point of a fully developed turbulent boundary layer in the pipe. The effect upon the average value of h for a long pipe would not be significant, but the pressure drop of the fluid flow might be increased. The average value of h for a short pipe could be effected. If the high local values of h in the initial laminar boundary layer region were reduced more than the overall gain from the initial increased turbulence, the average value of h for a short pipe could be reduced.

• *The Effect of the Entrance Configuration.* The pipe entrance configuration has an effect on the increased conductance of a short pipe due to the influence on the boundary layer growth. Typical pipe entrance geometries are illustrated in Figure 4-29 and the associated boundary layer growth patterns in Figure 4-30. The contoured entrance provides a smooth flow pattern with an increasing flow velocity until the pipe diameter is reached. This keeps the laminar boundary thin throughout the entrance section,

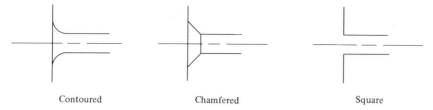

Contoured   Chamfered   Square

Figure 4-29   Typical pipe entrance geometries.

resulting in the highest local *h* values. A good contour is an ellipse with a 2:1 aspect ratio. The chamfered entrance is often easier to fabricate. Usually a 30° half-angle chamfer is used, with the large diameter approximately twice the pipe diameter. In this configuration, the point of flow separation as illustrated in Figure 4-30 is at a low velocity. The subsequent converging flow passage and acceleration of the flow thins and stabilizes the boundary layer. The fluid flow into a square entrance causes immediate separation followed by a turbulent boundary layer. The local value of *h* in the separation area is low, and a region of high conductance laminar boundary layer has been eliminated.

• *The Effect of Elbows and Turbulence Generators.* Elbows and turbulence generators, which increase the turbulence level over that of the fully developed turbulent boundary layer, cause an increase in the thermal conductance of the film. However, the associated pressure drop (loss) is relatively greater than that for an additional length of smooth straight pipe to provide the same increase in heat transfer.

• *The Effect of Coiling the Pipe.* The film conductance (as well as the pressure drop) is higher for coiled pipe than for straight pipe. In the absence of any directly applicable data, McAdams suggests using the multiplier for the *h* of long straight pipes developed by Jeschke for turbulent gas flows, as outlined on page 168.

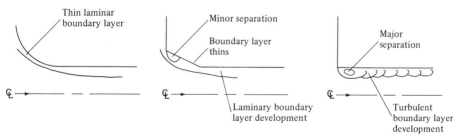

Figure 4-30   Typical boundary layer growth for three-pipe entrance configurations.

· *Turbulent Flow in Concentric Circular Tube Annuli.* The film coefficient h has been found to be different between heated inner and outer walls. In general, the h for the inner wall is higher, and smaller ratios between the inner and outer radii of the annulus result in higher values of h. The degree of eccentricity between the centerlines of the inner and outer cylinders also has an effect. Details of these effects and the values of the Nusselt number for various configurations are discussed by authors such as Kays[35] and Jakob.[40]

· *The Effect of Circumferential Heat Flux Variation.* There are applications where the heat flux and surface temperature are not uniform around the pipe. Authors such as Reynolds[41] have addressed this problem. Many conditions can be approximated by a simple cosine variation, because the thermal conductivity of the pipe wall tends to smooth out circumferential temperature variations.

$$(\dot{Q}/A)_\phi = (\dot{Q}/A)_o (1 + B\cos\phi) \quad [\text{W/m}^2] \tag{4-38}$$

Kays[35] gives the following equation and table for calculating the local film coefficient h_ϕ:

$$h_\phi = \frac{k}{D}\left[\frac{1+\cos\phi}{1/\text{Nu}+(S_1 b/2)\cos\phi}\right] \quad [\text{W/m}^2 \cdot \Delta_1 {}^\circ\text{C}] \tag{4-39}$$

where

$B=$ heat flux variation magnitude factor $\qquad [-]$
$\phi=$ circumferential angle from position of
  maximum heat flux $\qquad [-]$
$\text{Nu}=$ Nusselt number (hD/k) for uniform
  peripheral temperature and heat flux $\qquad [-]$
$S_1=$ circumferential heat flux function (from
  Table 4-7) $\qquad [-]$

· *Water Flow on Flat Surfaces.* When fluid flow passes over a flat plate aligned with the flow direction, a laminar boundary layer is initially formed, changing into a turbulent boundary layer, as discussed on page 159.

· *Laminar Flow.* An approximate equation for the value of h for laminar flow of water as given by Bailey, Owen, and Turner[30] is

$$h = 0.664\frac{k}{x}(\text{Pr})^{1/3}\text{Re}_L^{1/2} \quad [\text{W/m}^2 \cdot \Delta_1 {}^\circ\text{C}] \tag{4-40}$$

[40] M. Jakob, *Heat Transfer*, Vol. I, pp. 550–554. Wiley, New York, 1949.
[41] W. C. Reynolds, *Int. J. Heat Mass Transfer*, Vol. 6, pp. 445–454 (1963).

Table 4-7 CIRCUMFERENTIAL HEAT FLUX FUNCTIONS S_1 FOR FULLY DEVELOPED TURBULENT FLOW IN A CIRCULAR TUBE WITH CONSTANT HEAT RATE

Re Pr	LAMINAR	10^4	3×10^4	10^5	3×10^5	10^6
0.	1.000	1.000	1.000	1.000	1.000	1.000
0.001		1.000	1.000	0.999	0.974	0.901
0.003		0.999	0.994	0.957	0.831	0.473
0.01		0.991	0.952	0.733	0.409	0.161
0.03		0.923	0.699	0.348	0.145	0.0535
0.7		0.121	0.0490	0.0180	0.00721	0.00275
3		0.0448	0.0178	0.00629	0.00246	0.000902
10		0.0239	0.00931	0.00322	0.00123	0.000438
30		0.0151	0.00582	0.00199	0.00751	0.000166
100		0.00994	0.00383	0.00130	0.000486	0.000166
1000		0.00513	0.00198	0.000667	0.000248	0.0000841

SOURCE: From Kays.[35]

where

$h =$ the average film coefficient $\left[\mathrm{W}/\mathrm{m}^2{\cdot}\Delta_1{}^\circ\mathrm{C}\right]$

and

$$\mathrm{Re}_L = \frac{U_\infty \rho L}{\mu}$$

where

$U_\infty =$ free stream velocity   [m/s]
$L =$ length of plate along the direction
  of the fluid flow   [m]

and the fluid properties are at the average temperature between the free fluid stream and the plate surface.

· *Turbulent Flow.* A simplified equation for the value of h for turbulent flow of water as given by Johnson and Rubesin[8] is

$$h_x = 0.0292 \frac{k}{x}(\mathrm{Pr})^{1/3}(\mathrm{Re}_x)^{0.8}   \left[\mathrm{W}/\mathrm{m}^2{\cdot}\Delta_1{}^\circ\mathrm{C}\right] \tag{4-41}$$

This equation is based upon the fluid velocity profile following the 1/7th power law.

$$\frac{U}{U_\infty} = \left(\frac{y}{\delta}\right)^{1/7}   [-]$$

where

$$U = \text{fluid flow velocity} \qquad [\text{m/s}]$$
$$U_\infty = \text{fluid free stream velocity} \qquad [\text{m/s}]$$
$$y = \text{distance from flat plate surface} \qquad [\text{m}]$$
$$\delta = \text{distance from flat plate to the undisturbed}$$
$$\text{free stream fluid velocity} \qquad [\text{m}]$$

• *Combination of Laminar and Turbulent Flow.* An approximate equation for the average value of h_a for flow across a flat plate including both the initial laminar region and the turbulent region, as given by Bailey, Owen, and Turner[42] is

$$h_a = 0.0365 \frac{k}{L} (\text{Pr})^{1/3} \left[\text{Re}_L^{0.8} - \text{Re}_{\text{crit}}^{0.8} + 18.2 \text{Re}_{\text{crit}}^{0.5} \right] \quad \left[\text{W/m}^2 \cdot \Delta_1 \degree \text{C} \right]$$

(4-42)

where Re_{crit} is the critical Reynolds number. If the critical Reynolds number is not known, an approximation of $\text{Re}_{\text{crit}} = 40{,}000$ can be used.

Example Problem 4-9

Water at 25°C is flowing with a velocity of 4 m/s along a flat plate 2 m in length. If the temperature of the flat plate is 55°C, what is the average heat transfer film coefficient h_a?

SOLUTION

SKETCH

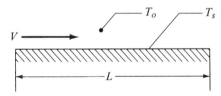

Figure 4-31

EQUATION

$$h_a = 0.0365 \frac{k}{L} (\text{Pr})^{1/3} \left(\text{Re}_L^{0.8} - \text{Re}_{\text{crit}}^{0.8} + 18.2 \text{Re}_{\text{crit}}^{0.5} \right)$$

(4-43)

[42] F. J. Bayley, J. M. Owen, and A. B. Turner, *Heat Transfer*, p. 242. Barnes & Noble Books, New York, 1972.

PARAMETERS

Evaluate k and Re at the average boundary layer temperature.

$T_{av} = (T_o + T_s)/2 \quad [°C]$

$T_o = 25°C \quad$ (given)

$T_s = 55°C \quad$ (given)

$T_{av} = (25 + 55)/2 \quad ([°C] + [°C])/2 = [°C]$

$T_{av} = 40°C$

$k = 0.633 \quad [W/m \cdot \Delta_1 °C] \quad$ (at 40°C, Table B-3, Appendix B)

$L = 2 \text{ m} \quad$ (given)

$Pr = 4.33 \quad [—] \quad$ (at 40°C, Table B-3, Appendix B)

$Pr^{1/3} = 1.629 \quad [—]$

$Re_L = LV\rho/\mu \quad [—]$

$L = 2 \text{ m} \quad$ (given)

$V = 4 \text{ m/s} \quad$ (given)

$\rho = 992.6 \quad [kg/m^3] \quad$ (at 40°C, Table B-3, Appendix B)

$\mu = 0.656 \times 10^{-3} \quad [kg/m \cdot s] \quad$ (at 40°C, Table B-3, Appendix B)

$Re_L = \dfrac{(2)(4)(992.6)}{0.656 \times 10^{-3}} \quad \dfrac{[m][m/s][kg/m^3]}{[kg/m \cdot s]} = [—]$

$Re_L = 12,104,878 \quad [—]$

$Re_L^{0.8} = 463,840 \quad [—]$

$Re_{crit} = 40,000 \quad [—] \quad$ (assume per text)

$Re_{crit}^{0.8} = 4804 \quad [—]$

$Re_{crit}^{0.5} = 200 \quad [—]$

SUBSTITUTION

$h_a = (0.0365)\dfrac{(0.633)}{(2)}(1.629)[463,840 - 4804 + (18.2)(200)]$

$\quad [—]\dfrac{[W/m \cdot \Delta_1 °C]}{[m]}[—][[—] - [—] + ([—][—])]$

$\quad = [W/m^2 \cdot \Delta_1 °C]$

$\quad = (0.01882)(462,676)$

ANSWER

$h_a = 8708 \quad [W/m^2 \cdot \Delta_1 °C]$

• *Flow Across a Circular Cylinder.* The basic characteristics of water flow across a cylinder are the same as for gas, as discussed on page 174.

• *The Average Value of h.* The average value of h for liquids flowing across a single tube normal to the tube axis for the range of Reynolds numbers from 50 to 10,000 can be calculated from the formulae given below.

McAdams[22] based upon the results of several authors recommends

$$\bar{h} = (0.35 + 0.56\text{Re}^{0.52})\frac{k}{D}\text{Pr}^{0.3} \quad \left[\text{W}/\text{m}^2 \cdot \Delta_1 {}^\circ\text{C}\right] \tag{4-43}$$

Jakobs and Hawkins (Reference 3, page 141) recommend

$$\bar{h} = 0.6\frac{k}{D}\text{Re}^{0.5}\text{Pr}^{0.31} \quad \left[\text{W}/\text{m}^2 \cdot \Delta_1 {}^\circ\text{C}\right] \tag{4-44}$$

These two equations give similar results with variations of as much as 10%.

• *Flow Through Tube Banks.* The majority of heat transfer equipment designs, wherein a liquid flow passes across the tubes in a tube bank, are of the nature of a multipass tube-in-shell heat exchanger. Because the fluid flow through such a configuration is relatively complex, designs are made on the basis of the overall heat transfer coefficient U, which is determined from tests of the overall performance of various configurations. Consequently, in contrast to gas flow, little experimental data are available for a uniform flow of liquid passing through a tube bank.

EQUATIONS FOR h FOR LIQUIDS SIGNIFICANTLY MORE VISCOUS THAN WATER

The effect of the liquid viscosity upon the heat transfer coefficient is manifested primarily in the boundary layer. When the viscous liquid is being heated, the viscosity in the boundary layer adjacent to the hot wall becomes less. This favors increased turbulence, resulting in a higher heat transfer. Conversely, when the viscosity in the boundary layer adjacent to a cold wall increases, a lower degree of flow turbulence and a decrease in the heat transfer results.

• *Flow Inside of Pipes—Laminar Flow.* For heat transfer under conditions of small values for the pipe diameter D and the temperature difference between the average fluid bulk and the wall surface, and when the buoyancy effects are negligible, Sieder and Tate[27] suggest the following formula for short tubes:

$$h = 1.86\frac{k}{D}\left(\frac{\mu}{\mu_s}\right)^{0.14}\left(\text{Re}\frac{D}{L}\right)^{1/3}(L \cdot \text{Pr}^{1/3}) \quad \left[\text{W}/\text{m}^2 \cdot \Delta_1 {}^\circ\text{C}\right] \tag{4-45}$$

where all properties, except the viscosity at the wall temperature, are evaluated at the average bulk temperature.

For laminar flow inside horizontal pipes heated by steam, Eubank and Proctor[43] obtained the following empirical formula for petroleum oils:

$$h = 1.75 \frac{k}{D} \left(\frac{\mu}{\mu_s} \right)^{0.14} \left[\frac{\pi}{4} \mathrm{Re} \mathrm{Pr} \left(\frac{D}{L} \right) + 0.04 \left(\frac{D}{L} \mathrm{Gr} \mathrm{Pr} \right)^{0.75} \right]^{1/3}$$

$$\left[W/m^2 \cdot \Delta_1 {}^\circ C \right] \tag{4-46}$$

where

D = pipe diameter		[m]
Gr = Grashof number evaluated at the wall surface temperature		[—]
L = pipe length		[m]
Pr = Prandtl number evaluated at the wall surface temperature		[—]
Re = Reynolds number evaluated at the average bulk temperature		[—]
μ = fluid viscosity evaluated at the average bulk temperature		[Pa·s]
μ_s = fluid viscosity evaluated at the average wall surface temperature		[Pa·s]

Equation (4-4) correlates within 30% over the range of (a) Prandtl numbers 140 to 15,200 and (b) (Gr)(Pr), 3.3×10^5 to 8.6×10^8.

• *Flow Inside of Pipes—Turbulent Flow.* For liquids with high viscosities, or for low viscosity liquids with substantial temperature differences the Sieder and Tate[44] equation is usually used.

$$h = 0.027 \frac{k}{D} \left(\frac{DV_\rho}{\mu} \right)^{0.8} \left(\frac{\mu c_p}{k} \right)^{1/3} \left(\frac{\mu}{\mu_s} \right)^{0.14} \quad \left[W/m^2 \cdot \Delta_1 {}^\circ C \right] \tag{4-47}$$

The correlation of this equation with test data is shown in Figure 4-32 by the straight line between Re of 10,000 to 150,000. Below 10,000 the L/D ratio of the tube has a significant effect.

The spread of the test data is significantly greater than that of the test data for water shown in Figure 4-23. Consequently, the accuracy of the h calculated for viscous liquids is more nearly ±30%.

[44] E. N. Sieder and G. E. Tate, Ind. Eng. Chem. Vol. 28, p. 1429 (1936).

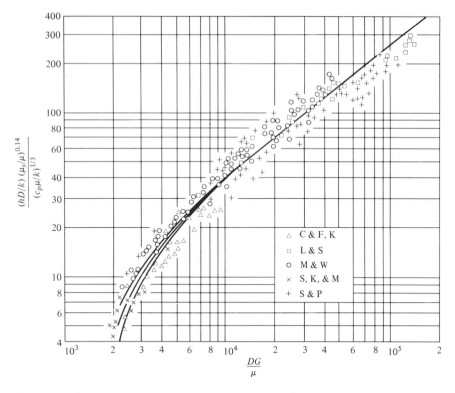

Figure 4-32 Heating and cooling of oils inside tubes. (SOURCE: *Heat Transmission*, 2d ed., Fig. 91, p. 195. McGraw-Hill Book Co., New York, 1942)

· *Other Configurations.* As configurations involving heat transfer between viscous fluids and flat plates or the outside of tubes is relatively uncommon, these configurations are not discussed in this text.

EQUATION FOR *h* FOR LIQUID METALS Pr<0.1

As liquid metals are characterized by a high thermal diffusivity, conduction (even in a highly turbulent stream) is the dominant heat transfer mechanism. Consequently, the empirical equations for gases and liquids with a Prandtl number greater than 0.7 do not apply, because turbulence, by contrast, becomes a dominant factor in the heat transfer mechanism in this Prandtl number region. In the analysis of heat transfer two basic heat transfer configurations are used, namely

- constant heat flow rate (heat flux)
- constant wall temperature

Theoretical analyses indicate a difference in the Nusselt number (Nu), and consequently a difference in *h* between these two configurations. This

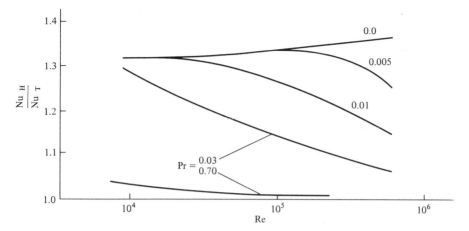

Figure 4-33 The relationship between the ratio of the Nusselt number for constant heat flow rate to the Nusselt number for constant surface temperature, the Prandtl number, and the Reynolds number. (SOURCE: C. A. Sleicher and M. Tribus, *Heat Transfer Fluid Mechanics, Inst.*, Fig. 12, p. 78. Stanford University Press, Stanford, CA, 1956)

difference is a function of the Prandtl number and the Reynolds number. The results of studies by Sleicher and Tribus[45] for fully developed flow conditions showing the relationship between the ratio of the Nusselt number for constant heat flow rate and the Nusselt number for constant surface temperature, the Prandtl number, and the Reynolds number are shown in Figure 4-33. For liquid metals with a Prandtl number <0.1 the difference between the two Nusselt numbers can be greater than 30%. For gases with a Prandtl number of 0.7 the difference is only a few percent. For water and viscous liquids there is essentially no difference between the two heat transfer configurations. Hence, for liquid metals usually two equations for the Nusselt number or *h* are given, one for constant heat flow rate and another for constant wall temperature. For gases and liquids with a higher Prandtl number only one equation is given, which is suitable for either configuration.

In the application of equations to liquid metal heat transfer it must be realized that significant discrepancies may be encountered between the actual and calculated performance because of effects such as

- impurities in the liquid metal
- degree of surface wetting

[45]C. A. Sleicher and M. Tribus, *Heat Transfer Fluid Mechanics Inot.*, p. 59. Stanford University Press, Stanford, CA, 1956.

Impurities can cause significant reductions in the thermal conductivities of metals as indicated on page 28. This can result in a lower actual Nusselt number and h than that calculated on the basis of the properties of the pure liquid metal. In systems with sodium, it should be noted that oxygen is soluble in sodium. The solubility increases rapidly as the temperature increases. Around 500°C the solubility approaches 0.1% by mass. Oxygen also reacts with sodium to form the oxide Na_2O, which deposits on heat transfer surfaces resulting in an added heat transfer resistance. Oxide contaminations can decrease the Nusselt number by as much as 50%.

Poor wetting characteristics are encountered with heavy liquid metals, such as mercury, lead, and bismuth and their alloys, resulting in higher surface resistances to heat flow. Consequently the heat transfer coefficients are lower than those for the lighter, more active metals such as sodium, magnesium, and potassium. The variation in the heat transfer coefficients may be as much as 50% between the heavier and lighter metals.

Test points in a series of test data, such as shown in Figure 4-34, may often contain a group of points giving significantly lower values of Nusselt number than the majority of the test data. This discrepancy is usually attributed to the liquid metal not wetting the heat transfer surface.

The factors affecting the wetting and nonwetting of the walls in liquid metal systems as well as the procedures to obtain and maintain wetting are relatively complex, and are not discussed in this text.

In many of the equations used with liquid metals, the product of the Reynolds number and the Prandtl number is encountered. This product, which is dimensionless, is called the Peclet number (Pe).

$$\text{Pe} = (\text{Re})(\text{Pr}) = \frac{DV\rho c_p}{k} \quad [-] \tag{4-48}$$

• *Flow Inside Circular Pipes.* In the following discussion, equations are presented for both the constant heat flux and the constant wall temperature conditions. One commonly used set of equations, based upon Martinelli's[46] analogy, simplified by Lyon,[47] is:

For uniform heat flux:

$$\text{Nu} = 7 + 0.025(\text{Pe})^{0.8} \quad [-] \tag{4-49}$$

For constant wall temperature

$$\text{Nu} = 5 + 0.025(\text{Pe})^{0.8} \quad [-] \tag{4-50}$$

[46] R. C. Martinelli, "Heat Transfer to Molten Metals," Trans. A.S.M.E., Vol. 69, p. 947 (1947).
[47] R. N. Lyon, *Liquid Metals Handbook*. Atomic Energy Commission and Department of the Navy, Washington, D.C., 1952.

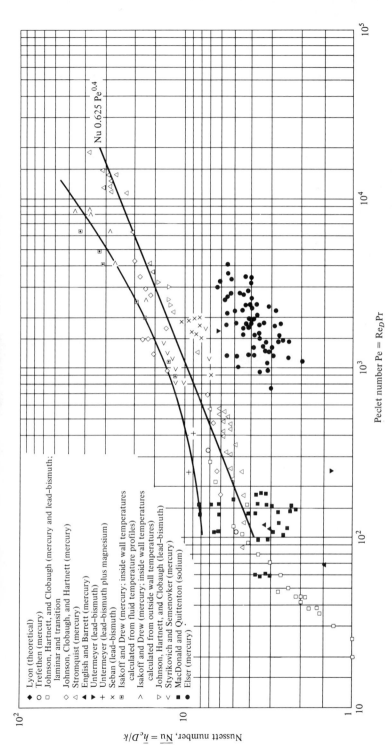

Figure 4-34 Comparison of measured and predicted Nusselt number for liquid metals heated in long tubes with constant heat input. (SOURCE: NASA, NACA TN 3336)

229

A similar set of equations recommended by Kays[28] is:

For uniform heat flux:

$$Nu = 6.3 + 0.003(Pe) \quad [-] \tag{4-51}$$

For constant wall temperature:

$$Nu = 4.8 + 0.003(Pe) \quad [-] \tag{4-52}$$

Skupinski et al.[48] developed the following equation for sodium/potassium (NaK) mixtures for uniform heat flux for the following conditions: $3600 < Re < 900,000$ and $100 < Pe < 10,000$.

$$Nu = 4.82 + 0.0185(Pe)^{0.827} \quad [-] \tag{4-53}$$

For most liquid metals, except mercury, Lubarsky and Kaufman[49] give the following equation for uniform heat flux for the following conditions: $100 < Pe < 10,000$ and $L/D > 60$.

$$Nu = 0.625(Pe)^{0.4} \quad [-] \tag{4-54}$$

• *Flow Across Rod Bundles*. Based upon experiments at Brookhaven National Laboratory,[50, 51] the following formula was developed for mercury flowing perpendicular to a rod bundle.

$$Nu_a = 4.03 + 0.228(Re_m Pr)^{0.67} \quad [-] \tag{4-55}$$

where

> Nu_a = average Nusselt number
> Re_m = Reynolds number based upon the maximum
> fluid velocity through the tube bundle
> Pr = Prandtl number taken with fluid properties
> at the average film temperature

All other properties are taken at the bulk temperature.

> The geometry of the tube bundle was as follows (refer to Figure 4-15):
> equilateral triangle arrangement $= 2Sn = S'p = 1.375D$
> tube diameter $= 1.27$ cm
> depth of array $= 10$ rows

[48] E. Skupinski, J. Tortel, and L. Vautrey, Intern. J. Heat Mass Transfer, Vol. 8, pp. 937–951 (1965).

[49] B. Lubarsky and S. S. Kaufman, "Review of Experimental Investigations of Liquid-Metal Heat Transfer," NACA TN 3336 (1955).

[50] R. J. Hoe, D. Dropkinn, and O. E. Dwyer, "Heat Transfer Rates for Cross Flowing Mercury in a Staggered Tube Bank—I," Trans. A.S.M.E., Vol. 79, pp. 899–918 (1957).

[51] C. L. Richards, O. E. Dwyer, and D. Dropkinn, "Heat Transfer Rates for Cross Flowing Mercury in a Staggered Tube Bank—II," Paper No. 57-Ht-11, ASME-AICHE Heat Conference (1957).

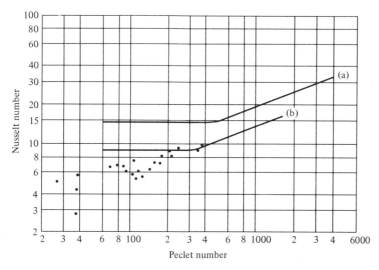

Figure 4-35   Heat transfer correlation for liquid metal flow parallel to rod bundles. $S'_p =$ (a), 1.5, (b), 1.2. [SOURCE: O. E. Dwyer, "Analytical Study of Heat Transfer for Liquid Metals Flowing In-Line Through Closely Packed Rod Bundles," Nuclear Science and Engineering, Vol. 25, No. 4 (August 1966)]

• *Flow Parallel to Rod Bundles.* The results of analytical work by Dwyer[52] are presented in Figure 4-35, which gives a relationship between the Nusselt number and the Peclet number for two rod bundle spacings on an equilateral triangle pattern. Experiments by Borishansky and Firsova[53] using sodium indicated a fair agreement with Dwyer's analytical curves over a range of Peclet numbers. The poor agreement at low Peclet numbers was attributed to oxide problems in the experiments.

• *Flow Between Parallel Plates.* Seban[54] developed the following equations for flow between two parallel plates. One equation is for constant heat flux through one wall only (with the other wall adiabatic); the other equation is for constant heat flux through both walls.

For constant flux through one wall with the other wall adiabatic the equation is

$$Nu = 5.8 + 0.02Pe^{0.8} \quad [—] \tag{4-56}$$

[52] O. E. Dwyer, "Analytical Study of Heat Transfer for Liquid Metals Flowing In-Line Through Closely Packed Rod Bundles," Nuclear Science and Engineering, Vol. 25, No. 4, pp. 343–358 (August 1966).

[53] V. M. Borishansky, and E. V. Firsova, "Heat Exchange in the Longitudinal Flow of Metallic Sodium Past a Tube Bank," Atomic Energy, Vol. 14, p. 584 (1963).

[54] R. A. Seban, "Heat Transfer to a Fluid Flowing Turbulently Between Parallel Walls with Asymmetric Wall Temperature," Trans. A.S.M.E., Vol. 72, p. 789.

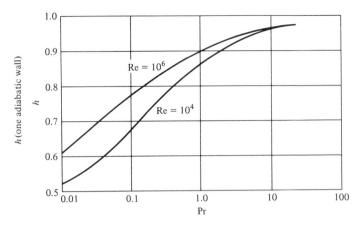

Figure 4-36   Conversion factor for constant heat flux through both walls of a parallel plate flow passage.

For constant heat flux through both walls the Nusselt number is more, being obtained by dividing the h obtained from Eq. (4-5) by the factor indicated by the plot in Figure 4-36.

· *Flow in Annuli.* In configurations where the ratio of the outside to inside diameters of the annuli are close to 1, the flow can be treated as flow between two parallel plates.

In configurations where the ratio of the outside to inside diameters of the annuli are greater than 1.4 significant error is encountered if the flow is treated as flow between two parallel plates. Bailey[55] and Werner et al.[56] suggest the following equation:

$$\text{Nu} = 5.25 + 0.0188 \text{Pe}^{0.8} \left(\frac{D_o}{D_i} \right)^{0.3} \quad [-] \tag{4-57}$$

4-3 FLUID FLOW MECHANICAL ENERGY LOSSES

In the design of both thermodynamic and heat transfer systems the mechanical energy loss in the fluid flow circuit is important. The mechanical pumping power required to overcome such losses can have a significant effect upon the performance of a thermodynamic system or the cost of operation of heat exchange equipment.

[55]R. V. Bailey, "Heat Transfer for Liquid Metals in Concentric Annuli," U.S. Atomic Energy Comm. Rep. ORNL-531, Oak Ridge National Laboratory (1950).
[56]R. C. Werner, E. C. King, and R. A. Tidball, "Forced Convection Heat Transfer with Liquid Metals," Paper presented at 42nd Annual Meeting, American Institute of Chemical Engineers, December 1949.

The energy loss, or pressure drop as it is often called, in a fluid flow system comprises the following elements:

- surface friction losses
     friction in straight pipes or ducts
     duct bend or elbow loss
- dynamic losses
     pipe or duct entrance loss
     pipe or duct enlargement loss
     exit loss

In many piping systems the surface friction losses are dominant and consequently the dynamic losses become relatively unimportant and often are neglected. However, in many heat exchanger designs relatively short lengths of flow passages are combined with flow passage entrances, exits, and manifolds of nonuniform flow cross sections. The dynamic losses can become important and must be considered. The surface friction and dynamic losses are discussed in the following sections.

4-3.1 Fluid Flow Mechanical Energy Loss Elements

SURFACE FRICTION LOSSES
Surface friction losses include both:

- friction in straight pipes or ducts
- duct bend or elbow losses

In a system comprising both straight pipe sections and bends or elbows, it is often convenient to characterize the friction of the bend or elbow as being equivalent to a certain length of straight pipe.

To calculate the friction loss of a system comprised of sections of straight pipe and bends or elbows, the system would be treated as a straight pipe with a length equal to the sum of the straight pipe sections plus the equivalent straight pipe lengths of the bends or elbows in the system.

The straight pipe friction loss and the bend or elbow losses are discussed in the following sections.

• *Straight Pipe Friction Loss.* The straight pipe friction loss is a function of three parameters, namely

- the friction factor $[f]$
- the pipe length (in terms of length/diameter) $[L/D]$
- the fluid flow kinetic energy $[V^2/2]$

The equation for the straight pipe friction loss is

$$\Delta P = f\frac{L}{D}\rho\frac{V^2}{2} \quad [\text{Pa}] \tag{4-58}$$

where

$$
\begin{aligned}
\Delta P &= \text{pressure loss} &&[\text{Pa}] \\
f &= \text{friction factor} &&[-] \\
L &= \text{pipe length} &&[\text{m}] \\
D &= \text{pipe diameter} &&[\text{m}] \\
\rho &= \text{fluid density} &&[\text{kg/m}^3] \\
V &= \text{fluid velocity} &&[\text{m/s}]
\end{aligned}
$$

The friction factor (f) is a function of

- Reynolds number (Re) of the fluid flow in the duct
- Relative surface roughness (ε/D) of the duct surface

The usual way to determine a value for f for flow inside circular ducts is to use the Moody diagram presented in Figure 4-37.[57] The coordinates of the graph are friction factor (f) versus Reynolds number (Re) with lines for various relative surface roughnesses (ε/D). The value of the friction factor (f) for a flow condition is determined by the intersection of the fluid flow Reynolds number and the relative surface roughness line.

The relative roughness of the pipe (ε/D) is the dimensionless ratio of the characteristic roughness (ε) of the surface to the diameter (D) of the pipe. Typical characteristic roughnesses of various surfaces are given in Table 4-8.

The friction factor f in Figure 4-37 is for isothermal flow in circular ducts. To modify the friction factor for heat transfer conditions (heating or cooling) the effect of the viscosity of the fluid is considered. When a viscous fluid is being heated, the viscosity at the wall temperature and in the boundary layer is less than the viscosity at the bulk temperature. The pressure drop is less than that for conditions of isothermal flow at the bulk temperature. Conversely, when a viscous fluid is being cooled, the pressure drop is more than that for conditions of isothermal flow at the bulk temperature.

When low viscosity fluids (similar to water) are involved and the temperature gradients are small (less than 10°C) the effects of the variation of viscosity in the boundary layer upon the pressure drop are sufficiently small that a modification of the friction factor determined using the fluid flow Reynolds number at the fluid flow bulk temperature is not necessary.

---

[57] L. F. Moody, "Friction Factors for Pipe Flows," Trans. A.S.M.E., Vol. 66, pp. 671–694 (1944).

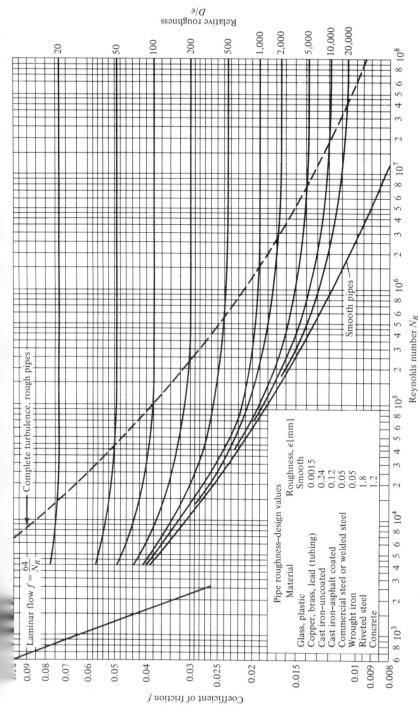

Figure 4-37 Friction factors for pipes. (SOURCE: L. F. Moody, "Friction Factors for Pipe Flows," Trans. A.S.M.E., Vol. 66, pp. 671–694 (1944)]

Table 4-8 TYPICAL CHARACTERISTIC ROUGHNESS
OF VARIOUS SURFACES

SURFACE	CHARACTERISTIC ROUGHNESS ε [mm]
Riveted steel	1–10
Concrete	0.3–3
Wood stave	0.18–1
Cast iron	0.25
Galvanized iron	0.15
Asphalted cast iron	0.12
Steel or wrought iron	0.05
Drawn tubing	0.0015

Experiments[58] have indicated that for most applications, a correction for the pressure drop of viscous fluids can be made by multiplying the basic pressure drop equation for surface friction losses [Eq. (4-58)] by the ratio of the wall temperature viscosity (μ_w) to the bulk temperature viscosity (μ_b) raised to the 0.14 power. The friction factor (f) is evaluated at the Reynolds number for the average (bulk) temperature of the fluid flow.

$$\Delta p = f \frac{L}{D} \rho \frac{V^2}{2} \left[\frac{\mu_w}{\mu_b} \right]^{0.14} \quad [\text{Pa}] \tag{4-59}$$

If the duct does not have a circular cross section, an equivalent diameter can be calculated for use in both the

- determination of Reynolds number for obtaining the friction factor (f) from Figure 4-34
- in Eqs. (4-59) and (4-60) for calculating Δp

The equivalent diameter is calculated from the cross-sectional area A of the duct (normal to the fluid flow) and the wetted perimeter of the duct P_w as follows:

$$D_e = \frac{4A}{P_w} \tag{4-60}$$

where

D_e = equivalent diameter of the duct   [m]
A = duct cross-sectional area (normal to the fluid flow)   [m^2]
P_w = duct wetted perimeter   [m]

This has been verified experimentally by investigations such as by Huebscher.[59]

[58] E. M. Sieder and G. E. Tate, Ind. Eng. Chem., Vol. 28, p. 1429 (1936).
[59] R. G. Huebscher, "Friction Equivalents for Round, Square, and Rectangular Ducts," Trans. A.S.H.V.E., Vol. 54, pp. 101–108 (1948).

Example Problem 4-10

Determine the equivalent diameter of a rectangular duct, with a side length of 0.2 m, and the other 0.1 m.

SOLUTION

SKETCH

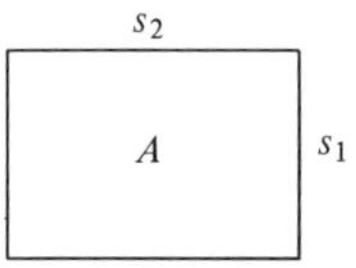

Figure 4-38

EQUATION

$$D_e = \frac{4A}{P_w} \quad [\text{m}] \quad [\text{Eq. (4-60)}]$$

PARAMETERS

$A = (s_1)(s_2) \quad [\text{m}^2]$

$s_1 = 0.1$ m

$s_2 = 0.2$ m

$A = (0.1)(0.2) = 0.02$ m^2

$P_w = s_1 + s_2 + s_1 + s_2 = 2(s_1 + s_2) \quad [\text{m}]$

$P_w = 2(0.1 + 0.2) = 0.6$ m

SUBSTITUTION

$$D_e = \frac{4 \times 0.02}{0.6} \quad [\text{m}^2]/[\text{m}] = [\text{m}]$$

ANSWER

$D_e = 0.133$ m

• *Bend or Elbow Loss.* When the fluid passes through a bend in the duct, added turbulence is introduced into the flow, as well as an additional length in the duct. The measurement of the loss introduced by the bend is expressed in terms of the equivalent length of straight pipe that has been added to the system. This equivalent length is given in terms of L/D. Typical values for common pipe fittings are given in Table 4-9.

Typical values for rectangular ducts are given in Table 4-10.

Table 4-9 RESISTANCE IN VALUES AND FITTINGS EXPRESSED AS EQUIVALENT LENGTH IN PIPE DIAMETERS, L_e/D

TYPE	EQUIVALENT LENGTH IN PIPE DIAMETERS, L_e/D
Globe valve—fully open	340
Angle valve—fully open	145
Gate valve—fully open	13
—$\frac{3}{4}$ open	35
—$\frac{1}{2}$ open	160
—$\frac{1}{4}$ open	900
Check valve—swing type	135
Check valve—ball type	150
Butterfly valve—fully open	40
90° standard elbow	30
90° long radius elbow	20
90° street elbow	50
45° standard elbow	16
45° street elbow	26
Close return bend	50
Standard tee—with flow through run	20
—with flow through branch	60

SOURCE: The Crane Co., New York, N.Y.

Table 4-10 EQUIVALENT LENGTH FOR 90° DUCT ELBOWS TO BE ADDED TO STRAIGHT RUNS OF DUCT MEASURED TO THE INTERSECTION OF THEIR CENTERLINES

ADDITION EQUIVALENT LENGTH IN TERMS OF WIDTH, L/W

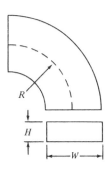

	CENTERLINE RADIUS RATIO, R/W					
ASPECT RATIO, H/W	0.50	0.75	1.00	1.25	1.50	2.00
0.15	20	10	6.4	4.4	3.2	2.0
0.25	24	12	7.2	4.5	3.5	2.1
0.50	33	15	8.7	5.6	3.9	2.3
1.0	46	19	10.6	6.6	4.4	2.4
2.0	65	26	13	7.8	5.1	2.6
4.0	—	35	17	9.5	6.0	2.9
6.0	—	42	20	10.6	6.5	3.0

SOURCE: From Reference 60.

[60]Aubrey I. Brown, and Salvatore M. Marco, *Introduction to Heat Transfer*, 3rd ed., p. 119. McGraw-Hill, New York 1958.

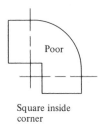

Square inside
corner

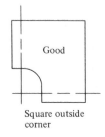

Square outside
corner

Figure 4-39   Duct elbow design modifications.

In rectangular ducts, the inside radius of an elbow (90° bend) is the most important dimension. A small radius increases the boundary layer separation turbulence. Two modifications of a constant width duct elbow design are often encountered. These are illustrated in Figure 4-39. The design with the small inside radius should be avoided.

In cases where large ducts must make sharp turns, vanes are often used. A typical design as illustrated in Figure 4-40 can be very efficient.

DYNAMIC LOSSES

• *Entrance Loss.* When the fluid flow is subject to a sudden flow cross-sectional area restriction, such as fluid flow leaving a large manifold to enter an array of tubes, an energy loss occurs. This loss is usually expressed in terms of the fluid flow velocity head in the smaller duct.

$$\Delta p = K \frac{\rho V^2}{2} \quad [\text{Pa}] \tag{4-61}$$

where

Δp = pressure loss	[Pa]
V = velocity in the smaller duct (average)	[m/s]
K = constant (see Table 4-11)	[—]
ρ = fluid density	[kg/m³]

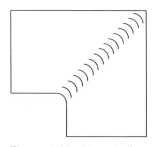

Figure 4-40   Vaned elbow.

Table 4-11 FRICTION COEFFICIENT FOR ABRUPT ENTRANCES

RATIO A_1/A_2	0.1	0.2	0.3	0.4	0.5	0.6	0.7	0.8	0.9	1.0
K		0.44	0.35	0.29	0.24	0.19	0.15	0.10	0.07	0.04	0.00

SOURCE: From Hughes and Safford.[61]

K is a function of the ratio of the cross-sectional area (A_1) of the smaller duct to the area (A_2) of the larger enclosure to which the smaller duct is connected. Both areas are taken normal to the fluid flow. Values of K have been obtained experimentally. Typical values are shown in Table 4-11.

Example Problem 4-11

Calculate the entrance loss sustained by a mass of water flowing from a larger pipe into a smaller pipe with a sudden flow area restriction. The reduction in the flow cross-sectional area (A_1/A_2) is 0.3. The water flow velocity in the smaller pipe is 5 m/s. If the density of the water is 1000 kg/m³, what is the entrance pressure loss (Δp) in Pa?

SOLUTION

SKETCH

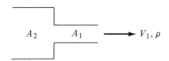

Figure 4-41

EQUATION

$$\Delta p = K \rho V^2 / 2 \quad [\text{Pa}] \qquad [\text{Eq. (4-61)}]$$

PARAMETERS

$$A_1/A_2 = 0.3 \qquad (\text{given})$$

Therefore

$$K = 0.29 \quad (\text{dimensionless}) \qquad (\text{Table 4-11})$$
$$\rho = 1000 \text{ kg/m}^3 \qquad (\text{given})$$
$$V = 5 \text{ m/s} \qquad (\text{given})$$

[61] H. C. Hughes and A. T. Safford, *Hydraulics*, Macmillan, New York, 1926.

SUBSTITUTION IN EQUATION

$$\Delta p = (0.29)(1000)(5^2)/(2) \quad [-][kg/m^3][m/s][-]$$
$$= [kg/ms^2] = [Pa]$$

ANSWER

$$\Delta p = 3625 \quad [Pa]$$

• *Enlargement Loss.* When the cross-sectional area of a fluid duct is increased in size by a sudden enlargement, such as shown in Figure 4-42, the fluid flow is subject to an energy loss. This results from the flow separation and resulting turbulence in the boundary layer at the sudden enlargement. Experimental measurements show the following expression is a reasonable measurement of the sudden expansion loss:

$$\Delta p = \frac{\rho(V_1^2 - V_2^2)}{2} \quad [Pa] \qquad (4\text{-}62)$$

Typical examples of sudden enlargements in fluid flow ducts are (1) the exit from a bank of tubes into a manifold, and (2) tees where the flow is split into two branches, each of which is often of the same size as the initial duct.

To minimize enlargement losses, the duct expansion can be made with a gradual divergence. The minimum losses are obtained with a duct wall divergence angle (wall surface to fluid stream axis) of approximately 3°. Steeper divergences result in higher losses from turbulence. Shallower angles reduce the turbulence losses, but increase the equivalent straight pipe friction loss associated with the expansion.

Example Problem 4-12

Calculate the sudden expansion loss sustained by a mass of water flowing from a smaller pipe into a larger pipe with a sudden flow area expansion of a 2:1 area ratio. The water flow velocity (V_1) in the smaller tube is 10 m/s. If the density (ρ) of the water is 1000 kg/m^3, what is the sudden expansion pressure loss (Δp) in Pa?

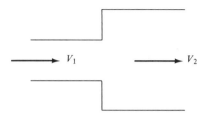

Figure 4-42   Sudden enlargement of a duct.

SOLUTION

SKETCH

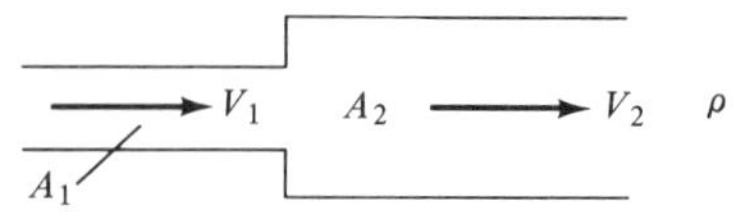

Figure 4-43

EQUATION

$$\Delta p = \frac{\rho \left[V_1^2 - V_2^2 \right]}{2} \quad [\text{Pa}] \quad [\text{Eq. (4-62)}]$$

PARAMETERS

$\rho = 1000 \text{ kg/m}^3$ \quad (given)

$V_1 = 10 \text{ m/s}$ \quad (given)

$V_2 = V_1(A_1/A_2)$ \quad (continuity equation)

$A_1/A_2 = \frac{1}{2}$ \quad (given)

$V_2 = (10)(\frac{1}{2})$

$V_2 = 5 \text{ m/s}$

SUBSTITUTION IN EQUATION

$$\Delta p = (1000)(10^2 - 5^2)/2 \quad \left[\text{kg/m}^3\right]\left[\text{m/s}\right]^2[-] = \left[\text{kg/ms}^2\right] = [\text{Pa}]$$

ANSWER

$$\Delta p = 37,500 \quad [\text{Pa}] \rightarrow 37.5 \quad [\text{kPa}]$$

· *Exit Loss.* If the fluid in a heat exchange system is not continuously recirculated in a duct system, but is discharged into a tank or reservoir, an exit loss is incurred. This loss is basically the same as a sudden expansion loss. Usually the expanded velocity is 0, so the loss becomes

$$\Delta p = \rho \frac{V^2}{2} \quad [\text{Pa}] \tag{4-63}$$

Example Problem 4-13

Calculate the exit loss sustained by a mass of water flowing out of a pipe into a large reservoir. The water flow velocity leaving the pipe is 3 m/s. If the density of the water is 1000 kg/m³, what is the exit pressure loss (Δp) in Pa?

SOLUTION

SKETCH

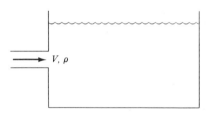

Figure 4-44

EQUATION

$$\Delta p = \rho V^2 / 2 \quad [\text{Pa}] \qquad [\text{Eq. (4-63)}]$$

PARAMETERS

$$\rho = 1000 \quad [\text{kg/m}^3] \qquad (\text{given})$$
$$V = 3 \quad [\text{m/s}] \qquad (\text{given})$$

SUBSTITUTION IN EQUATION

$$\Delta p = (1000)(3^2)/(2) \quad [\text{kg/m}^3][\text{m/s}]^2 [-] = [\text{kg/ms}^2] = [\text{Pa}]$$

ANSWER

$$\Delta p = 4500 \quad [\text{Pa}]$$

4-3.2 Elevation Change

If the system is closed, as illustrated in Figure 4-45, there is no elevation change effect to consider because the fluid does not leave the system.

In an open system, as illustrated in Figure 4-46, the fluid can enter the system at one elevation, and leave the system at another. If the fluid leaves at a higher elevation, the energy required to lift the fluid to the higher elevation must be provided to support the fluid flow. If the fluid leaves the system at a lower elevation, as shown in Figure 4-46, the energy associated with the drop in elevation is consumed by the fluid flow energy losses, and reduces the amount of energy that must be supplied to the fluid flow system.

The contribution to the total system pressure drop by the elevation change between the fluid intake and discharge from the system is

$$\Delta p = \rho g Z \quad [\text{Pa}] \tag{4-64}$$

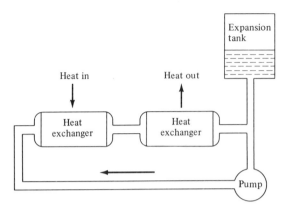

Figure 4-45   A closed fluid system.

where

Δp = pressure   [Pa]
ρ = fluid density   [kg/m^3]
g = acceleration of gravity   (9.81 m/s^2 at the earth's
                                                  surface)
Z = elevation change, $+$up, $-$down   [m]

If the elevation of the fluid discharged from the system were discharged at a higher elevation, the total system pressure drop (loss) would be increased by the amount of the elevation pressure; if discharged at a lower elevation, the total system pressure drop (loss) would be decreased by the amount of the elevation pressure.

Example Problem 4-14

In an open flow system, the surface of the reservoir into which the water flow discharges is 75 m lower in elevation (Z) than the surface of the intake reservoir. How much will this elevation head (pressure Δp) reduce the pressure that must be generated by the pump to equal the pressure

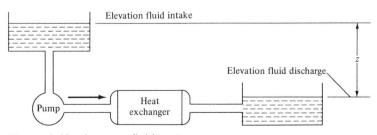

Figure 4-46   An open fluid system.

drop (Δp_f) of the piping system, if the temperature of the water is 15°C and the local acceleration of gravity is 9.81 m/s²?

SOLUTION

SKETCH

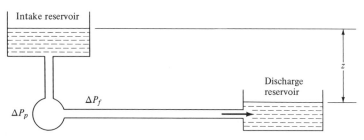

Figure 4-47

EQUATIONS

$$\Delta p_p = \Delta p_f + \Delta p$$

$$\Delta p = \rho g Z \quad [\text{Pa}] \qquad [\text{Eq. (4-64)}]$$

PARAMETERS

$$\rho = 999 \quad [\text{kg/m}^3] \qquad (\text{Appendix B-3, water at 15°C})$$

$$g = 9.81 \quad [\text{m/s}^2] \qquad (\text{given})$$

$$Z = -75 \quad [\text{m}] \qquad (\text{given})$$

SUBSTITUTE IN EQUATION

$$\Delta p = (999)(9.81)(-75) \quad [\text{kg/m}^3][\text{m/s}^2][\text{m}] = [\text{kg/ms}^2] = [\text{Pa}]$$

ANSWER

$$\Delta p = -735{,}000 \quad [\text{kPa}] \rightarrow 735 \quad [\text{kPa}]$$

4-3.3 Pressure Measurements

A flowing fluid stream has basically two energy components, which are

- static pressure (potential energy)
- velocity energy (kinetic energy)

Pressure is a measurement of this energy. Often these two components are measured in a height of fluid column that exerts an equal pressure. This height term is called "head," and is expressed in meters. The two energy components are then "static" head and "velocity" head, and the sum of the two is called the "total" head.

STATIC HEAD

The static head (static pressure) in a fluid system is based upon a selected elevation datum in the fluid system. At any point in a fluid system, the magnitude of the static pressure measurement is a function of the elevation at which the static pressure reading was taken. This is illustrated by Figure 4-48, which shows two pressure gauges connected to the same static pressure tap in the pipe line. One gauge is at the elevation of the pipe centerline and reads a pressure p. The second pressure gauge is at some other location at an elevation Z above the pipe centerline. It reads a reduced pressure p'.

$$p' = p - \rho g Z \quad [\text{Pa}] \tag{4-65}$$

where

$\rho =$ the density of the fluid in the tube connecting
  the pressure gauge to the pipe static pressure tap $\quad [\text{kg/m}^3]$
$g =$ acceleration of gravity $\quad (9.81 \text{ m/s}^2 \text{ at earth's surface})$
$Z =$ elevation change, $+$up, $-$down $\quad [\text{m}]$

The magnitude of the static head can be presented in various units, such as head (meters) of a fluid of a specified density or force per unit area (pascals), and must be referred to the selected static pressure elevation datum.

Example Problem 4-15

A pressure gauge is located 3 m above a pipe to which a static pressure tap is connected. If the fluid in the pipe is water at 20°C with a pressure of

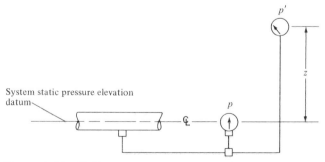

Figure 4-48   Static pressure measurement.

200 kPa at the pipe centerline, determine the reading on the gauge in kPa. (Assume $g = 9.81$ m/s².)

SOLUTION

SKETCH

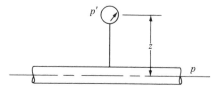

Figure 4-49

EQUATION

$$p' = p - \frac{\rho g Z}{1000} \quad [\text{kPa}] \qquad [\text{Eq. (4-65)}]$$

PARAMETERS

$p = 200$ kPa   (given)

$\rho = 998$ kg/m²   (from Appendix B-3, water at 20°C)

$g = 9.81$ m/s²   (given)

$Z = 3$ m   (given)

SUBSTITUTION

$$p' = 200 - \frac{(998)(9.81)(3)}{1000} \quad [\text{kPa}] - [\text{kg/m}^2][\text{m/s}^2][\text{m}][\text{kPa/Pa}]$$

$$= 200 - 29.4 \quad [\text{kPa}] - [\text{kPa}] = [\text{kPa}]$$

Therefore

$$p' = 170.6 \text{ kPa}$$

• *Velocity Head.* The dynamic (kinetic) energy per unit volume of the fluid can be related directly to pressure. The pressure can be measured in units of pascals or in units of head (meters of a fluid).

By substituting the mass of a cubic meter of the fluid (ρ) with units of kg/m³ for the mass (M) in the basic kinetic energy equation from physics,

$$KE = \tfrac{1}{2}MV^2 \quad [\text{kg} \cdot \text{m}^2/\text{s}^2] \tag{4-66}$$

the· following equation is obtained:

$$KE/m^3 = \frac{\rho V^2}{2} \quad [\text{Pa}] \tag{4-67}$$

To obtain the pressure in terms of head (Z) of a fluid, divide Eq. (4-67) by ρg where ρ is the density of the fluid in units of kg/m^3, and g is the acceleration of gravity. This results in

$$Z = \frac{V^2}{2g} \quad [\text{m}] \tag{4-68}$$

The energy in the fluid stream can be converted from static head to dynamic head, and vice versa. With no conversion losses, the sum of the velocity head and the static head of a fluid stream flow, based upon the same elevation datum, is constant. In actual practice, the conversion of static head to velocity head can be accomplished with relatively little loss. However, the reverse conversion of velocity head to static head is relatively inefficient.

4-3.4 Power Required to Move Fluids

The power required to move the fluid in a system is equal to the total mechanical energy loss. This power (work per unit time) is the product of the total pressure loss and the mass flow rate divided by the fluid density.

$$\dot{W}k = \frac{\Delta p \dot{M}}{\rho} \tag{4-69}$$

where

$$
\begin{aligned}
\dot{W}k &= \text{power required} & [\text{W}] \\
\Delta p &= \text{total pressure drop} & [\text{Pa}] \\
\dot{M} &= \text{mass flow rate of fluid} & [\text{kg/s}] \\
\rho &= \text{fluid density} & [\text{kg/m}^3]
\end{aligned}
$$

The total pressure loss is the sum of (as applicable)

- straight pipe friction loss
- bend or elbow loss
- entrance loss
- enlargement loss
- exit loss
- elevation change

The input power to the pumping equipment (pump or blower) will be higher than the power required to move the fluid in the system because the pumping equipment is not 100% efficient. The input power to the pump or blower is

$$P_i = P_L / \eta_m \tag{4-70}$$

where

P_i = input power [W]
P_L = power required to move the fluid in the system [W]
η_m = mechanical efficiency of pump and driver [—]

Example Problem 4-16

A pump motor is rated at 50 kW. If the mechanical efficiency of the pump and driver (pump motor) is 80%, determine the power available to pump the fluid.

SOLUTION

SKETCH

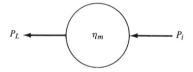

Figure 4-50

EQUATION

$$P_L = \eta_m P_i \quad [\text{kW}] \qquad [\text{Eq. (4-71)}]$$

PARAMETERS

$\eta_m = 80\% = 0.80$ (given)

$P_i = 50 \text{ kW}$ (given)

SUBSTITUTION

$$P_L = (0.80)(50)$$

ANSWER

$$P_L = 40 \text{ kW}$$

PROBLEMS

4-1. Describe or define forced convection. What forces control the fluid motion?

4-2. Give a general equation for the rate of heat transfer ($\dot{Q}$) by forced convection. How does this equation differ from the general equation for the rate of heat transfer by free convection?

4-3. What factors affect the value of the forced convection heat transfer film coefficient (h)?

4-4. What are the three major fluid flow patterns? Describe each. What is critical flow? Why should critical flow be avoided in heat exchange designs?

4-5. What is the boundary layer?

4-6. Describe the boundary layer for flow inside a round pipe for laminar and fully turbulent flow.

4-7. Describe the boundary layer development along a flat surface.

4-8. Describe the boundary layer for flow across a tube in laminar and turbulent flow.

4-9. What dimensionless numbers are used in forced convection heat transfer?

4-10. Give the basic equation for forced convection that was developed from dimensional analysis. How is this used in correlating test data?

4-11. What is the Reynold's analogy and what are its limitations to evaluating the convection heat transfer film coefficient (h)?

4-12. What is the basic equation for forced convection heat transfer for gases? Why is it different from Eq. (4-2)?

4-13. Calculate the average value for h for atmospheric pressure air at 25°C flowing through a 2-cm diameter tube 1 m long at a velocity of 10 m/s, with a parabolic velocity profile.

4-14. If the gas were hydrogen at 2 atm pressure flowing at the same temperature and velocity as the air in Problem 4-13 in the tube, what would be the average value for h?

4-15. How does the Reynolds number (Re) used to characterize fluid flow over a flat plate differ from the Reynolds number (Re) used to characterize fluid flow inside a circular tube? What significant length dimension is used in evaluating Re for flow over flat plates?

4-16. Atmospheric pressure air at 50°C is flowing over a 30°C flat plate 0.5 m long at a velocity of 5 m/s. What is the Reynolds number range along the flat plate? Is the boundary layer laminar or turbulent? What is your criteria for laminar/turbulent boundary layer? What is the average heat transfer film coefficient ($\bar{h}$) over the flat surface under these flow conditions?

4-17. Atmospheric pressure air at 50°C is flowing over a 30°C flat plate 4 m long at a velocity of 15 m/s. Calculate the value of the average heat transfer film coefficient ($\bar{h}$), assuming a critical Reynolds number (Re_{crit}) of 300,000.

4-18. How does the Reynolds number (Re) used to characterize fluid flow over a cylindrical surface (at right angles to the axis of the cylinder) differ from the Reynolds number (Re) used to characterize fluid flow inside a circular tube? What significant length dimension is used in evaluating Re for flow over a cylinder?

4-19. Give the basic equation for forced convection fluid flow over a cylindrical surface. What properties at what state are used to evaluate Re?

4-20. Air is flowing across a 2.5-cm diameter tube at right angles to the axis of the tube with a velocity of 5 m/s. The temperature of the air is 50°C and its pressure is 3 atm. The tube is cooled by water flowing through the tube keeping the tube surface at 30°C. What is the average value of the convection film coefficient $\bar{h}$? How much heat is transferred per meter length of the tube?

4-21. If the gas in Problem 4-20 were carbon dioxide instead of air, what would be the amount of heat transferred per meter length of the tube?

4-22. If the tube in Problem 4-20 were square 2.5 cm on a side with the diagonal parallel to the undisturbed air stream velocity, what would be the amount of heat transferred per meter length of the tube?

4-23. How does the local value of h_o vary around the surface of a long cylinder normal to the flow? How are the variations usually explained?

4-24. Atmospheric pressure air at 0°C is flowing through a staggered tube bank with 12 rows of 2.0-cm diameter tubes arranged in a pattern (see Figure 4-15) of $S_n/D_o = 1.5$ and $S_p/D_o = 1.25$. The temperature of the tube surface is 100°C. The air approach velocity to the tube bank is 5 m/s. Calculate the average heat transfer film coefficient $\bar{h}$.

4-25. What would be the average heat transfer film coefficient $\bar{h}$ if the tube bank in Problem 4-24 were in-line with $S_n/D_o = 1.5$ and $S_p/D_o = 1.25$ (see Figure 4-14)?

4-26. What would be the average heat transfer film coefficient $\bar{h}$ if the gas in Problem 4-24 were carbon dioxide?

4-27. What are the three categories of liquids used in heat transfer? Why are liquids divided into these groups?

4-28. Water flowing inside a vertical pipe 6 cm in diameter and 2.5 m long has a flow velocity of 60 m/h. The tube wall is 100°C and the water temperature is 50°C. Calculate the film coefficient h using the King correlation [Eq. (4-28)].

4-29. Compare the value of h obtained in Problem 4-28 with the value obtained using the Bayley correlation for a fluid with constant properties with a constant heat flux [Eq. (4-23)].

4-30. For precise evaluation of h in laminar flow, why should the buoyancy forces be considered? What is the mechanism of their effect?

4-31. In turbulent flow equations for h for water are often separated into two categories. What are these two categories and why is the distinction made?

4-32. How can noncircular ducts be treated to apply formulae for circular ducts?

4-33. What would be the equivalent diameter of: (1) a rectangular cross section duct with an aspect ratio of 2:1, and (2) an annulus between an inner cylinder with a diameter D and a concentric outer cylinder with a diameter $2D$?

4-34. What factors affect the value of h for water flowing inside a circular pipe?

4-35. Water with a bulk temperature of 40°C is flowing with a velocity of 10 m/s inside a 2.5-cm condenser tube 3 m long whose surface is 50°C. Calculate the convective surface coefficient h.

4-36. If the tube surface temperature in Problem 4-35 were 100°C, calculate the convective surface coefficient h.

4-37. If the tube in Problem 4-35 were 0.25 m long, what difference in the value of h would be expected?

4-38. What effect on the value of h in a short pipe can the entrance geometry to the pipe have? What three entrance configuration types are common?

4-39. How do the general equations for the value of h for water flow along flat plates differ from the equations for water flow inside circular tubes?

4-40. Calculate the value of h for water with a free stream velocity of 60 m/h flowing over a flat plate 0.3 m long when the plate temperature is 35°C and the water temperature is 25°C.

4-41. Calculate the value of h_x for water with a free stream velocity of 2 m/s flowing over a flat plate at a point 4 m from the leading edge of the plate when the plate temperature is 35°C and the water temperature is 25°C.

4-42. Calculate the average value of h_a for the entire 4-m length of the flat plate in Problem 4-41.

4-43. How do the general equations for the value of h for water flowing across a single tube normal to the tube axis differ from the equations for water flow inside circular tubes?

4-44. Calculate the average value for $\bar{h}$ for water with a velocity of 3 m/s and a temperature of 80°C flowing across a tube 2.5 cm in diameter with a surface temperature of 100°C.

4-45. How do the general equations for the value of h for liquids significantly more viscous than water differ from the equations for the value of h for water?

4-46. Engine oil at 40°C is flowing inside a 1-cm diameter horizontal pipe with a velocity of 20 m/min. The pipe wall temperature is 100°C. What is the value of h?

4-47. Ethylene glycol at $40°C$ is flowing at 4 m/s, in a 2.5-cm diameter tube. The temperature of the tube wall is $100°C$. Calculate the value of h.

4-48. How do the general equations for the value of h for liquid metals differ from the general equations for the value of h for water?

4-49. What factors significantly affect the value of h for liquid metals that are not usually involved with, and/or have little affect, if any, upon the value of h for water?

4-50. NaK is flowing in a 2.0-cm diameter circular duct through a nuclear reactor with a velocity of 1 m/s, where it is subject to a uniform heat flux. If the temperature of the NaK is $200°C$, what is the value of h?

4-51. In heat exchange equipment, the energy loss in pumping the fluids through the heat exchange passages and plumbing systems can be divided into two categories. What are these and identify several components of each.

4-52. Calculate the straight pipe friction loss of water flowing at 10 m/s inside a 6 m long smooth drawn condenser tube 2 cm in diameter when the bulk water temperature is $30°C$.

4-53. Calculate the friction loss of a plumbing system comprised of the following:

30-m, 2.5-cm i.d. galvanized pipe .
6—90° standard elbows
1 ball type check valve
1 globe valve fully open

4-54. $50°C$ water flowing in a manifold enters a tube bundle (flowing inside the tubes) with a square edged entrance. If the flow cross-sectional area of the tube bundle is 20%, the manifold flow cross-sectional area and the water flow velocity in the manifold is 3 m/s, what would be the entrance loss indicated by Hughes and Safford?

4-55. What would be the loss when the water leaves the tubes of Problem 4-54 and enters an exit manifold the same size as the entrance manifold?

4-56. If a water cooling system discharges $40°C$ water through spray nozzles in a cooling tower at a velocity of 10 m/s, what would be the exit loss from the water circulating system?

4-57. In a cooling water system, water is taken from a tank with a water level 10 m above the surface of a reservoir into which the water is discharged. When the average temperature of the water is $30°C$, how much will this change in elevation reduce the pumping pressure that must be generated by the pump?

4-58. A pressure gauge on a control room panel is 5 m below the centerline of the pipe whose pressure is being measured. If the tubing connecting the pressure gauge to the pipe is filled with water at 25°C, what correction must be applied to the gauge reading to obtain the pipe pressure?

4-59. If 100 kg/s of water at 30°C were being pumped through a system with a total pressure drop of 100 kPa, what pumping power would be required? If the pump were 80% efficient and the electric motor 90% efficient, how many kW of electrical power input would be required?

Chapter 5
Boiling

5-1 GENERAL

Boiling is one of the most commonly used heat transfer processes in our modern technology. Its applications include:

- production of steam for uses such as:
    power generation
    industrial processes and cooking
    space heating
- absorption (removal) of heat in air conditioning and refrigeration systems
- distillation and refining of liquids
- concentration, dehydration, and drying of foods and materials

Boiling is a heat transfer process that involves a phase change from liquid to vapor through heat input into the liquid. Under low heat transfer flux rates, boiling involves liquid convective flow currents that circulate heated fluid to a liquid/vapor interface surface at which liquid evaporation occurs. Under higher heat transfer fluxes, vapor bubbles are formed on the heat transfer surface, and can cause vigorous circulation flow patterns in the boiling liquid capable of effecting high rates of heat transfer. Both conduction and convection enter into the boiling process. However, because of the relatively large difference in the density of the liquid and vapor phases as well as the relative volumes that may be involved, the details of the transformation from the liquid to the vapor phase has a strong influence upon the heat transfer characteristics. Consequently, factors concerned with the phase change from liquid to vapor must be considered in addition to the liquid properties. Factors that affect the boiling heat transfer include

- Specific heat of the saturate liquid
- Latent heat of vaporization
- Temperature difference between the wall and saturation
- Prandtl number of the saturated liquid
- Liquid viscosity
- Surface tension of the liquid-vapor interface
- Surface wetability factor
- Density of saturated liquid
- Density of saturated vapor
- Gravitational acceleration

The two basic types of boiling are analogous to free and forced convection in single-phase heat transfer. These two basic types of boiling heat transfer are called "pool" and "forced convection."

Pool boiling occurs when liquid is only subjected to free convection forces. Consequently, there can be a significant difference between pool boiling and a saturated liquid (wherein the vapor bubbles rise to the top of the liquid) and in a subcooled liquid (wherein the vapor bubbles condense shortly after being formed).

Forced convective boiling occurs when the fluid is pumped and forced to flow across surfaces in a controlled manner.

The basic boiling heat transfer equation has the same general form as the convective heat transfer equation.

$$\dot{Q} = hA \, \Delta T \quad [\text{W}] \tag{5-1}$$

where

$\dot{Q}$ = heat flow rate $\qquad$ [W]
h = boiling film coefficient $\qquad$ [$\text{W}/\text{m}^2{\cdot}\Delta_1{}^\circ\text{C}$]
A = surface area $\qquad$ [m^2]
ΔT = temperature excess $(T_w - T_{\text{sat}})$ $\quad$ [$\Delta{}^\circ\text{C}$]

Whether boiling occurs in pool boiling or in forced convection boiling, there are six definite regimes of boiling associated with progressively

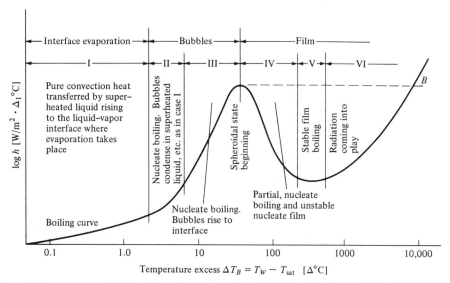

Figure 5-1 Boiling regimes. Heat flux data from an electrically heated platinum wire. [SOURCE: E. A. Farber and E. L. Scorah, Trans. A.S.M.E., Vol. 70, p. 369 (1948)]

increasing heat fluxes. What is termed the "critical heat flux" is encountered at the boundary between boiling regimes III and IV. At the critical heat flux (discussed in Section 5-2.2) a large step increase in the wall temperature can be encountered as illustrated in Figure 5-1 by moving from point (a) between Regimes III and IV to point (b) in Regime VI. The resulting increase in the boiler wall temperature may cause a burnout failure of the heat transfer surface.

The six regimes of boiling are identified in Figure 5-1.[1] When the slope of the curve is rising, indicating an increasing h with increasing temperature excess, the boiling regime is stable. When it decreases, the boiling regime is unstable, and should be avoided in design.

Illustrations from Kreith[2] of nucleate boiling on a wire under pool boiling conditions are presented in Figures 5-2 and 5-3.

The mechanism and the cycle of bubble formation is illustrated in Figure 5-4.

At (a) the liquid next to the heated wall has become superheated.

At (b) a vapor nucleus of sufficient size to initiate the formation of a bubble has been created.

[1] E. A. Farber, and R. L. Scorah, "Heat Transfer to Water Boiling under Pressure," Trans. A.S.M.E., Vol. 79, 1948.
[2] F. Kreith, *Principles of Heat Transfer*, 3rd ed., p. 499. Harper & Row, New York, 1973.

Figure 5-2 Photograph showing nucleate boiling on a wire in water. (SOURCE: From J. T. Castles in F. Kreith and W. Black, *Basic Heat Transfer*, p. 466, Harper & Row Publishers, New York, 1980.)

At (c) the bubble grows in size, pushing the layer of superheated liquid away from the wall.

At (d) the top of the bubble has come in contact with cooler liquid. The rate of growth of the bubble begins to slow down.

At (e) the inertia of the liquid and the bubble has caused the bubble to grow to a size and position in the cooler fluid so it loses more

Figure 5-3 Photograph showing film boiling on a wire in water. (SOURCE: From J. T. Castles in F. Kreith and W. Black, *Basic Heat Transfer*, p.466, Harper & Row Publishers, New York, 1980.)

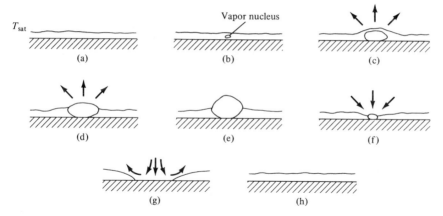

Figure 5-4 Sketches illustrating the flow pattern induced by a bubble in a subcooled liquid boiling on a horizontal surface. (SOURCE: F. Kreith, *Principles of Heat Transfer*, 1st ed., p. 404. International Textbook Co., Scranton, Pa., 1958)

heat to the cooler fluid than it gains by conduction from the heating surface, causing the bubble to begin to collapse.

At (f) the bubble is rapidly collapsing with the cooler fluid gaining velocity filling in the bubble volume.

At (g) the bubble has completely collapsed, and the inertia of the cooler fluid brings the cooler fluid into contact with the heating surface.

At (h) the cooler fluid has been heated above saturation temperature, placing the fluid and bubble formation cycle to condition (a) for the start of another bubble cycle.

The effect of the surface wetability on the bubble contact angle is illustrated in Figure 5-5. The greater the wetability, the greater the heat transfer.

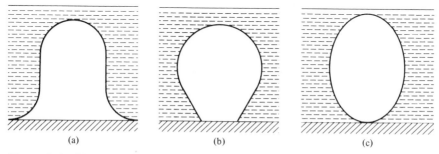

Figure 5-5 Sketches illustrating the effect of surface wetability on the bubble contact angle. (a) Not wetted; (b) partially wetted; (c) totally wetted. (SOURCE: F. Kreith, *Principles of Heat Transfer*, 1st ed., p. 407. International Textbook Co., Scranton, Pa., 1958)

In forced convection boiling, the boundary layer shear and turbulence accelerates the separation of the vapor bubble from the heat transfer surface in the nucleate boiling regimes. In the film boiling regimes the pattern of the vapor film along the surface is significantly altered from that encountered in pool boiling.

5-2 THE EQUATION FOR h

To correlate experimental data in the nucleate boiling regime the basic expression

$$Nu = \phi(\text{Re})\psi(\text{Pr})$$

(where ϕ and ψ are appropriate constants) is modified. The Reynolds number is replaced by a modulus significant of the turbulence and mixing motion for the boiling process. Such a modulus can be made by combining the average bubble diameter (D_b), the mass velocity of the bubbles per unit area (G_b), and the liquid viscosity (μ_l) to form the dimensionless modulus

$$\text{Re}_b = \frac{D_b G_b}{\mu_l} \quad [-]$$

The Prandtl number is also modified using the bubble diameter and the thermal conductivity of the liquid, resulting in

$$\text{Pr}_b = \frac{D_b}{k_l} \quad [-]$$

5-2.1 Nucleate Pool Boiling

The equation Rohsenow[3] found most convenient for the reduction and correlation of experimental nucleate pool boiling experimental data is

$$\frac{c_p \Delta T_B}{r \cdot \text{Pr}^{1.7}} = C_B \left[\frac{\dot{Q}/A}{\mu \cdot r} \sqrt{\frac{\sigma}{g(\rho_f - \rho_g)}} \right]^{0.33} \quad [-]$$

Solving this equation for the boiling heat flux $(\dot{Q}/A)$ gives:

$$\dot{Q}/A = C_B' \cdot \mu \cdot r \left[\frac{g(\rho_f - \rho_g)}{\sigma} \right]^{0.5} \left[\frac{c_p \Delta T_B}{r \cdot \text{Pr}^{1.7}} \right]^3 \quad [\text{kW}/\text{m}^2] \tag{5-2}$$

[3]W. M. Rohsenow, "A Method of Correlating Heat Transfer Data from Surface Boiling Liquids," Trans. A.S.M.E., Vol. 74, 1952.

where

μ = liquid absolute viscosity $[Pa \cdot s] = [kg/m \cdot s]$
r = enthalpy of vaporization $[kJ/kg]$
g = acceleration of gravity $[m/s^2]$
ρ_f = density of saturate liquid $[kg/m^3]$
ρ_g = density of saturate vapor $[kg/m^3]$
σ = surface tension of liquid-vapor interface $[N/m]$
c_p = specific heat of saturate liquid $[kJ/kg \cdot \Delta_1 \degree C]$
ΔT_B = temperature difference between surface
     and saturation $[\Delta \degree C]$
Pr = Prandtl number of saturated liquid $[-]$
C_B' = constant (determined from experimental
     data)—some values given in Table 5-1 $[-]$

Table 5-1   VALUES OF THE COEFFICIENT C_B' IN EQ. (5-2) FOR VARIOUS LIQUID-SURFACE COMBINATIONS

FLUID-HEATING SURFACE COMBINATION	C_B' $[-]$
Water-copper[4]	17
Carbon tetrachloride-copper[4]	17
35† K_2CO_3-copper[4]	185
n-Butyl alcohol-copper[4]	327
50% K_2CO_3-copper[4]	364
Isopropyl alcohol-copper[4]	444
n-Pentane-chromium[5]	67
Water-platinum[6]	77
Benzene-chromium[5]	100
Water-brass[7]	152
Ethyl alcohol-chromium[5]	370
Dowtherm A-steel	90

Adapted from Kreith.[8]

[4] E. L. Piret and H. S. Isbin, "Natural Circulation Evaporation Two-phase Heat Transfer," Chem. Eng. Progress, Vol. 50, p. 305 (1954).
[5] M. T. Cichelli and C. F. Bonilla, "Heat Transfer to Liquids Boiling under Pressure," Trans. AIChE, Vol. 41, pp. 755–787 (1945).
[6] J. N. Addoms, "Heat Transfer at High Rates to Water Boiling Outside Cylinders," D.Sr. Thesis, Dept. of Chem. Eng., Massachusetts Institute of Technology, 1948.
[7] D. S. Cryder and A. C. Finalbargo, "Heat Transmission from Metal Surfaces to Boiling Liquids: Effect of Temperature of the Liquid on Film Coefficients," Trans. AIChE, Vol. 33, pp. 346–362 (1937).
[8] F. Kreith, *Principles of Heat Transfer*, p. 409. International Textbook Company, Scranton, Pa., 1958.

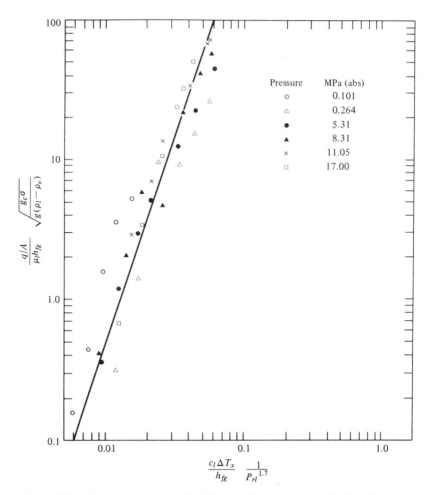

Figure 5-6 Correlation of pool-boiling heat-transfer data by the Rohsenow method. [SOURCE: W. M. Rohsenow, "A Method of Correlating Heat Transfer Data from Surface Boiling Liquids," Trans. A.S.M.E., Vol. 74 (1952)]

Values for the coefficient C'_B for various liquid-surface combinations are given in Table 5-1.

Correlations of Eq. (5-2) with pool boiling of water on a 0.6-mm diameter platinum wire for several conditions, as presented by Kreith,[9] is shown in Figure 5-6.

[9] F. Kreith, *Principles of Heat Transfer*, p. 408. International Textbook Company, Scranton, Pa., 1958.

NUCLEATE POOL BOILING OF WATER AT ATMOSPHERE PRESSURE

• *Boiling on a Horizontal Flat Plate.* Nucleate boiling on a flat plate at low heat fluxes (boiling regimes I and II) follows the heat transfer characteristics of single-phase fluid free convection quite closely. In this region Jakob[10] suggests correlation of boiling data at atmospheric pressure by

$$Nu = 0.16(Gr \cdot Pr)^{1/3} \quad [-]$$

The 1/3 exponent for the product of the Grashof and Prandtl numbers eliminates the characteristic length from the resulting equation for the boiling film coefficient (h).

$$h = 0.16 \left[\frac{k^2 \rho^2 \beta g \, \Delta T c_p}{\rho^2} \right]^{1/3} \quad [W/m^2 \cdot \Delta_1 °C] \qquad (5\text{-}3)$$

The absence of a characteristic length would be expected because the only difference between boiling on a large or small horizontal flat plate would be the effect of the edge.

As the boiling heat flux ($\dot{Q}/A$) increases (to boiling regime III) a radical departure from the correlation of Eq. (5-3) is encountered. In the heat flux range of 15,000 to 250,000 W/m² Fritz[11] recommends the following equation for water boiling on a horizontal surface:

$$h = 1.54(\dot{Q}/A)^{0.75} = 5.58 \Delta T_B^{\,3} \quad [W/m^2 \cdot \Delta_1 °C] \qquad (5\text{-}4)$$

where

$$\dot{Q}/A = \text{heat flow rate} \qquad [W/m^2]$$
$$\Delta T_B = \text{temperature difference between}$$
$$\text{surface and saturation} \qquad [\Delta °C]$$

It should be noted that the numerical values in Eq. (5-4) have the following dimensions:

$$1.54 \, [1/\Delta_1 °C][W/m^2]^{1/4}$$
$$5.58 \, [W/m^2 \cdot \Delta_1 °C^4]$$

[10] M. Jakob, *Heat Transfer*, Vol. I, p. 640. Wiley, New York, 1949.
[11] W. Fritz, Zeitschr. d. Ver deutsch. Ing., Beiheft "Verfahrenstechnik," No. 5, p. 149 (1937).

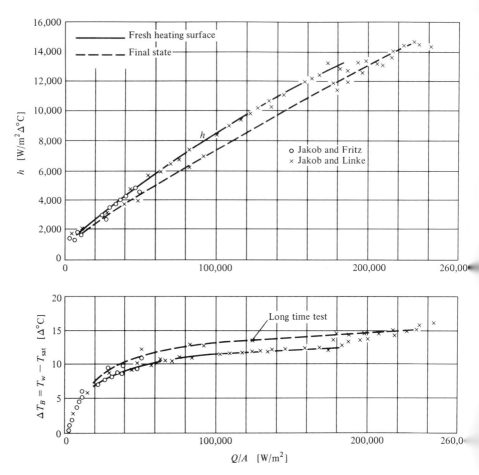

Figure 5-7   Coefficient of heat transfer (h) and temperature difference between wall and saturation (ΔT_B) versus heat transfer flux ($\dot{Q}/A$) in boiling water on a smooth horizontal surface. (SOURCE: M. Jakob, *Heat Transfer*, Vol. I. Wiley, New York, 1949)

Equation (5-4) is based on the experimental points of Jakob and Fritz[12] (1931), Jakob and Linke[13] (1933), Church and Cobb[14] (1922), and Dunn and Vincent.[15] To apply Eq. (5-4) to other liquids, appropriate numerical values must be substituted for those in Eq. (5-4).

[12] M. Jakob and W. Fritz, Forschung a.d. Geb. d. Ingenieurwes., Vol. 2, p. 435 (1931).
[13] M. Jakob and W. Linke, Forschung a.d. Geb. d. Ingenieurwes., Vol. 4, p. 75 (1933).
[14] J. W. Church and H. B. Cobb, Thesis, Massachusetts Institute of Technology, 1922.
[15] P. S. Dunn and A. D. Vincent, Thesis, Massachusetts Institute of Technology, 1931.

A curve for h and ΔT_B as a function of the heat flux $\dot{Q}/A$ is shown in Figure 5-7 as presented by Jakob.[16]

Example Problem 5-1

Estimate the $\dot{Q}/A$ for water boiling on a horizontal flat surface in pool boiling when the temperature difference (ΔT_B) between the surface and saturation is $\Delta 10°C$ using Eq. (5-4) and comparing with Figure 5-8.

SOLUTION

PLAN

(A) Calculate $\dot{Q}/A = h\,\Delta T_B$, determining h by Eq. (5-4).
(B) Calculate $\dot{Q}/A = (h/1.72)^{4/3}$ using the value of h from (a).
(C) Use Figure 5-8 entering at ΔT_B to determine $\dot{Q}/A$ and h.

EQUATION FOR (A)

$$\dot{Q}/A = h\,\Delta T_B \quad \left[\text{W}/\text{m}^2\right]$$

PARAMETERS FOR (A)

$$h = 5.58\Delta T_B^3 \quad \left[\text{W}/\text{m}^2 \cdot \Delta_1°\text{C}\right] \quad \left[\text{Eq. (5-4)}\right]$$

$$\Delta T_B = 10\Delta°C \quad \text{(given)}$$

$$h = (5.58)(10^3)$$

$$h = 5580 \quad \left[\text{W}/\text{m}^2 \cdot \Delta_1°\text{C}\right]$$

SUBSTITUTION FOR (A)

$$\dot{Q}/A = (5580)(10) \quad \left[\text{W}/\text{m}^2 \cdot \Delta_1°\text{C}\right]\left[\Delta°\text{C}\right]$$

ANSWER FOR (A)

$$\dot{Q}/A = 55{,}800 \quad \left[\text{W}/\text{m}^2\right] = 55.8 \quad \left[\text{kW}/\text{m}^2\right]$$

EQUATION FOR (B)

$$\dot{Q}/A = (h/1.72)^{4/3} \quad \left[\text{W}/\text{m}^2\right] \quad \left[\text{Eq. (5-8)}\right]$$

$$h = 5580 \quad \left[\text{W}/\text{m}^2 \cdot \Delta_1°\text{C}\right] \quad \left[\text{from (a)}\right]$$

$$\dot{Q}/A = (5580/1.72)^{4/3}$$

[16] M. Jakob, *Heat Transfer*, Vol. I, p. 639. Wiley, New York, 1949.

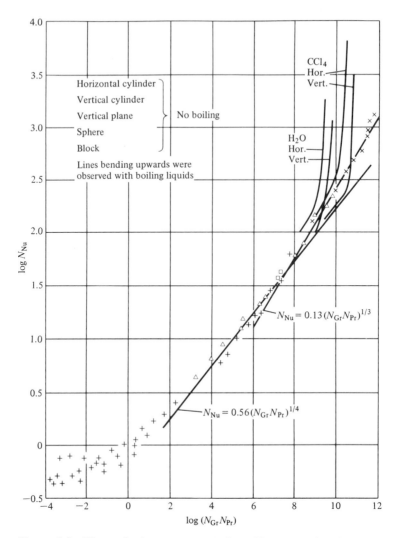

Figure 5-8 Dimensionless representation of heat transfer. (SOURCE: M. Jakob, *Heat Transfer*, Vol. I. Wiley, New York, 1949)

ANSWER FOR (B)

$$\dot{Q}/A = 48{,}010 \quad \left[\text{W}/\text{m}^2\right]$$

ANSWER FOR (C)
Using Figure 5-4 entering at $\Delta T_B = 10\Delta°C$, $\dot{Q}/A$ ranges from 60,000 W/m² for new surfaces to 35,000 W/m² for old surfaces, with a corresponding range in h from 6000 W/m²·Δ_1°C for new surfaces to 3500 W/m²·Δ_1°C for old surfaces.

• *Boiling on a Vertical Flat Plate.* Nucleate boiling on a vertical flat plate at low heat fluxes (boiling regimes I and II) follows the heat transfer characteristics of single-phase fluid free convection quite closely. In this region Jakob[10] suggests correlation of boiling data at atmospheric pressure by

$$Nu = 0.61(Gr \times Pr)^{0.25}$$

The 0.25 exponent for the product of the Grashof and Prandtl numbers introduces a characteristic length (L) into the resulting equation for the boiling film coefficient (h).

$$h = 0.61 \left[\frac{k^3 \rho^2 \beta g \, \Delta T}{L \mu} \right]^{0.25} \quad \left[W/m^2 \cdot \Delta_1{}^\circ C \right] \tag{5-5}$$

The characteristic length (L) is the vertical height of the flat plate (in consistent units).

At a low heat flux the influence of the formation and motion of vapor bubbles upon the heat transfer film coefficient is almost negligible compared with the heat transfer for the nonboiling liquid. At a high heat flux, the vapor bubbles rising in more or less permanent columns, exert a significant circulating effect upon the liquid which results in large increases in the heat transfer film coefficient.

As the boiling heat flux ($\dot{Q}/A$) increases (to boiling regime III) a radical departure from the correlation of Eq. (5-5) is encountered, as illustrated in Figure 5-8 which is Jakob's presentation[10] of King's data.[17] Note the figure includes several shapes (which have a characteristic height dimension), and that King uses a slightly different numerical constant for his correlation equations: 0.56 versus 0.61 for vertical plates and 0.13 versus 0.16 for horizontal plates.

Jakob[16] presents a relationship of the boiling film coefficient (h) as a function of the heat flux ($\dot{Q}/A$) for water boiling at atmospheric pressure on a smooth vertical surface, as shown in Figure 5-9.

• *Boiling Inside Vertical Tubes.* The process of boiling in free convection in a vertical tube heated from the outside is a complicated process. The bubbles rise in a relatively narrow space, becoming more and more crowded and displacing more liquid. Hence, the fluid changes in composition density and velocity.

Data of different investigators collected by Fritz[11] on water boiling at atmospheric pressure in tubes from 0.5 to 2 m in length, can be approximately expressed by the empirical formula

$$h = 1.973(\dot{Q}/A)^{0.75} = 15.11(\Delta T_B)^3 \quad \left[W/m^2 \cdot \Delta_1{}^\circ C \right] \tag{5-6}$$

[17] W. J. King, *Refrigerating Engineering*, Vol. 25, p. 83 (1933).

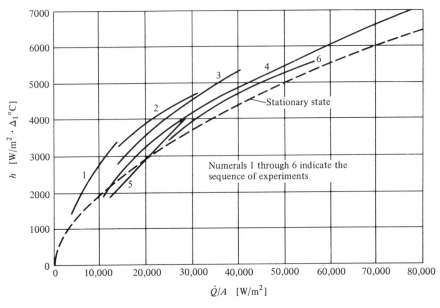

Figure 5-9   Coefficient of heat transfer (h) on a smooth vertical heating surface in boiling water at atmospheric pressure. (SOURCE: M. Jakob, *Heat Transfer*, Vol. I. Wiley, New York, 1949)

It should be noted that the numerical values in Eq. (5-6) have the following dimensions:

$$1.973 \quad [1/\Delta_1°C][W/m^2]^{1/4}$$
$$15.11 \quad [W/m^2 \cdot \Delta_1°C^4]$$

The film coefficient for boiling inside a tube with forced convection, where the fraction of evaporation is so small that the formation and motion of bubbles do not appreciably disturb the turbulent liquid flow pattern, is similar to single-phase liquid forced convection. Boarts[18, 19] obtained an approximate correlation of his test data with water with the equation

$$Nu = C \, Re^{0.8} Pr^{0.4}$$

For a range of Reynolds numbers from 65,000 to 300,000. C was 0.028 [—]. Solving this equation for h results in

$$h = 0.028 \frac{k}{D} Re^{0.8} Pr^{0.4} \quad [W/m^2 \cdot \Delta_1°C] \tag{5-7}$$

[18] R. M. Boarts, W. L. Badger, and S. J. Meisenburg, Trans. Am. Inst. Chem. Engrs., Vol. 33, p. 363 (1937).
[19] R. M. Boarts, W. L. Badger, and S. J. Meisenburg, Ind. Eng. Chem., Vol. 29, p. 912 (1937).

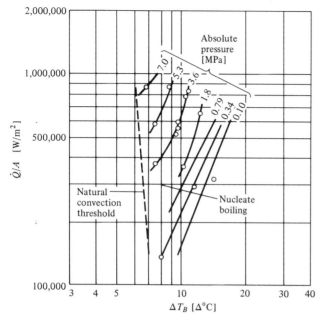

Figure 5-10 Heat flux versus ΔT_B for boiling water on a surface with a slight scale deposit. [SOURCE: M. T. Cichelli and C. F. Bonilla, "Heat Transfer to Liquids Boiling Under Pressure," Trans. Amer. Inst. Chem. Eng., Vol. 41, p. 155 (1945)]

This equation is identical to the McAdams equation (see page 207) for turbulent liquid forced convection, with the exception of the value of the numerical constant (0.028 vs 0.023).

Investigators have found that in sections with vehement boiling significant increases in the value of h can be experienced. However, if the vapor volume ratio in the flow inside the tube becomes too large adequate heat fluxes between the tube wall and the boiling liquid inside the tube cannot be maintained. This results in overheating and perhaps failure of the tube. Consequently, in the design of water tube boilers one criteria is that limiting the vapor volume ratio at the exit of the tube.

BOILING OF VARIOUS FLUIDS

A careful experimental investigation with water, carbon tetrachloride, isopropanol, and a 40% sucrose solution was performed by Insinger and Bliss[20] (1940). They used a vertical heating cylinder, having almost the same length (15.24 cm), diameter (3.18 cm), and hydraulic starting and trailing sections as Jakob and Linke's[13] heater. The result of these experi-

[20] Th. H. Insinger, Jr., and H. Bliss, Trans. Am. Inst. Chem. Engr., Vol. 36, p. 491 (1940).

ments have been subjected to various correlations as discussed by Jakob.[21] These correlations not only satisfied Jakob's experimental results, but also those of Jakob and Linke[13] on water and carbon tetrachloride, Linden and Montillon[22] and Dunn and Vincent[15] on water, and Akin and McAdams[23] on water, isopropanol, isobutanol, and n-butanol. Furthermore, as was done by Jakob and Linke[24] in some experiments, a wetting agent (called Triton W-30) was added to water, 0.207% by volume of which reduced the surface tension by 33% and increased h by 27% (for $\dot{Q}/A = 40{,}300$ W/m^2), but without sensible influence upon the dimensionless representation.

EFFECT OF PRESSURE ON NUCLEATE BOILING

Pressure has a significant effect upon nucleate boiling. Extension of the performance at one pressure to another usually is relatively inaccurate. Illustrative curves of the effect of pressure for water, ethyl alcohol, and propane are given in Figures 5-10, 5-11, and 5-12 from Cichelli and Bonilla.[25]

Example Problem 5-2

Water is boiling at a rate of 100 kg/h in a boiler at a pressure of 3.6 MPa with a temperature difference (ΔT_B) between the metal surface and the saturation temperature of $10\Delta°C$. If the pressure were reduced to 0.79 MPa and the same ΔT_B were maintained, what would be the boiling rate?

SOLUTION

PLAN
Use the general heat transfer equation $\dot{M} = \dot{Q}/r$ and (assuming the same heat input per kg of water to effect boiling) Figure 5-10 for values of $\dot{Q}/A$.

EQUATIONS

$$\dot{M}_2 = \frac{\dot{Q}_2}{r} \quad [kg/h]$$

$$\dot{Q}_2 = \left(\frac{\dot{Q}_2}{A}\right)(A) \quad [W]$$

---

[21] M. Jakob, *Heat Transfer*, Vol. 1, p. 647. Wiley, New York, 1956.
[22] C. M. Linden and G. A. Montillon, Ind. Engr. Chem., Vol. 22, p. 708 (1930).
[23] G. A. Akin and W. H. McAdams, Trans. Am. Inst. Chem. Engrs., Vol. 41, p. 155 (1945).
[24] M. Jakob and W. Linke, Physik. Zeitschr. Vol. 36, p. 267 (1935).
[25] M. T. Cichelli and C. F. Bonilla, Trans. Am. Inst. Chem. Engrs., Vol. 41, p. 155 (1945).

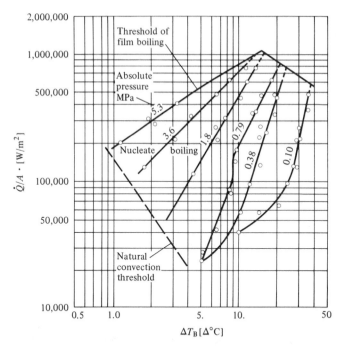

Figure 5-11 Boiling of pure ethyl alcohol from a clean horizontal surface. [SOURCE: M. T. Cichelli and C. F. Bonilla, "Heat Transfer to Liquids Boiling Under Pressure," Trans. Amer. Inst. Chem. Eng., Vol. 41, p. 155 (1945)]

$$A = \frac{\dot{M}_1 r}{\left(\dot{Q}_1/A\right)}$$

$$\dot{Q}_2 = \left(\frac{\dot{Q}_2}{A}\right)\left(\frac{\dot{M}_1 r}{\left(\dot{Q}_1/A\right)}\right)$$

$$\dot{M}_2 = \frac{\dot{M}_1\left(\dot{Q}_2/A\right)}{\left(\dot{Q}_1/A\right)} \quad [\text{kg/h}]$$

PARAMETERS

$$\dot{M}_1 = 100 \quad [\text{kg/h}] \qquad \text{(given)}$$

$$\frac{\dot{Q}_2}{A} = 285{,}000 \quad \left[\text{W/m}^2\right] \qquad \text{(Figure 5-10 at 0.79 MPa, } 10\,\Delta°\text{C)}$$

$$\frac{\dot{Q}_1}{A} = 650{,}000 \quad \left[\text{W/m}^2\right] \qquad \text{(Figure 5-10 at 3.6 MPa, } 10\,\Delta°\text{C)}$$

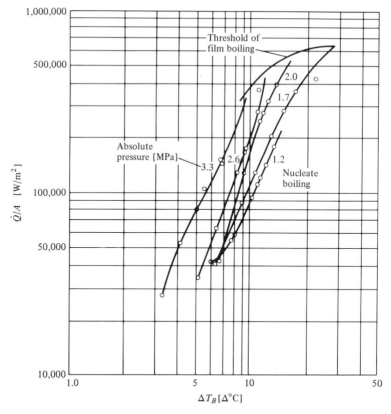

Figure 5-12 Boiling of propane (99% pure) from a clean horizontal surface. [SOURCE: M. T. Cichelli and C. F. Bonilla, "Heat Transfer to Liquids Boiling Under Pressure," Trans. Amer. Inst. Chem. Eng., Vol. 41, p. 155 (1945)]

SUBSTITUTION

$$\dot{M} = (100)\frac{(285,000)}{(650,000)} \quad [\text{kg/h}][\text{W/m}^2]/[\text{W/m}^2]$$

ANSWER

$$\dot{M} = 43.85 \quad [\text{kg/h}]$$

SIMPLIFIED RELATIONSHIPS FOR BOILING WATER

Simplified relationships for boiling water have been developed by various authors from correlations of empirical data. Holman[26] suggests the values given in Table 5-2 for pool boiling of water at atmospheric pressure. For increased pressures the values in Figure 5-3 should be increased by the

Table 5-2 SIMPLIFIED RELATIONSHIPS FOR POOL BOILING WATER AT ATMOSPHERIC PRESSURE

SURFACE ATTITUDE	$\dot{Q}/A$ [kW/m^2]	h [W/m$^2 \cdot \Delta_1$°C]
Horizontal	$\dot{Q}/A < 16$	$1042(\Delta T_B)^{1/3}$
	$16 < \dot{Q}/A < 240$	$5.56(\Delta T_B)^3$
Vertical	$\dot{Q}/A < 3$	$537(\Delta T_B)^{1/7}$
	$3 < \dot{Q}/A < 63$	$7.96(\Delta T_B)^3$

SOURCE: From Holman.[26]

factor $(p/p_o)^{0.4}$, where

> p = the boiling pressure
>
> p_o = atmospheric pressure

The forced-convection local boiling inside vertical tubes, the following equation was recommended[26] for the pressure range of 0.5 to 17 MPa:

$$h = 2.54(\Delta T_B)^3 e^{P/1.551} \quad \left[\text{W/m}^2 \cdot \Delta_1\text{°C}\right] \tag{5-7}$$

where

> ΔT_B = the temperature difference between
> the surface of the tube and saturation   [Δ°C]
> p = the saturation pressure   [MPa]

5-2.2 Critical Heat Flux

As the temperature difference between the wall and the boiling fluid is increased in the nucleate boiling regime, the heat flux is increased, as indicated by the curve in Figure 5-1. The heat flux reaches a maximum at the end of the nucleate boiling regime, as indicated by point a. The heat flux corresponding to the heat flux at point a is called the "critical heat flux." Any further increase in the heat flux results in a shift in the boiling regime to point b. At point b there is a significant increase in the temperature difference between the wall temperature and the fluid temperature. This results in a marked increase in the wall temperature. If this temperature exceeds the temperature limit of the wall material, "burnout" of the wall (i.e., structural failure) will occur. If this temperature does not exceed the temperature limit of the wall material, boiling will continue under the conditions of the higher wall temperature.

If the boiling is occurring with a constant heat flux being applied (such as by an electrical resistance, rocket combustion gases, an atomic

[26] J. P. Holman, *Heat Transfer*, 4th ed., pp. 376, 377. McGraw-Hill, New York, 1976.

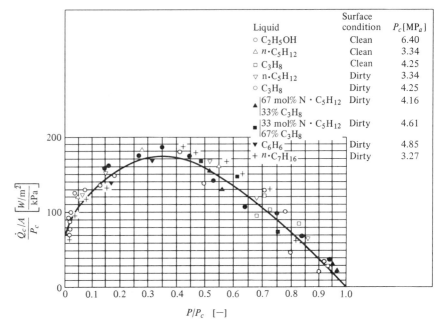

Figure 5-13 Ratio of critical heat flux/thermodynamic critical pressure versus ratio of boiling pressure/thermodynamic critical pressure for various liquids in pool boiling. [SOURCE: M. T. Cichelli and C. F. Bonilla, "Heat Transfer to Liquids Boiling Under Pressure," Trans. Amer. Inst. Chem. Eng., Vol. 41, p. 155 (1945)]

reactor, etc.), burnout usually results when the critical heat flux is reached. However, in other configurations with an increased thermal resistance to heat being transferred to the wall, the increased wall temperature will result in a reduced heat flux. The boiling will then occur at a reduced heat flux and wall temperature which may be below the burnout limits of the wall material. In this case burnout may not occur.

CRITICAL HEAT FLUX CHARACTERISTICS

The primary factor affecting the magnitude of the critical heat flux is pressure. Pressure affects both the nucleate boiling regime and $\dot{Q}_c/A$ because it affects the vapor density, the latent heat of vaporization, and, because it changes the boiling temperature and the surface tension. Experiments by Cichelli and Bonilla[25] on pool boiling of water and a variety of organic liquids and experiments by Kazakova[27] on water, showed a shift in the nucleate boiling regime and showed that $\dot{Q}_c/A$ increased with pressure up to a maximum, at some optimum pressure, and then

[27] E. A. Kazakova, "Maximum Heat Transfer to Boiling Water at High Pressures," Engrs. Dig., Vol. 12, p. 81 (1951).

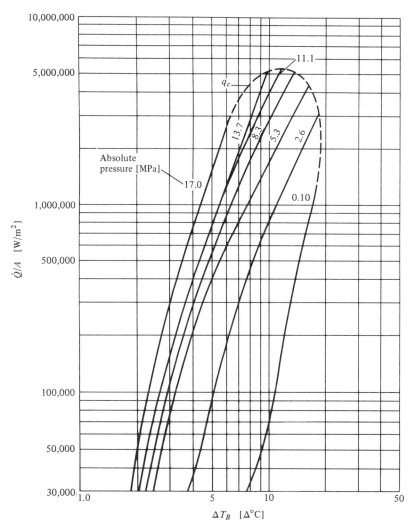

Figure 5-14 Heat flux versus T_B for water in pool boiling, showing nucleate boiling lines at various pressures, and the critical heat flux locus data. [SOURCE: F. Kreith, *Principles of Heat Transfer*, p. 407 International Textbook Co., Scranton, Pa., (1958) as extracted from W. M. Rohsenow, "A Method of Correlating Heat-Transfer Data from Surface Boiling Liquids," Trans. A.S.M.E., Vol. 74 (1952).

decreased. This optimum pressure was in all cases approximately equal to $33\frac{1}{3}$ to 40% of the critical pressure of the liquid in question. Figure 5-13 shows the results of some of the above experiments. The solid line in Figure 5-13 represents the average for clean surfaces. For contaminated surfaces $\dot{Q}_d/A$ is to be increased by 15% from the values indicated by the curve in Figure 5-13.

The value of $\dot{Q}_c/A$ has also been found to increase with certain additives, particularly those with molecules much heavier than the boiling fluid, with ultrasonic and electrostatic fields, but to decrease with the presence of dissolved gases and surface agents, which reduce the surface tension.

The critical heat flux for water is shown in Figure 5-14 as a function of pressure by the dashed line. The solid lines are the nucleate boiling lines for different pressures. These lines terminate at the critical heat flux line.

If the maximum heat flux is desired in a pool boiling system, the boiling should be done at a pressure level giving an adequately high critical heat flux.

For saturated pool boiling Rohsenow and Griffith[28] proposed the following correlation for the critical heat flux ($\dot{Q}_c/A$) based in part on theoretical analysis and in part on experimental data:

$$\dot{Q}_c/A = 0.00374\, r\rho_v \left(\frac{\rho_l - \rho_v}{\rho_v}\right)^{0.6} \left(\frac{g}{g_c}\right)^{0.25} \quad [\text{W}/\text{m}^2] \tag{5-8}$$

where

r = latent heat of vaporization at system pressure [J/Kg]
ρ_l = density of saturated liquid at system pressure [kg/m³]
ρ_v = density of saturated vapor at system pressure [kg/m³]
g = gravitational acceleration [m/s²]
g_c = gravitational constant (9.80665 m/s²)

Equation (5-8) is applicable for water and a variety of organic liquids.

5-2.3 Forced Convective Boiling

The heat transfer rates ($\dot{Q}/A$) in forced convective boiling are considerably higher than corresponding pooling boiling heat transfer rates, and are significantly affected by the geometry of the boiling flow. The performance of each forced convective boiling configuration is relatively specialized and designs should be made by extrapolation of test data of applicable configurations. Consequently, this text does not discuss forced convective boiling performance calculations.

PROBLEMS

5-1. What is boiling? When does it occur?

5-2. List 10 factors that affect boiling heat transfer.

---

[28]W. Rohsenow and P. Griffith, "Correlation of Maximum Heat Flux Data for Boiling of Saturated Liquids," (Reprint, Heat Transfer Symposium, Am. Inst. Chem. Engrs., Louisville, Ky., March 1955).

5-3. Explain or describe pool boiling.

5-4. How does forced corrective boiling differ from pool boiling?

5-5. Describe the six boiling regimes described in this text.

5-6. Would you expect pool boiling $\dot{Q}/A$ to be different for boiling on a horizontal surface and on a vertical surface? Explain your reasoning.

5-7. Estimate the $\dot{Q}/A$ for pool boiling water with a temperature difference (ΔT_B) between the surface and saturation of $8\Delta°C$. (a) on a horizontal surface; (b) on a vertical surface.

5-8. To effect pool boiling on a horizontal surface with water at atmospheric pressure at a heat flux $(\dot{Q}/A)$ of 62 kW/m², what surface temperature is required?

5-9. If the boiling in Problem 5-8 were on a vertical flat plate, what surface temperature would be required?

5-10. If the boiling in Problem 5-8 were in a vertical tube to which Eq. (5-6) applied, what surface temperature would be required?

5-11. If water were boiling at atmospheric pressure on a horizontal surface with a ΔT_B of 12 $\Delta°C$, what would be the boiling heat flux $(\dot{Q}/A)$?

5-12. If the surface in Problem 5-11 were boiling ethyl alcohol with the same ΔT_B, what would be the boiling heat flux $(\dot{Q}/A)$ at atmospheric pressure?

5-13. What effect does a change in pressure have on the boiling $\dot{Q}/A$ at a constant temperature difference (ΔT_B) between the surface temperature and the fluid saturation temperature? Why do you think this occurs?

5-14. What effect does a change in the temperature difference (ΔT_B) between the surface temperature and the fluid saturation temperature have on the boiling Q/A at constant pressure? Why do you think this occurs?

5-15. If the pressure of pool boiling water on a horizontal surface were increased from 0.1 to 3.6 MPa and the temperature difference (ΔT_B) were maintained at 10 $\Delta°C$, what change in the boiling heat flux $(\dot{Q}/A)$ could be expected?

5-16. Using the simplified relationships for boiling water, (a) estimate the boiling heat flux $(\dot{Q}/A)$ for boiling on a horizontal surface under atmospheric pressure with ΔT_B of 8 $\Delta°C$; (b) compare with Fig. 5-10; (c) compare with Eq. (5-4).

5-17. Using the simplified relationships estimate the boiling heat flux for the conditions of Problem 5-16, but with a pressure of 5.3 MPa. Compare with Figure 5-10.

5-18. What are the critical heat flux? Of what importance is it?

5-19. What parameters affect the value of the critical heat flux? What are the mechanics of their effect?

Chapter 6
Condensation

Condensation is another commonly used heat transfer process in our modern technology. It plays an important role in steam power plants, refrigeration and air conditioning systems, and many other applications. Condensation is a heat transfer process wherein fluid in a gaseous or vapor state is changed to a liquid state by the transfer of heat from the gas or vapor. Condensing heat transfer can occur in various configurations. In "jet condensers" a liquid is sprayed into the gas or vapor to effect condensation by heat transfer between the liquid and the vapor. In fractionating distillation the vapor may be bubbled through a liquid maintained at a controlled temperature so as to only condense the desired fraction of the vapor. In a wide variety of condensers, the heat transfer is between the gas or vapor and a solid surface. This latter type of condensation is widely used, and will be the condensation process discussed in this chapter. Both conduction and convection enter into the condensing process. Heat is transferred from the gas or vapor at the heat transfer surface, resulting in the fluid state change to liquid. For the heat to be removed from the condensing surface, the heat must be conducted through the condensate film, if any, on the solid surface. As the vapor or gas is condensed, the surrounding vapor moves to fill the volume vacated by the condensing gas or vapor. Condensing heat transfer has been characterized by two types of condensation, namely

- film condensation
- dropwise condensation

In film condensation, which is the normal stable type of condensation, the condensate wets the solid cooling surface. This results in the formation of a layer (or film) of condensate covering the solid cooling surface. This layer or film of liquid condensate on the solid cooling surface imposes a resistance to the transfer of heat between the vapor and the solid cooling surface. The thickness (and thermal resistance) of the condensate film stabilizes by reaching an equilibrium between the rate of addition of liquid from the vapor condensation and the rate of removal of the liquid condensate by flowing along and off the cooling surface under the forces of gravity and/or the vapor flow. Consequently, the condensation heat transfer coefficients for film condensation are considerably lower than the coefficients for dropwise condensation (wherein there is no thermal resistance of a surface film).

In dropwise condensation, the condensate does not wet the solid cooling surface. Discrete drops form on the cooling surface. These drops grow in volume because vapor is condensed until the drops are removed from the cooling surface by the forces of gravity and/or the vapor flow.

Dropwise condensation has only been obtained by contaminating the condensation surface to prevent wetting. The effect of the currently known contaminants wears away in use, resulting in a reversion to film condensation with its lower condensation heat transfer coefficients. Practical condenser design should be based upon the lower values of film condensation heat transfer coefficients because dropwise condensation is unstable with time, and reverts to the film condensation mode.

In general there is little difference in the magnitude of the condensation heat transfer coefficient between the condensation of saturated vapor and superheated vapor because the amount of heat in the superheat is only a small fraction of the total amount of the heat transferred to effect condensation. Consequently, the equations for saturated vapor condensation are usually used. When superheated vapor is involved, the vapor saturation temperature (not the superheat temperature) is used for the temperature difference as well as for the properties of the condensate fluid.

6-1 FILM CONDENSATION

Film condensation occurs when the condensate wets the solid cooling surface and forms a continuous film. The heat being transferred must pass through this liquid film as it is transferred from the vapor to the cooling

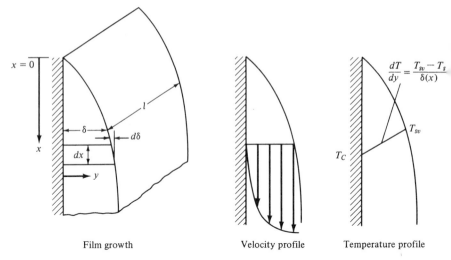

Figure 6-1 Filmwise condensation on a vertical surface-film growth, temperature distribution, and velocity profile. (SOURCE: F. Kreith, *Principles of Heat Transfer*, 3rd ed., p. 524. Harper & Row, New York 1973)

wall. Because the thermal conductivity of liquids is low, the liquid film, thin as it may be, presents a significant resistance to the flow of the heat.

The character of the condensate film on the solid condensing surface can range from laminar to highly turbulent. Turbulent condensate films can result in increased condensation heat transfer coefficients compared to laminar films. Because turbulent films are usually encountered in special cases involving high vapor velocities and substantial surface lengths, the more generally applicable case of laminar condensation films is used as the basis for the equations presented in the following sections.

The general characteristics of the liquid film forming on a vertical flat plate is illustrated in Figure 6-1.

Representative approximate values of condensing film coefficients for several fluids are given below in Table 6-1.

Table 6-1 TYPICAL VALUES OF HEAT TRANSFER COEFFICIENTS FOR FILM CONDENSATION

FLUID (VAPOR)	FILM COEFFICIENT $[\text{W}/\text{m}^2 \cdot \Delta_1 {}^{\circ}\text{C}]$
Steam condensing inside tubes	6,800
Steam condensing on horizontal tubes	11,400
Ammonia	9,200
Benzene	2,000
Carbon tetrachloride	1,700
Diphenyl vapor	1,700

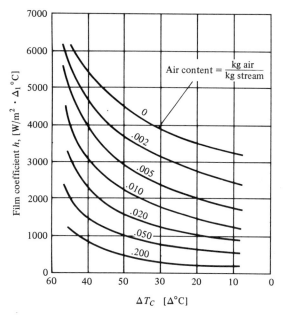

Figure 6-2 Condensation film coefficient for steam at 20 kPa as a function of air content and ΔT_c. (SOURCE: M. Jakob, *Heat Transfer*, Vol. I. Wiley, New York, 1949)

Even small amounts of noncondensible gases decrease the rate of condensation. This is attributed to the noncondensible gases forming a gaseous insulating layer over the cooling surfaces. Typical reductions in the condensation film coefficient for steam as a function of the air content in the steam and the temperature difference between the condensation surface and the steam saturation (ΔT_c) is shown in Figure 6-2.

6-2 THE EQUATION FOR h FILM CONDENSATION

In usual condensers, the condensation of the vapor occurs on one of the following configurations:

1. on the outside of horizontal tubes
2. on the surface of flat plates
3. inside tubes

Condensation film coefficient of these configurations are different, and equations for each are given in the following sections. In the equations for the condensation film coefficient, the temperature of the surface of the liquid condensate film in contact with the cooling wall is assumed to be equal to the temperature of the wall surface. The surface of the film in contact with the vapor is assumed to be at the saturation temperature. By combining the laws of laminar flow of the fluid and of heat conduction through it, Nusselt[1] arrived at the following relation:

$$\text{Nu} = \frac{hL}{k} = C\left(\frac{g\rho^2 r L^3}{\mu k\,\Delta T_c}\right)^{0.25} \quad [-] \tag{6-1}$$

wherein, using meters, seconds, watts, kilograms, joules, and degrees Celsius

h = the average coefficient of heat transfer, taken for
the whole condensing surface $\qquad$ [W/m^2·Δ_1·°C]
L = the height of a vertical wall or the outer diameter of
a horizontal tube on which the condensation occurs $\qquad$ [m]
g = the acceleration of gravity $\qquad$ [m/s^2]
ρ, μ, and k = respectively, the density, dynamic viscosity,
and thermal conductivity of the condensate at its
average temperature
r = the latent heat of evaporation $\qquad$ [J/kg]
ΔT_c = temperature difference between vapor
saturation and the tube surface $\qquad$ [Δ°C]

6-2.1 Heat Transfer Coefficient for Film Condensation on the Outside Surface of Tubes

SINGLE HORIZONTAL TUBE
To calculate the average heat transfer coefficient for the condensation of a pure saturated vapor on the outside of a single horizontal tube, the C in Nusselt's equation [Eq. (6-1)] is usually evaluated as 0.725. When the Nusselt's equation is rearranged by solving for h and g is evaluated as 9.807 m/s^2 the following equation results:

$$h_c = 1.283\left(\frac{k^3\rho^2 r}{D\mu\,\Delta T_c}\right)^{0.25} \quad [\text{W/m}^2\text{·}\Delta_1\text{°C}] \tag{6-2}$$

where

k = thermal conductivity of the liquid condensate
film $\qquad$ [W/m·Δ_1°C]
ρ = density of the liquid condensate film $\qquad$ [kg/m^3]
r = vapor latent heat of condensation $\qquad$ [J/kg]
D = diameter of tube $\qquad$ [m]

[1]W. Nusselt, Z. Ver. Deut. Ing., Vol. 60, pp. 541, 569 (1916).

μ = absolute viscosity $\qquad\qquad$ [Pa·s]

ΔT_c = temperature difference between vapor
$\qquad$ saturation and the tube surface $\qquad$ [Δ°C]

Note: The numerical value in the equation (1.283) is the product of the dimensionless coefficient C in Eq. (6-1) and the acceleration of gravity (g) to the 0.25 power. Hence the numerical value has units of $[m/s^2]^{0.25}$. The properties of the liquid film are evaluated at the arithmetic average temperature between the temperature of the condensing surface and the vapor saturation temperature.

Example Problem 6-1

Saturate steam at 90°C is condensing on a single horizontal tube 2 cm in diameter. With a surface temperature of 70°C, calculate the average heat transfer film coefficient.

SOLUTION

SKETCH

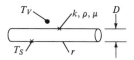

Figure 6-3

PLAN
Use Eq. (6-2). Evaluate fluid properties at the average temperature (T_{av}) between the saturated steam (T_v) and the tube surface (T_s). Evaluate the latent heat of condensation (r) at the saturated steam temperature.

EQUATION

$$h_c = 1.283 \left(\frac{k^3 \rho^2 r}{D\mu\,\Delta T_c} \right)^{0.25} \quad [\text{W/m}^2 \cdot \Delta_1 {}^\circ\text{C}] \qquad [\text{Eq. (6-2)}]$$

PARAMETERS
For liquid property evaluation at T_a

$$T_{av} = \frac{T_v + T_s}{2} \quad [^\circ\text{C}]$$

$T_v = 90°C \qquad$ (given)

$T_s = 70°C \qquad$ (given)

$$T_{av} = \frac{90+70}{2} \quad [°C] + [°C] = [°C]$$

$$T_{av} = 80°C$$

$$k = 0.671 \quad [W/m \cdot \Delta_1 °C] \qquad (\text{Table B-3, Appendix B at } 80°C)$$

$$\rho = 971.8 \quad [kg/m^3] \qquad (\text{Table B-3, Appendix B at } 80°C)$$

$$r = 2283 \quad [kJ/kg] \rightarrow 2.283 \times 10^6 \quad [J/kg]$$

$$(\text{Table B-3, Appendix B at } 90°C)$$

$$D = 2 \text{ cm} \rightarrow 0.020 \text{ m} \qquad (\text{given})$$

$$\mu = 0.359 \times 10^{-3} \quad [kg/m \cdot s] \qquad (\text{Table B-3, Appendix B at } 80°C)$$

$$\Delta T_c = T_v - T_s \quad [\Delta°C]$$

$$T_v = 90°C \qquad (\text{given})$$

$$T_s = 70°C \qquad (\text{given})$$

$$\Delta T_c = 90 - 70 \quad [°C] - [°C] = [\Delta°C]$$

$$\Delta T_c = 20 \quad [\Delta°C]$$

SUBSTITUTION

$$h_c = 1.283 \left[\frac{(0.671)^3 (971.8)^2 (2.283 \times 10^6)}{(0.020)(0.359 \times 10^{-3})(20)} \right]^{0.25}$$

$$\left[\frac{m}{s^2} \right]^{0.25} \left[\frac{[W/m \cdot \Delta_1 °C]^3 [kg/m^3]^2 [J/kg]}{[m][kg/m \cdot s][\Delta°C]} \right]^{0.25}$$

$$= \left[\frac{m}{s^2} \right]^{0.25} \left[\frac{W^3 \cdot J \cdot s}{m^9 \cdot \Delta_1 °C^4} \right]$$

$$= \left[\frac{[W^3][J/s]}{[m^8][\Delta_1 °C]} \right]^{0.25} = \left[\frac{W^4}{m^8 \cdot \Delta_1 °C^4} \right]^{0.25} = [W/m^2 \cdot \Delta_1 °C]$$

$$h_c = 1.283 \left[\frac{(0.3021)(0.9444 \times 10^6)(2.283 \times 10^6)}{0.1436 \times 10^{-3}} \right]^{0.25}$$

$$= 1.283 (0.4536 \times 10^{16})^{0.25} = (1.283)(0.821 \times 10^4)$$

ANSWER

$$h_c = 10,530 \quad [W/m^2 \cdot \Delta_1 °C]$$

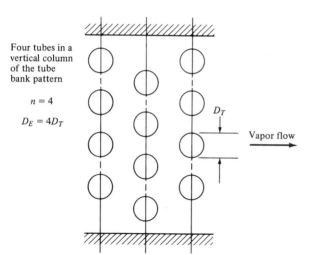

Four tubes in a vertical column of the tube bank pattern

$n = 4$

$D_E = 4D_T$

D_T

Vapor flow

Figure 6-4   The equivalent tube diameter of a horizontal tube bank.

BANKS OF HORIZONTAL TUBES

For horizontal tube banks, the average heat transfer coefficient is calculated using Eq. (6-2), with the substitution of an "equivalent" tube diameter (D_E) for the single tube diameter (D) used for the case of the single horizontal tube. The equivalent tube diameter for calculating the average heat transfer coefficient is the sum of the outside tube diameters of all the tubes in a vertical column of the tube pattern:

$$D_E = nD_T \quad [\text{m}] \tag{6-3}$$

where

D_E = the equivalent tube diameter of the tube bank   [m]
n = the number of tubes in a vertical column of the tube bank pattern
D_T = the diameter of a single tube in the bank   [m]

A horizontal tube bank arrangement is illustrated in Figure 6-4 to show the equivalent tube diameter for the tube bank.

Example Problem 6-2

If steam were condensing under the conditions of Example Problem 6-1, but on a vertical array of eight tubes in a vertical column, what would be the average heat transfer film coefficient?

SOLUTION

SKETCH

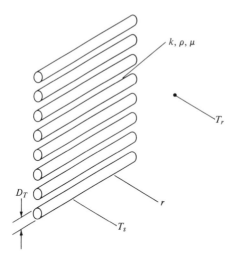

Figure 6-5

PLAN
Use Eq. (6-2), but use an equivalent tube diameter D_E. Evaluate fluid properties as in Example Problem 6-1.

EQUATION
$$h_c = 1.283 \frac{k^3 \rho^2 r}{D_E \mu \, \Delta T} \quad \left[\text{W/m}^2 \cdot \Delta_1 \text{°C}\right] \quad \left[\text{Eq. (6-2)}\right]$$

PARAMETERS
For liquid property evaluation at T_{av} use T_{av} from Example Problem 6-1, $T_{av} = 80\text{°C}$.

$k = 0.671 \quad \left[\text{W/m} \cdot \Delta_1 \text{°C}\right] \quad$ (Table B-3, Appendix B at 80°C)

$\rho = 971.8 \quad \left[\text{kg/m}^2\right] \quad$ (Table B-3, Appendix B at 80°C)

$r = 2283 \quad \left[\text{kJ/kg}\right] \rightarrow 2.283 \times 10^6 \quad \left[\text{J/kg}\right]$
$\qquad\qquad\qquad\qquad\qquad$ (Table B-3, Appendix B at 90°C)

$D_E = n D_T \quad \left[\text{m}\right] \quad \left[\text{Eq. (6-3)}\right]$

$n = 8 \quad \left[-\right] \quad$ (given)

$D_T = 2 \text{ cm} \rightarrow 0.020 \text{ m} \quad$ (given)

$D_E = (8)(0.020) \quad \left[-\right]\left[\text{m}\right] = \left[\text{m}\right]$

$D_E = 0.160 \quad \left[\text{m}\right]$

$\mu = 0.359 \times 10^{-3}$ [kg/m·s]   (Table B-3, Appendix B at 80°C)

$\Delta T = T_v - T_s$ [Δ°C]

$T_v = 90°C$ (given)

$T_s = 70°C$ (given)

$\Delta T = 90 - 70$ [°C] − [°C] = [Δ°C]

$\Delta T = 20$ [Δ°C]

SUBSTITUTION

$$h_c = 1.283 \left[\frac{(0.671^3)(971.8)^2(2.283 \times 10^6)}{(0.160)(0.359 \times 10^{-3})(20)} \right]^{0.25}$$

$$\left[\frac{m}{s^2} \right]^{0.25} \left[\frac{[W/m \cdot \Delta_1 °C]^3 [kg/m^3]^2 [J/kg]}{[m][kg/m \cdot s][\Delta °C]} \right]^{0.25}$$

$$= \left[\frac{m}{s^2} \right]^{0.25} \left[\frac{W^3 \cdot J \cdot s}{m^9 \cdot \Delta_1 °C^4} \right]^{0.25}$$

$$= \left[\frac{[W^3][J/s]}{[m^8][\Delta_1 °C^4]} \right]^{0.25} = \left[\frac{W^4}{m^8 \cdot \Delta_1 °C^4} \right]^{0.25} = [W/m^2 \cdot \Delta_1 °C]$$

$$h_c = 1.283 \left[\frac{(0.3021)(0.9444 \times 10^6)(2.283 \times 10^6)}{1.149 \times 10^{-3}} \right]^{0.25}$$

$$1.283(0.05669 \times 10^{16})^{0.25} = (1.282)(0.488 \times 10^4)$$

ANSWER

$$h_c = 6260 \quad [W/m^2 \cdot \Delta_1 °C]$$

VERTICAL TUBES

Whenever the tube diameter is large relative to the thickness of the condensate liquid film, the heat transfer coefficients for the condensation on a vertical tube are similar to those of a vertical flat surface. Equation (6-4) should be used. The average heat transfer coefficient for a vertical tube bank will be the same as for a single vertical tube if the tube bank pattern does not restrict the vapor flow causing significant vapor turbulence.

In usual condenser designs vertical tubes are seldom used, because the average heat transfer coefficients are lower than for horizontal tubes. In the usual case the length (height) of the vertical tube more than offsets the increased coefficient in the Nusselt's equation. When the length of a tube is greater than 2.85 times its diameter, the average heat transfer coefficient for a single tube in the vertical position will be less than the average heat transfer coefficient for the tube in a horizontal position.

6-2.2 Heat Transfer Coefficient for Film Condensation on a Flat Surface

VERTICAL FLAT SURFACES

To calculate the average heat transfer coefficient for the condensation of a pure saturated vapor on a flat vertical surface, the C in Nusselt's equation [Eq. (6-1)] is usually evaluated at 0.943. When the Nusselt's equation is rearranged by solving for h and g is evaluated as 9.807 m/s^2, the following equation results:

$$h_c = 1.669 \left(\frac{k^3 \rho^2 r}{L \mu \, \Delta T_c} \right)^{0.25} \quad \left[W/m^2 \cdot \Delta_1 \,^\circ C \right] \tag{6-4}$$

where the units of the parameters are as listed for Eq. (6-2) and

$L =$ height of the vertical surface   [m]

Note the numerical value in the equation (1.669) is the product of the dimensionless coefficient C in Eq. (6-1) and the acceleration of gravity (g) to the 0.25 power. Hence the numerical value has units of $[m/s^2]^{0.25}$.

Example Problem 6-3

If steam were condensing under the conditions of Example Problem 6-1, but on a vertical flat plate 1.5 m high, what would be the average heat transfer film coefficient?

SOLUTION

SKETCH

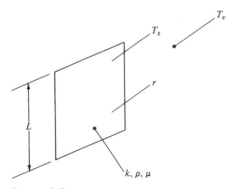

Figure 6-6

PLAN
Use Eq. (6-4). Evaluate fluid properties as in Example Problem 6-1.

EQUATION

$$h_c = 1.669 \left(\frac{k^3 \rho^2 r}{L \mu \, \Delta T_c} \right)^{0.25} \quad \left[W/m^2 \cdot \Delta_1 °C \right] \quad \left[\text{Eq. (6-4)} \right]$$

PARAMETERS
For liquid property evaluation at T_{av} use T_{av} from Example Problem 6-1, $T_{av} = 80°C$.

$k = 0.671 \quad \left[W/m \cdot \Delta_1 °C \right] \qquad$ (Table B-3, Appendix B at 80°C)

$\rho = 971.8 \quad \left[kg/m^3 \right] \qquad$ (Table B-3, Appendix B at 80°C)

$r = 2283 \quad \left[kJ/kg \right] \rightarrow 2.282 \times 10^6 \quad \left[J/kg \right]$

(Table B-3, Appendix B at 90°C)

$L = 1.5 \quad \left[m \right] \qquad$ (given)

$\mu = 0.359 \times 10^{-3} \quad \left[kg/m \cdot s \right] \qquad$ (Table B-3, Appendix B at 80°C)

$\Delta T_c = T_v - T_s \quad \left[\Delta °C \right]$

$T_v = 90 \quad \left[°C \right] \qquad$ (given)

$T_s = 70 \quad \left[°C \right] \qquad$ (given)

$\Delta T_c = 90 - 70 \quad \left[°C \right] - \left[°C \right] = \left[\Delta °C \right]$

$\Delta T_c = 20 \quad \left[\Delta °C \right]$

SUBSTITUTION

$$h_c = 1.669 \left[\frac{(0.671^3)(971.8^3)(2.283 \times 10^6)}{(1.5)(0.359 \times 10^{-3})(20)} \right]^{0.25}$$

$$\left[\frac{m}{s^2} \right]^{0.25} \left[\frac{\left[W/m \cdot \Delta_1 °C \right]^3 \left[kg/m^3 \right]^2 \left[J/kg \right]}{\left[m \right] \left[kg/m \cdot s \right] \left[\Delta °C \right]} \right]^{0.25}$$

$$= \left[\frac{m}{s^2} \right]^{0.25} \left[\frac{W^3 \cdot J \cdot s}{m^9 \cdot \Delta_1 °C^4} \right]^{0.25}$$

$$= \left[\frac{\left[W^3 \right] \left[J/s \right]}{\left[m^8 \right] \left[\Delta_1 °C^4 \right]} \right]^{0.25} = \left[\frac{W^4}{m^8 \cdot \Delta_1 °C^4} \right]^{0.25} = \left[W/m^2 \cdot \Delta_1 °C \right]$$

$$h_c = 1.669 \left[\frac{(0.3021)(0.9444 \times 10^6)(2.283 \times 10^6)}{10.77 \times 10^{-3}} \right]^{0.25}$$

$$= 1.669 (60.48 \times 10^{12})^{0.25} = (1.669)(2.789 \times 10^3)$$

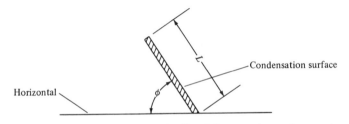

Figure 6-7 Geometry of an inclined flat condensation surface.

ANSWER

$$h_c = 4655 \quad \left[\text{W}/\text{m}^2 \cdot \Delta_1 °\text{C}\right]$$

INCLINED FLAT SURFACES
For inclined flat surfaces, the equation for vertical surfaces [Eq. (6-4)] is modified by introducing the sine of the inclination angle (from the horizontal) as indicated by Eq. (6-5) and illustrated by the sketch shown in Figure 6-7. The accuracy of Eq. (6-5) diminishes as the angle of inclination approaches the horizontal.

$$h_c = 1.669 \left(\frac{k^3 \rho^2 r \sin \phi}{L \mu \, \Delta T_c} \right)^{0.25} \quad \left[\text{W}/\text{m}^2 \cdot \Delta_1 °\text{C}\right] \tag{6-5}$$

Example Problem 6-4

Steam at atmospheric pressure is condensing on a flat plate positioned 60° from horizontal with a length of 2 m, and a surface temperature of 80°C. Estimate the condensing film coefficient (h_c).

SOLUTION

SKETCH   FIG. 6-8

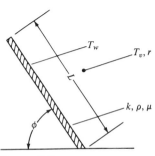

Figure 6-8

EQUATION

$$h_c = 1.669 \left(\frac{k^3 \rho^2 r \sin \phi}{L \mu \, \Delta T_c} \right)^{0.25} \quad \left[\text{W/m}^2 \cdot \Delta_1 {}^\circ \text{C} \right] \quad \left[\text{Eq. (6-5)} \right]$$

PARAMETERS

Properties are for the liquid at the average film temperature (T_{av}).

$T_{av} = T_v + T_w / 2 \quad [^\circ \text{C}]$

$T_v = 100 ^\circ \text{C} \qquad \text{(saturation at 1 atm pressure)}$

$T_w = 80 ^\circ \text{C} \qquad \text{(given)}$

$T_{av} = 100 + 80 = 90 ^\circ \text{C}$

$k = 0.676 \text{ W/m} \cdot {}^\circ \text{C} \qquad \text{(Table B-3, Appendix B at 90°C)}$

$\rho = 965.4 \text{ kg/m}^3 \qquad \text{(Table B-3, Appendix B at 90°C)}$

$r = 2.257 \times 10^6 \text{ J/kg} \qquad \text{(Table B-3, Appendix B at 100°C)}$

$L = 2 \text{ m} \qquad \text{(given)}$

$\mu = 0.318 \times 10^{-3} \text{ kg/m} \cdot \text{s} \qquad \text{(Table B-3, Appendix B at 90°C)}$

$\Delta T_c = T_v - T_w \quad [\Delta ^\circ \text{C}]$

$\Delta T_c = 100 - 80 = 20 \, \Delta ^\circ \text{C}$

$\phi = 60 ^\circ \qquad \text{(given)}$

$\sin \phi = 0.866 \quad [-] \qquad \text{(trig. tables)}$

SUBSTITUTION

$$h_c = 1.669 \left[\frac{(0.676)^3 (965.4)^2 (2.257 \times 10^6)(0.866)}{(2)(0.318 \times 10^{-3})(20)} \right]^{0.25}$$

$$\left[\frac{\text{m}}{\text{s}^2} \right]^{0.25} \left[\frac{\left[\text{W/m} \cdot \Delta_1 {}^\circ \text{C} \right] \left[\text{kg/m}^3 \right]^2 \left[\text{J/kg} \right] [-]}{[\text{m}][\text{kg/m} \cdot \text{s}][\Delta ^\circ \text{C}]} \right]^{0.25}$$

$$\left[\frac{\text{m}}{\text{s}^2} \right]^{0.25} \left[\frac{\left[\text{W}^3/\text{m}^3 \cdot \Delta ^\circ \text{C}^3 \right] \left[\text{kg}^2/\text{m}^6 \right] \left[\text{J/kg} \right]}{[\text{m}][\text{kg/m} \cdot \text{s}][\Delta ^\circ \text{C}]} \right]^{0.25}$$

$$\left[\frac{\text{m}}{\text{s}^2} \right]^{0.25} \left[\frac{\text{W}^3 \cdot \text{J} \cdot \text{s}}{\text{m}^9 \cdot \Delta_1 {}^\circ \text{C}^4} \right]^{0.25} = \left[\frac{\text{W}^3 \cdot \text{J/s}}{\text{m}^8 \cdot \Delta_1 {}^\circ \text{C}^4} \right]$$

$$\left[\frac{\text{W}^4}{\text{m}^8 \cdot \Delta_1 {}^\circ \text{C}^4} \right]^{0.25} = \left[\text{W/m}^2 \cdot \Delta_1 {}^\circ \text{C} \right]$$

$$h_c = 1.669 \left[\frac{(0.3089)(0.932 \times 10^6)(2.257 \times 10^6)(0.866)}{0.0127} \right]^{0.25}$$

$$1.669(44.31 \times 10^{12})^{0.25} = (1.669)(2.580 \times 10^3)$$

ANSWER

$$h_c = 4306 \ \text{W/m}^2 \cdot \Delta_1 °\text{C}$$

6-2.3 Heat Transfer Coefficient for Film Condensation on the Inside Surface of Tubes

In condensation on the inside surface of tubes, the vapor flow in the tube has an effect upon the condensate film character and thickness, thereby affecting the heat transfer coefficient. The effect of the vapor flow will be different between horizontal and vertical tubes, as well as between upward and downward vapor flow of the same velocity in vertical tubes.

Investigations have been made by authors such as Jakob.[2] Measurements presented by Jakob for saturated steam are presented in Figure 6-9, and for superheated steam in Figure 6-10. Empirical equations recommended by Jakob are as follows.

For saturated steam:

$$\frac{\dot{Q}}{A} = (3951 + 116 \ V_{v,o})(\Delta T_c)\left(\frac{1.21}{L}\right)^{1/3} \quad [\text{W/m}^2] \tag{6-6}$$

For superheated steam:

$$\frac{\dot{Q}}{A} = (4068 + 59V_{v,o})(\Delta T_c)\left(\frac{1.21}{L}\right)^{1/3} \quad [\text{W/m}^2] \tag{6-7}$$

where

$$
\begin{aligned}
&L = \text{the cooled length} && [\text{m}] \\
&V_{v,o} = \text{the entrance velocity} && [\text{m/s}] \\
&\Delta T_c = \text{the temperature difference between vapor} \\
&\qquad \text{saturation and the tube surface} && [\Delta°\text{C}]
\end{aligned}
$$

6-2.4 Dehumidification

When a mixture of a condensible vapor and a noncondensible gas flows in contact with a surface colder than the dew point of the vapor-gas mixture, some condensation occurs. Figure 6-11 illustrates the flow pattern adjacent to the cold solid wall. A liquid film forms on the solid wall, and a

---

[2]M. Jakob, *Heat Transfer*, Vol. I, pp. 682–689. Wiley, New York, 1956.

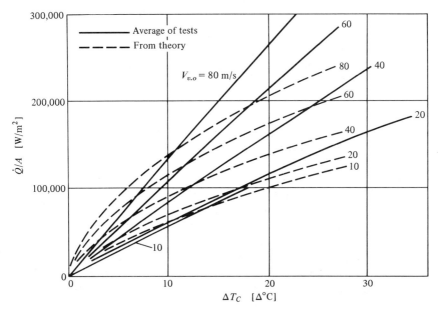

Figure 6-9   Heat flux by condensation of saturate steam at atmospheric pressure. (SOURCE: M. Jakob, *Heat Transfer*, Vol. I. Wiley, New York, 1949)

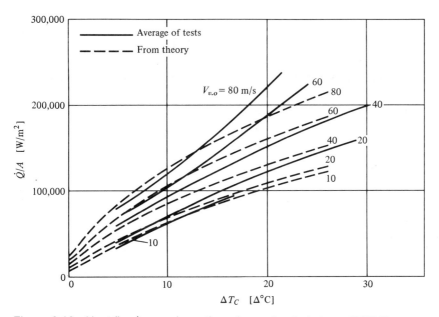

Figure 6-10   Heat flux by condensation of superheated steam (325°C) at atmospheric pressure. (SOURCE: M. Jakob, *Heat Transfer*, Vol. I. Wiley, New York, 1949)

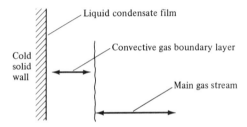

Figure 6-11   Flow pattern adjacent to the cold wall in dehumidifying flow.

convective gas boundary layer forms between the gas side surface of the liquid film and the main gas flow. Because of the higher partial pressure of the condensible vapor in the main gas stream than at the liquid film-gas interface, the vapor diffuses from the main body through the gas boundary layer to liquify on the liquid film-gas boundary layer interface.

The heat transferred is a combination of sensible heat from the gas vapor steam and latent heat from the vapor condensation. The rate of sensible heat transmission is governed by the usual laws of convective and conductive heat transfer from the main stream, through the gas boundary layer, and through the liquid condensate film to the solid surface. The rate of condensation is governed by the laws of diffusion of vapor from the main gas stream, through the gas boundary layer to the surface of the liquid film.

This text does not provide a detailed discussion of dehumidification heat transfer performance. In the design of any dehumidification equipment, it is recommended that test performance data of the configurations of interest be obtained for the planned operating conditions. Extrapolations of test data from one operating condition to another or from one configuration to another may not be accurate due to the relative complexity of the physical phenomena involved.

6-3 DROPWISE CONDENSATION

When the condition of the cooling surface is such that the condensate does not wet the surface, the vapor will condense in drops rather than in a continuous film. Condensation under these conditions is known as drop-wise condensation. A part of the condensation surface is directly exposed to the vapor without an insulating film of condensate liquid, as illustrated in Figure 6-12.

Figure 6-12 shows, for example, steam approaching a cooling surface. The adiabatic compression of the deceleration of the steam velocity causes a thin layer of superheated steam on the cooling surface. The steam in contact with a liquid drop condenses, causing two circulation patterns. One causes a flow velocity along the cooling surface. The other causes a

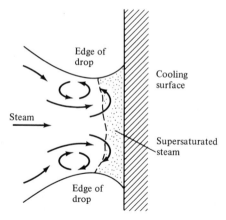

Figure 6-12   Dropwise condensation. (SOURCE: M. Jakob, *Heat Transfer*, Vol. I. Wiley, New York, 1949)

flow over the surface of a portion of the drop. Both are effective heat transfer systems, especially that of the vapor along the cooling surface without any liquid layer in the boundary layer to introduce additional thermal resistance. The droplets grow, break away, and then run down the condensation surface knocking off other droplets as they go. The result is a substantial increase in dropwise condensation heat transfer coefficients over those of film condensation.

Heat transfer fluxes on the order of 750 kW/m² have been obtained with dropwise condensation. For the same vapor conditions and cooling surface temperatures, dropwise condensation can result in an increased heat transfer which may be as much as 20 times that obtained with film condensation. Normally the stable cooling surface condition with clean condensing vapors results in film condition. To obtain dropwise condensation the cooling surface must be treated to prevent wetting. Because of the potential performance gain with dropwise condensation, dropwise condensation has been the subject of considerable interest and investigations.

Dropwise condensation is most readily obtained with steam. Numerous substances while actually on the cooling surface can make the cooling surface nonwetable by the condensate; however, only those substances that are strongly absorbed or otherwise firmly held on the surface are significant drop promoters in a condenser. Some contaminants, such as mercaptans, are specifically effective on certain metals (e.g., mercaptans on copper alloys). Other contaminants, such as fatty acids, are generally effective. The character of the cooling surface also has an effect. Smooth surfaces are generally more conducive to dropwise condensation than rough surfaces. Hampson[3] summarizes the conditions that seem to control

[3] H. Hampson, "General Discussion on Heat Transfer," I.M.E. and A.S.M.E., London (1951).

the length of the effective period of various surface contaminants (promoters) and surface combinations.

Because dropwise condensation can be expected only under carefully controlled conditions, which can not always be maintained in practice, it is recommended that condensers not be designed on the basis of dropwise condensation.

As a matter of interest, the following brief summary of experimental results in dropwise condensation is presented.

Schmidt, Schurig, and Sellschopp[4] (1930) who probably were the first to study dropwise condensation systematically, used a polished or chromium-plated copper plate, the rear side of which was vigorously cooled by a jet of water. They claim to have obtained the most stable drop condensation with clean polished surfaces; treatment of the surface with oil or fat had no effect at all, and treatment with kerosene a temporary effect only. The mean coefficients h_m of heat transfer which they found were between 34 and 45 kW/m^2·Δ_1°C.

Nagle, Bays, Blenderman, and Drew[5] (1935), using oleic acid on a chromium-plated copper tube, found values of $h_m \approx 80$ kW/m^2·Δ_1°C with deviations of only ±10%, whereas $\dot{Q}/A$ was changed from 243 to 535 kW/m^2. These results were obtained with filtered steam, whereas steam containing rust gave much smaller values.

Fitzpatrick, Baum, and McAdams[6] (1939) who used benzyl mercaptan as promoter on a vertical cooper tube, obtained values of $h_m = 39$ to 93 kW/m^2·Δ_1°C.

PROBLEMS

6-1. What is condensation? When does it occur?

6-2. What are the two types of condensation? Describe each and show how they are different. Which type has the highest heat transfer film coefficient (h_c) and explain why this is so.

6-3. In the design of condensers, which of the two types of condensation is usually selected? Explain why.

6-4. In the equations for evaluating the condensing heat transfer film coefficient (h_c) are the fluid properties involved evaluated for the gas phase? If not, how are they evaluated and explain why.

[4] E. Schmidt, W. Schurig, and W. Sellschopp, Tech. Mech. v. Thermodynamik, Vol. I, p. 53 (1930).

[5] W. M. Nagle, G. S. Bays, L. M. Blenderman, and T. B. Drew, "Heat Transfer Coefficients During Dropwise Condensation of Steam," Trans. Am. Inst. Chem. Engrs., Vol. 31, p. 593 (1935).

[6] J. P. Fitzpatrick, S. Baum, and W. H. McAdams, Trans. Am. Inst. Chem. Engrs., Vol. 35, p. 97 (1939).

6-5. If the outside surface of a tube were to be used for condensing a vapor, does it make any difference if the axis of the tube were positioned vertically or horizontally? Explain your answer.

6-6. It is desired to use a 2.5-cm o.d. tube that is 1 m long to condense water vapor at a saturation temperature of 50°C. The surface of the tube will be maintained at 40°C. What would be the condensing film coefficient for film condensation under these conditions if the tube axis were positioned horizontally?

6-7. What would be the condensate mass flow rate obtained from the tube in Problem 6-6?

6-8. If the tube in Problem 6-6 were positioned with its axis vertically, what would be the condensing film coefficient under the conditions of Problem 6-6?

6-9. What would be the condensate mass flow rate obtained from the tube in Problem 6-8?

6-10. If the condensation rate of Problem 6-7 needed to be increased by 25% and this were to be accomplished by increasing the ΔT between the tube surface and the saturation pressure, and if the saturation pressure were to be held constant, what would be the required temperature of the tube surface?

6-11. If the condensation rate of Problem 6-7 needed to be increased 25% but the temperature of the tube surface could not be reduced, what would be the saturation temperature of the resultant condensation?

6-12. If the condensate rate of Problem 6-7 were needed to be increased by a factor of 4, and it was decided to use four condensing tubes (just like the original one), would it make a difference if they were positioned in bank of:

BANK ROWS

	VERTICAL	HORIZONTAL
(a)	1	4
(b)	2	2
(c)	4	1

If you think it would make a difference, explain why and which configuration would give the most condensate, and which would give the least.

6-13. Calculate the condensation film coefficient for the tube bank configurations (b) and (c) in Problem 6-12 using Eq. (6-2) modified for tube banks.

6-14. If four times the condensate flow rate of Problem 6-7 were to be obtained from four tubes (instead of one) at the same steam saturation temperature, what would be the required tube surface temperature for the tube bank configurations (b) and (c) of Problem 6-12?

6-15. If the tube surface temperature could not be changed, what would be the condensing pressure for:
   (a) tube bank configuration (b) of Problem 6-12?
   (b) tube bank configuration (c) of Problem 6-12?

6-16. What factors cause the condensation film coefficient for condensation on the inside of a tube to be different from the condensation on the outside of a tube? What is the primary controlling factor?

Chapter 7
Radiation Heat Transfer

In the modes of heat transfer discussed thus far, such as conduction, convection, boiling, and condensation, the heat has been transferred by direct molecular contact, transmitting the increased energy of molecular motion progressively along a temperature gradient extending through the solid, liquid, or gaseous media involved. Heat may also be transmitted by thermal radiations between two bodies without an intermediate material to act as a carrier of energy. This mode of heat transfer is called radiation heat transfer. The thermal radiant energy is transmitted in a wavelength bandwidth ($\log \lambda$ [m] -4 to -7) of the electromagnetic radiation spectrum as indicated in Figure 7-1. The wavelength of thermal radiations is a function of the temperature of the body. The long wavelength edge of the thermal radiation bandwidth is limited by low temperatures (absolute zero), and the short wavelength edge shown in Figure 7-1 is established by the temperature of the sun. The wavelength bandwidth of thermal radiations associated with temperatures from 20°C to 2500°C, for example, lie largely between $\log \lambda$ of -5 to -6, extending into visible light at the high temperature end. Thermal radiations exhibit characteristics similar to those of visible light, and follow optical laws. They travel in a straight line at the speed of light. They can be reflected, refracted, and are subject to scattering and absorption when they pass through a media.

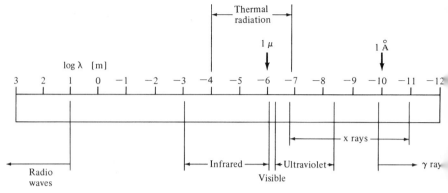

Figure 7-1 Electromagnetic spectrum.

The basic mechanics of radiant heat transfer is simple. A body radiates energy in accordance with the temperature and the surface characteristics of the body. The body absorbs radiant energy incident upon the surface of the body to varying degrees in accordance with the characteristics of the body. The balance of the incident radiant energy that is not absorbed is transmitted through the body and/or reflected from the surface of the body. The net amount of radiant energy transfer to the body is the difference between the energy radiated (energy loss) and the incident energy absorbed (energy gain). When the body radiates more energy than it absorbs, the net heat transfer is away from the body. When the body absorbs more energy than it radiates, the net heat transfer is to the body.

The following discussion of radiation heat transfer is divided into three parts. The first part discusses the radiant energy emissions from a body; the second part discusses the response of a body to incoming incident radiant energy; and the third part discusses the net radiant energy interchange (heat transfer) between two bodies.

7-1 RADIANT ENERGY EMISSIONS BY A BODY

A body, whether it be a solid, liquid, or a gas, radiates energy in accordance with

- the fourth power of its absolute temperature [K]
- the emissivity of the body (ε)

7-1.1 The Emitted Radiant Energy Flux Density

The emitted radiant energy flux density (W) is given by the following equation:

$$W = \varepsilon \sigma T^4 \quad [\mathrm{W/m^2}] \tag{7-1}$$

where

ε = the emissivity of the body $[-]$

σ = the Stefan-Boltzmann constant 5.670×10^{-8} $[\text{W/m}^2 \cdot \text{K}^4]$

T = the temperature of the body $[\text{K}]$

The *emissivity* of a body (ε) is the ratio of the emitted radiant energy flux density (W) to the emitted radiant energy flux density of a blackbody at the same temperature. Values for emissivities range from 0.0 to 1.0.

A *blackbody* is a concept in radiation heat transfer that has the following "ideal" radiation heat transfer characteristics:

- a blackbody radiates the maximum energy possible at the temperature of the body
- a blackbody absorbs all of the incident radiation energy regardless of wavelength

Under certain conditions real bodies can closely approach the characteristics of blackbodies.

The Stefan-Boltzmann constant was obtained by the integration of the monochromatic emissive power (the radiant emissive energy per unit wavelength) over the entire wavelength bandwidth from 0 to ∞. The wavelengths of the energy emissions from a blackbody cover a wide spectrum. The monochromatic emissive power (the radiant emissive energy per unit wavelength) varies across the wavelength spectrum. For any blackbody temperature, the monochromatic emissive power peaks at a particular wavelength.

The monochromatic emissive power decreases progressively at longer or shorter wavelengths, finally reaching energy levels approaching zero, as illustrated in Figure 7-2.

The wavelength at which the monochromatic emissive power peaks, is related to the blackbody temperature by the formula (Wien's law)

$$\lambda_{\text{max}} = \frac{2898}{T} \quad [\mu\text{m}] \tag{7-2}$$

where

λ_{max} = wavelength of maximum monochromatic emissive power $[\mu\text{m}]$

T = blackbody temperature $[\text{K}]$

The monochromatic emissive power at any wavelength in the blackbody radiant emissions can be calculated using the following formula developed by Planck. This formula agrees well with experimental data.

$$W_{b\lambda} = \frac{2\pi h c^2 \lambda^{-5} \times 10^{24}}{e^{(ch/k\lambda T) \times 10^6} - 1} \quad [\text{W/m}^2 \cdot \mu\text{m}] \tag{7-3}$$

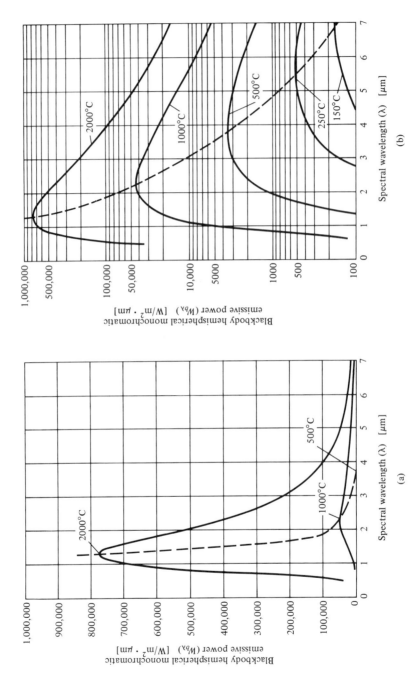

Figure 7-2 Blackbody monochromatic emissive power versus wavelength.

where

$h =$ Planck constant, 0.66236×10^{-33} $[\text{J} \cdot \text{s}]$

$c =$ velocity of light in a vacuum $= 299.79 \times 10^{6}$ $[\text{m/s}]$

$\lambda =$ wavelength $[\mu\text{m}]$

$k =$ Boltzmann constant, 13.802×10^{-24} $[\text{J/K}]$

$T =$ temperature $[\text{K}]$

The symbol $W_{b\lambda}$ denotes the monochromatic emissive power and is defined as the energy emitted per unit surface area at wavelength λ per unit wavelength interval around λ. That is, the rate of energy emission in the interval $d\lambda$ is equal to $W_{b\lambda} \, d\lambda$.

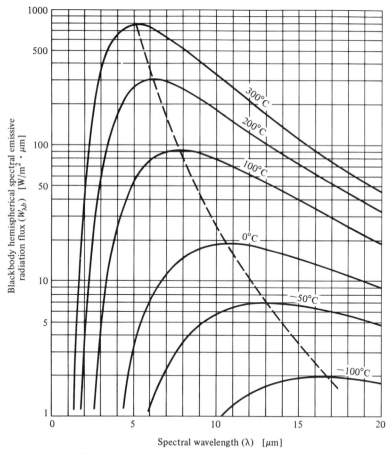

Figure 7-3 Blackbody monochromatic emissive power curves for temperatures below 300°C.

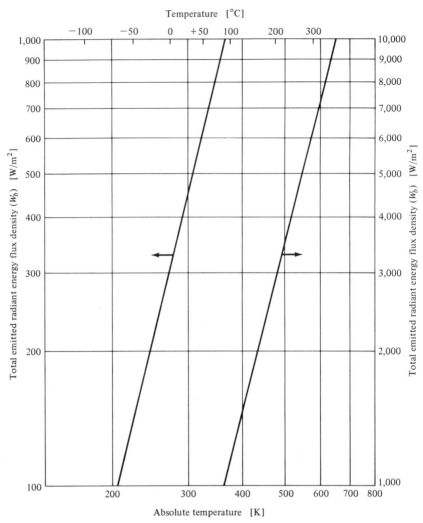

Figure 7-4   Blackbody total emissive power per unit area (radiant flux density)($W_{b\lambda}$) versus absolute temperature.

The variation (or distribution) of the monochromatic emissive power with wavelength is shown in Figures 7-2 and 7-3. Figure 7-2(a) shows on a linear scale the blackbody monochromatic emissive power for three temperatures. By the use of a logarithmic scale Figure 7-2(b) extends the blackbody monochromatic emissive power to lower temperatures. Figure 7-3 presents a plot of the blackbody monochromatic emissive powers for temperatures below 300°C. The dashed line in these figures indicates the wavelength of the peak monochromatic emissive power as a function of temperature.

The total emissive power per unit area (radiant energy flux density) radiating from a blackbody is the total area under the monochromatic emissive power versus wavelength curve for the blackbody temperature. The total emissive power per unit area can be calculated using Eq. (7-4), and is plotted versus temperature in Figure 7-4.

$$W_b = \int_0^\infty W_{b\lambda} \, d\lambda = \frac{2\pi^5 k^4}{15c^2 h^3} T^4 = \sigma T^4 \quad [W] \tag{7-4}$$

where c, h, and k were defined for Eq. (7-3),

$T =$ the blackbody temperature $\quad [K]$

$\sigma =$ the Stefan-Boltzmann constant $5.670 \times 10^{-8} \quad \left[W/m^2 \cdot K^4\right]$

Example Problem 7-1

What is the radiant energy flux density (W_b) from a blackbody at 727°C?

SOLUTION

PLAN
Use Eq. (7-1).

EQUATION

$$W_b = \varepsilon \sigma T^4 \quad \left[W/m^2\right] \qquad \left[\text{Eq. (7-1)}\right]$$

PARAMETERS

$\varepsilon = 1.0 \quad [-] \qquad$ (definition of a blackbody)

$\sigma = 5.670 \times 10^{-8} \quad \left[W/m^2 \cdot K^4\right]$

$T = °C + 273 \quad [K]$

$727 + 273 = 1000 = 10^3 \quad [K] \qquad$ (given)

$T^4 = 10^{12} \quad \left[K^4\right]$

SUBSTITUTION

$$W_b = (1.0)(5.670 \times 10^{-8}) \times 10^{12} \quad [-]\left[W/m^2 \cdot K^4\right]\left[K^4\right] = \left[W/m^2\right]$$

ANSWER

$$W_b = 5.670 \times 10^4 \quad \left[W/m^2\right] \quad \text{or} \quad 56.7 \quad \left[kW/m^2\right]$$

7-1.2 The Emitted Radiant Energy

The emitted radiant energy (E) is the product of the area (A) of the radiating body and the emitted radiant energy flux density (W).

$$E = AW \quad [W] \tag{7-5}$$

which can be written

$$E = A\varepsilon\sigma T^4 \quad [\text{W}]$$ (7-6)

where

$E =$ radiant energy $\quad[\text{W}]$

$A =$ emitting surface area $\quad[\text{m}^2]$

$W =$ radiant energy flux density $\quad[\text{W/m}^2]$

$\varepsilon =$ emissivity of the body $\quad[-]$

$\sigma =$ Stefan-Boltzmann constant $\quad[\text{W/m}^2 \cdot \text{K}^4]$

$T =$ temperature $\quad[\text{K}]$

Example Problem 7-2

What is the total radiant energy emitted by a blackbody in the form of a cube 1 m long on the side when the temperature of the blackbody is 1000 K?

SOLUTION

SKETCH

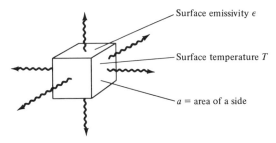

Figure 7-5

EQUATION

$$E = A\varepsilon\sigma T^4 \quad [\text{W}] \qquad [\text{Eq. (7-5)}]$$

PARAMETERS

$A = 6a \quad [\text{m}^2]$

$\quad a = 1 \times 1 = 1 \quad [\text{m}^2] \qquad \text{(given)}$

$A = (6)(1) = 6 \quad [\text{m}^2]$

$\varepsilon = 1.0 \quad [-] \qquad \text{(definition of a blackbody)}$

$\sigma = 5.670 \times 10^{-8} \quad [\text{W/m}^2 \cdot \text{K}^4]$

$T = 1000 \quad [\text{K}] \qquad \text{(given)}$

SUBSTITUTION

$$E = (6)(1.0)(5.67 \times 10^{-8})(1000^4) \quad [\text{m}^2][—][\text{W}/\text{m}^2 \cdot \text{K}^4][\text{K}^4] = [\text{W}]$$

ANSWER

$$E = 3.402 \times 10^5 \quad [\text{W}] \quad \text{or} \quad 340.2 \quad [\text{kW}]$$

7-1.3 Emissivities of Real Bodies

The following discussion of the emissivities of real bodies is divided into two parts. The first part discusses solid surfaces, and the second part gas bodies.

EMISSIVITIES OF SOLID SURFACES

The emissivity of an opaque solid surface is a function of the surface rather than of the material. Most real surfaces do not have a constant value of monochromatic emissivity (ε_λ) with respect to wavelength, as illustrated in Figure 7-6.

The monochromatic emissive power (radiant flux density) (W_λ) of a real surface can vary significantly from the blackbody monochromatic emissive power, as illustrated in Figure 7-7. This is due to the fact that the monochromatic emissive power at any wavelength is the product of the monochromatic emissivity of the real surface and the blackbody monochromatic emissive power. Because the value of the real surface monochromatic emissivity is a characteristic of the surface and not the temperature of the surface, the area under the real surface monochromatic emissive power curve will not be a constant fraction of the area under the blackbody monochromatic emissive power curve as the temperature changes. Thus the effective emissivity of a real surface varies with temperature.

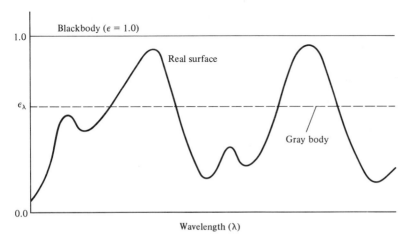

Figure 7-6   Comparison between real surface and blackbody monochromatic emissivities.

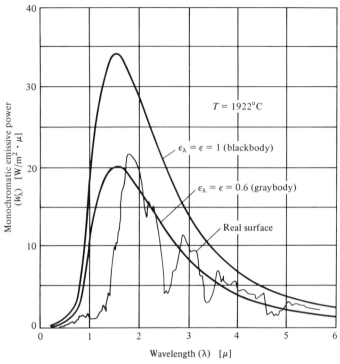

Figure 7-7 Comparison of emissive power of a blackbody, ideal gray body, and real surface.

The emissivity of a surface is the ratio of the area under the mono-chromatic emissive power—spectral wavelength plot of the surface to that of a blackbody. Emissivity data are readily available in reference books. Tables of emissivities for a variety of materials are included in Appendix B-9 of this text. When the value of the emissivity of a material changes with temperature, tables usually give a series of emissivity values with associated temperatures over the temperature range normally encountered with the material.

When the emissivity of the material does not change with tempera-ture, it is called a "gray" body. A gray body has a constant monochromatic emissivity (ε_λ) with respect to wavelength, as illustrated in Figure 7-5. Consequently, a gray body has a characteristic emissivity value (which must be less than 1.0) which does not vary with temperature.

Example Problem 7-3

How much heat radiates from an aluminum-surfaced roof 4 by 8 m in size, when the temperature of the roof surface is at 43°C?

SOLUTION

PLAN
Use Eq. (7-5). Obtain value of ε from Appendix B-8.

SKETCH

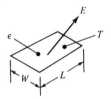

Figure 7-8

EQUATION

$$E = A\varepsilon\sigma T^4 \quad [\text{W}] \qquad [\text{Eq. (7-5)}]$$

PARAMETERS

$$A = LW \quad [\text{m}^2]$$
$$L = 8 \quad [\text{m}] \qquad \text{(given)}$$
$$W = 4 \quad [\text{m}] \qquad \text{(given)}$$
$$A = (8)(4) \quad [\text{m}][\text{m}] = [\text{m}^2]$$
$$A = 32 \quad [\text{m}^2]$$
$$\varepsilon = 0.216 \quad [-] \qquad \text{(from Appendix B-9, Al surfaced roofing at } 43°\text{C)}$$
$$\sigma = 5.670 \times 10^{-8} \quad [\text{W/m}^2 \cdot \text{K}^4]$$
$$T = °\text{C} + 273 \quad [\text{K}]$$
$$T = 43 + 273 = 316 \quad [\text{K}] \qquad \text{(given)}$$
$$T^4 = 99.71 \times 10^8 \quad [\text{K}^4]$$

SUBSTITUTION

$$E = (32)(0.216)(5.670 \times 10^{-8})(99.71 \times 10^8)$$
$$[\text{m}^2][-][\text{W/m}^2 \cdot \text{K}^4][\text{K}^4] = [\text{W}]$$

ANSWER

$$E = 3910 \quad [\text{W}]$$

EMISSIVITIES OF GASES AND FLAMES

The magnitude of the radiant energy flux density from a body of gas at a temperature T is a function of the mass of gas involved as well as other characteristics which are different for (1) gases and (2) luminous flames.

Emissivities of gases emit radiations at discrete wavelengths, depending upon the molecular structure of the gas. Many of the common gases such as O_2, N_2, and H_2 are practically transparent to thermal radiations, neither emitting nor absorbing significant amounts of radiant energy at commonly encountered temperatures. These gases have symmetrical molecules. Gases with nonsymmetrical molecules, such as CO_2, H_2O, SO_2, CO, NH_3, HCl, and so forth, can radiate and absorb significant amounts of energy. Gases emit and absorb radiation only in relatively narrow regions of wavelength, which are called bands. Each gas has its characteristic bands, which are a result of the molecular structure of the gas. Assuming blackbody radiation in the emission bandwidths, the ratio of the total emission (E_g) from these gases to the total emission of a blackbody (E_b) is given in Table 7-1.

The reduction in the ratio E_g/E_b is due to the shift of the intensities of the blackbody monochromatic radiant flux densities to shorter wavelengths as the temperature increases. This is shown by the plots of monochromatic emissive power versus wavelength over the temperature range usually encountered with combustion and flue gases presented in Figure 7-9. In the case of CO_2 and H_2O this reduces the percentage of the blackbody radiant energy falling within the emission bands of these two gases.

As gas radiations are emitted and absorbed throughout the gas volume, the magnitude of the monochromatic radiation emissions and absorption are a function of

- radiation path length(s) through the gas
- monochromatic absorption coefficient (α_λ)

The rate of absorption of radiant energy (I_λ) passing through a gas is

$$dI_\lambda = - I_\lambda \alpha_\lambda \, ds \quad [\text{W}/\text{m}] \tag{7-7}$$

The integration of this equation gives

$$I_{\lambda s} = I_{\lambda o} e^{-\alpha_\lambda s} \quad [\text{W}] \tag{7-8}$$

Table 7-1

TEMPERATURE	E_g/E_b	
	CO_2	H_2O
200°C	0.12	0.67
1000°C	0.105	0.46
2000°C	0.06	0.24

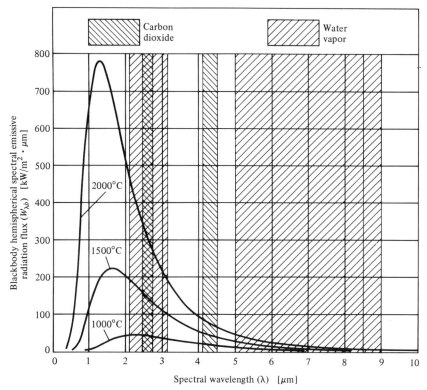

Figure 7-9 Emission bands of carbon dioxide and water vapor and blackbody spectral emissive radiation flux.

where

$I_{\lambda s}$ = monochromatic radiation energy at the end of
the path lengths [W]

$I_{\lambda o}$ = initial monochromatic radiation energy [W]

α_λ = monochromatic absorption coefficient [—]

s = path length [m]

The monochromatic transmissivity (T_λ) is

$$T_\lambda = e^{-\alpha_\lambda s} \quad [-] \tag{7-9}$$

Assuming the reflectivity of gas-to-gas interfaces along the path length are negligible $(\rho = 0)$, the monochromatic absorptivity (α_λ) is

$$\alpha_\lambda = 1 - T_\lambda = 1 - e^{-\alpha_\lambda s} \quad [-] \tag{7-10}$$

According to Kirchhoff's law $(\varepsilon = \alpha)$ the monochromatic emissivity of the gas body is

$$\varepsilon_\lambda = 1 - e^{-\alpha_\lambda s} = \frac{I_{\lambda s}}{I_{b\lambda}} \quad [-] \tag{7-11}$$

where

$I_{b\lambda}$ = monochromatic blackbody radiation energy   [W]

The effective emissivity of a gas body for determining the radiant energy exchange between the gas mass and an element of the boundary surface is obtained by integrating all of the radiations infringing upon the boundary surface element from all directions.

To determine the total radiant heat transferred, a second integration must be performed over the entire boundary surface involved. In general these calculations are complex.

Gases radiating in only a few monochromatic bands do not radiate significant amounts of heat for consideration in usual radiation heat transfer. Gases that do radiate over a sufficiently wide band of wavelengths to warrant consideration in usual engineering calculations of radiant heat transfer include

ammonia
carbon dioxide
carbon monoxide
hydrocarbons
hydrogen chloride
sulfur dioxide
water vapor

Hottel[1] studied the radiation from nonluminous gases, and presented the radiant flux (I_g) emitted from a gas mass to a unit area of its bounding surface in terms of

1. The gas temperature.
2. The product term PL where

P = the partial pressure of the radiating gas [atm]

L = the effective length of the radiant path through the gas volume [m]

The value of L is dependent upon the shape of the gas volume. The effective radiant path length factors to apply to a characteristic dimension of the gas volume are given in Table 7-2.

The radiant energy flux density ($I = [W/m^2]$) at the boundary surface of a volume of gas is given for carbon dioxide in Figure 7-11, water vapor in Figure 7-12, and for a typical flue gas containing 0.8 mol H_2O/mol CO_2

[1] H. C. Hottel, *Trans. Am. Inst. Chem. Eng.* 19, 173 (1927).

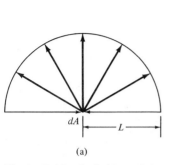
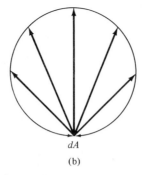

(a)   (b)

Figure 7-10   Definition of the equivalent radius of a radiating gas body.
(SOURCE: H. C. Hottel and E. Eckert, *Heat Transmission*, 3rd ed., Chapter
4, (W. H. McAdams, ed.), McGraw-Hill, New York, 1954.

Table 7-2   THE EFFECTIVE RADIATION PATH LENGTH OF GAS VOLUMES OF
DIFFERENT SHAPES (ACCORDING TO H. C. HOTTEL AND E. ECKERT)

SHAPE CONFIGURATION	EFFECTIVE PATH LENGTH
Circular cylinder: height$=$diameter$=d$	
Irradiation into the center of the base	$0.77d$
Circular cylinder: height$=\infty$, diameter$=d$	
Irradiation into the convex surface	$0.95d$
Cylinder: height$=\infty$, base semicircle with radius r	
Irradiation into the center of the plane rectangular surface	$1.26r$
Sphere: diameter$=d$	
Irradiation into the surface	$0.65d$
Volume between two infinite planes separated by distance l	
Irradiation into the plane	$1.8l$
Circular cylinder: height$=\infty$, diameter$=d$	
Irradiation into the center of the base	$0.9d$
Tube bundle:*	
In triangular arrangement	
$s=2d$	$3.0(s-d)$
$s=3d$	$3.8(s-d)$
In square arrangement	
$s=2d$	$3.5(s-d)$
Cube: length$=a$	
Irradiation into each surface plane	$0.66a$

SOURCE: H. C. Hottel and E. Eckert, Chapter 4, *Heat Transmission* (W. H. McAdams, ed.), 3rd ed.,
McGraw-Hill, New York, 1954.
*$s=$distance between tube centers.
 $d=$tube diameter.

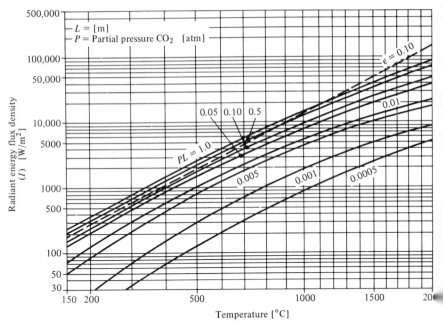

Figure 7-11 Total radiation due to carbon dioxide. (SOURCE: H. C. Hottel and R. S. Egbert, "Radiant Heat Transmission from Water Vapor," *AIChE Trans.* 38 (1942).)

in Figure 7-13. These figures are plots of radiant energy flux density versus gas temperatures for various values of radiating gas masses described by the pL factor. The dashed line is the radiant energy flux density for a gray body with $\varepsilon = 0.10$, which is included as a reference.

EMISSIVITY OF LUMINOUS FLAMES

Luminous flames are those containing solid particles, such as soot or powdered coal suspended in the flame. The radiations are the sum of those from the gas and the solid particles, and can be characterized by an effective emissivity (ε_F). The effective emissivity of a luminous flame is different for flames wherein luminous soot is formed by the combustion process and for powdered coal flames containing luminous unburnt coal and ash particles.

The radiation flux density from a luminous flame follows the Stefan-Boltzmann law using an effective emissivity (ε_F).

$$W = \varepsilon_F \sigma T^4 \quad [\text{W}/\text{m}^2] \tag{7-12}$$

In using this formula to determine the radiant energy flux density from luminous flames, the difference between luminous flames with soot and luminous flames with burning powdered coal and ash is in the

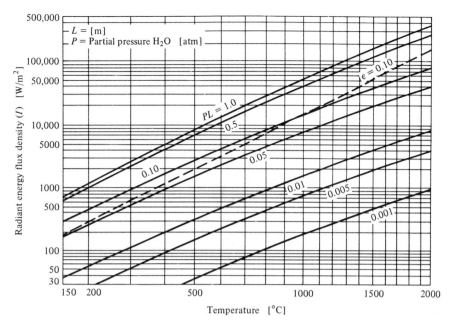

Figure 7-12   Total radiation due to water vapor. (SOURCE: H. C. Hottel and
R. S. Egbert, "Radiant Heat Transmission from Water Vapor," *AIChE
Trans.* 38 (1942).)

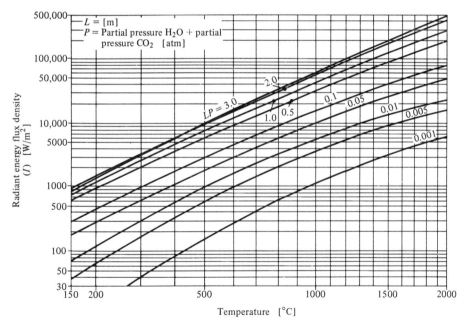

Figure 7-13   Total radiation due to water vapor and carbon dioxide, for
a flue gas containing 0.8 mol of H_2O per mole of CO_2. (SOURCE: H. C.
Hottel and R. S. Egbert, "Radiant Heat Transmission from Water Vapor,"
AIChE Trans. 38 (1942).)

evaluation of ε_F for the furnace size and combustion temperatures involved. A method for evaluating the ε_F for these two types of flames is discussed in the following sections.

LUMINOUS FLAMES WITH SOOT

To determine the emissivity of luminous flames with soot, Hottel and Broughton[2] developed the charts presented in Figures 7-14 and 7-14A. To use these charts the green brightness temperature and the red brightness temperature must be obtained from sighting an optical pyrometer equipped with green and red lights into the furnace flame or must be given to define the problem. By using the red brightness temperature [K] and the difference between the red brightness and the green brightness temperatures in Figure 7-14, the true flame temperature and the flame absorptive strength can be determined from the chart. These two values are applied to Figure 7-14A to obtain the emissivity due to luminosity, which is used in Eq. (7-12) to obtain the radiation flux density from the luminous flame.

LUMINOUS POWDERED COAL FLAMES

To determine the emissivity of luminous flames with powdered coal, the following equation has been developed:

$$\varepsilon_F = 1 - e^{-X} \quad [-] \tag{7-13}$$

where

$$X = \frac{176L\left[(1-V)\rho_o/\rho_1\right]^{2/3}}{G_B\rho_o d_o T_F} \quad [-] \tag{7-14}$$

In this expression

L = effective flame thickness (radiation path length), see Table 7-2 [m]

V = fraction of volatile matter plus moisture in the coal as fired [—]

ρ_o and ρ_1 = initial densities of the coal and of the coked coal, respectively [kg/m^3]

G_B = kilograms of combustion products per kilogram of coal

d_o = initial average particle diameter [m]

T_F = average temperature of the flame [K]

A graphical solution of this equation for two typical types of coal was developed by Hottel[3] and is presented in Figure 7-15. This figure gives a

---

[2] H. C. Hottel, and Broughton, F. P., *Ind. Eng. Chem. Anal. Ed.* 4, 166 (1932).
[3] H. C. Hottel, *Mech. Eng.* 52, 703 (1930).

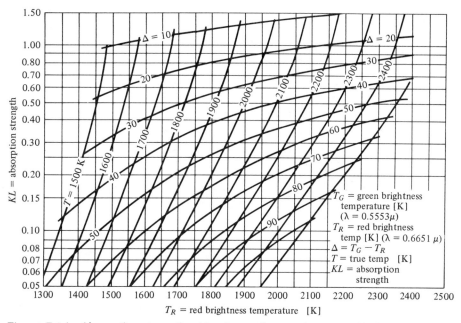

Figure 7-14 Absorptive strength of luminous flames. (SOURCE: H. C. Hottel and F. P. Broughton, *Ind. Eng. Chem., Anal. Ed.* 4, 166 (1932).)

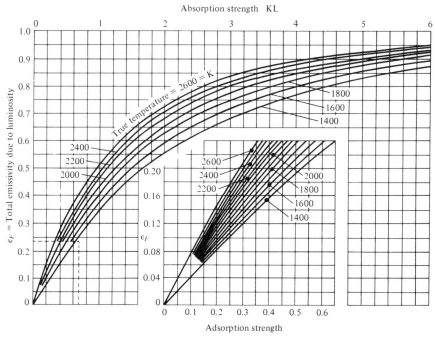

Figure 7-14A Emissivity of luminous flame. (SOURCE: H. C. Hottel and F. P. Broughton, *Ind. Eng. Chem., Anal. Ed.* 4, 166 (1932).)

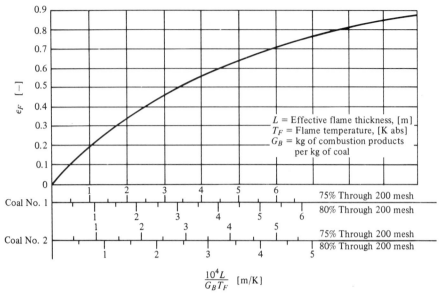

Figure 7-15 Emissivity (ε_F) of powdered-coal flames. (SOURCE: H. C. Hottel, *Mech. Eng.* 52, 703 (1930).)

plot of the emissivity of the flame (ε_F) versus the parameter $10^4 L / G_B T_F$ where

> $L =$ the effective flame thickness (radiation path length). The relationship between the volume shape and the effective radiation path is given in Table 7-2. [m]
>
> $G_B =$ ratio of the weight of the combustion products to the weight of the coal [—]
>
> $T_F =$ the true flame temperature as determined from Figure 7-13 or from other sources [K]

Because the flame emissivity is a function of the type of coal as well as the mesh size of the powdered coal, separate scales are provided for two different types of coal and two typical mesh sizes for an example. The characteristics of these two coals are given in Table 7-3.

Table 7-3 RADIATION FROM POWDERED COAL FLAMES

COAL NUMBER	TYPE	HEATING VALUE [MJ/kg]	V	ρ_0 [kg/m^3]	ρ_1 [kg/m^3]
1	Illinois bituminous, Saline County	29.7	0.384	1345	1201
2	McDowell, West Virginia	33.7	0.160	1442	1201

SOURCE: H. C. Hottel, *Mech. Eng.* 52, 703 (1930).

Example Problem 7-4

Determine the effective emissivity of a flame obtained from burning Illinois bituminous (Saline County) coal ground so 75% passes through a 200 mesh screen when burning in a boiler furnace that is a cube 3 m on a side with a true flame temperature of 1900 K with the ratio of combustion products to the weight of coal of 14:1.

SOLUTION

PLAN
Use Figure 7-15. To use this figure to obtain the effective emissivity (ε_F) of the flame, the parameter $10^4 L / G_B T_F$ must be determined.

SKETCH

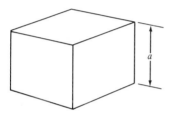

Figure 7-16

PARAMETERS

$$L = 0.66a \quad [\text{m}] \qquad (\text{Table 7-2})$$
$$a = 3 \text{ m} \qquad (\text{given})$$
$$L = (0.66)(3) = 1.98 \text{ m}$$
$$G_B = 14 \quad [-] \qquad (\text{given})$$
$$T_F = 1900 \text{ K} \qquad (\text{given})$$
$$10^4 L / G_B T_F = (10^4)(1.98)/(14)(1900) \quad [-][\text{m}]/[-][\text{K}] = [\text{m}/\text{K}]$$
$$= 0.744 \quad [\text{m}/\text{K}]$$

ANSWER
Entering Figure 7-15 at $10^4 L / G_B T_F = 0.744 \qquad [\text{m}/\text{K}]$
$$\varepsilon_F = 0.14 \quad [-]$$

7-2 RADIANT ENERGY INCIDENT UPON A SURFACE

The following discussion of radiations incident upon a body consider
- the incident radiation flux
- the reaction between the body and the incident radiations.

7-2.1 The Incident Radiation Flux

The radiated energy incident upon a body is emitted by another body. Only a portion of the total emissive power radiating from a body will be incident upon another body because the emissive power radiates in all directions from the radiating surface. The portion of the energy emitted by one body incident on a unit area of another body is a function of the geometry of the relative disposition of the two bodies.

Considering two flat surfaces, the radiant energy E_{1-2} emitted from a flat diffuse surface A_1 incident upon a flat surface A_2 is the product of the solid angle (ψ) subtended by A_2 and the intensity ($\bar{I}$) of the emitting surface A_1.

THE SUBTENDED SOLID ANGLE
When A_2 is normal to the line of propagation of the radiation between the two surfaces, the subtended solid angle is illustrated in Figure 7-17.

The solid angle (ψ) subtended by A_{2n} is defined by geometry as

$$\psi = \frac{A_{2n}}{r^2} \quad [-] \tag{7-15}$$

where

$A_{2n} =$ the projection of the incident surface A_2 normal to the line of propagation $[m^2]$

$r =$ length of the line of propagation between the two surfaces $[m]$

When the incident surface A_2 is not normal to the line of propagation, the projection of the surface normal to the line of propagation is

$$A_{2n} = A_2 \cos \theta \quad [m^2] \tag{7-16}$$

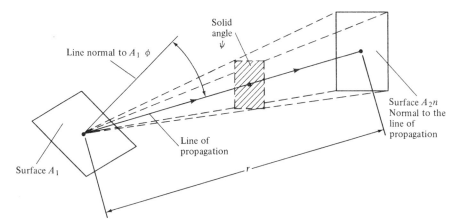

Figure 7-17   The geometry of the relative disposition of two surfaces.

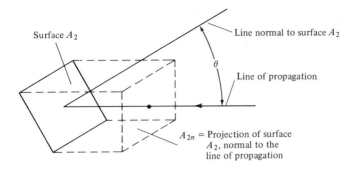

$$A_{2n} = A_2 \cos \theta$$

Figure 7-18   Projection of an incident flat surface normal to the line of propagation.

where

$A_2 =$ the area of the incident surface $[m^2]$
$\theta =$ the angle between the normal to the surface of A_2 and the line of propagation $[rad]$

This is illustrated in Figure 7-18.

When the incident surface A is a sphere, the projection of the surface normal to the line of propagation is the silhouette disk of the sphere, which is a circle of the diameter of the sphere. This is illustrated in Figure 7-19.

THE INTENSITY OF A RADIATING SURFACE

The intensity of a radiating surface is a measure of the strength or magnitude of the emitted radiant energy. It is analogous to the wattage of an electric light bulb. A 100-W light emits four times the light energy of a 25-W light.

The intensity $\bar{I}$ of an emitting surface is the product of the radiation flux intensity I and the area A of the emitting surface.

$$\bar{I} = IA \quad [W/st] \tag{7-17}$$

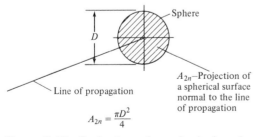

Figure 7-19   Projection of a spherical surface normal to the line of propagation.

The units of the radiation flux intensity (I) of an emitting surface are W/m^2 per unit solid angle. The unit of measurement of the solid angle is called a steradian [st], and it is dimensionless. Therefore the units of

$$I = [W/m^2 \cdot st]$$

and

$$\bar{I} = [W/st]$$

THE INTENSITY OF AN EMITTING BODY

The radiant energy emitted by a body radiates in all directions. For a diffuse emitting surface (a surface whose emissivity does not vary with the direction of the emission), the intensity is a maximum along a ray normal to the emitting surface. To characterize an emitting body an intensity parameter $\bar{I}$ is used. The intensity $\bar{I}$ of an emitting surface is defined as the energy radiating along a ray normal to the emitting surface, evaluated in terms of watts per unit solid angle [W/st]. The equation for evaluating $\bar{I}$ is

$$\bar{I} = \frac{A \varepsilon \sigma T^4}{\pi} \quad [W/st] \tag{7-18}$$

where

$A =$ the area of the emitting surface $\quad [m^2]$

$\varepsilon =$ the emissivity of the emitting surface $\quad [-]$

$\sigma =$ the Stefan-Boltzmann constant (5.670×10^{-8}) $\quad [W/m^2 \cdot K^4]$

$T =$ the temperature of the emitting surface $\quad [K]$

The amount of emitted radiant energy from a body that is incident upon another surface (E_i) is simply the product of (1) the intensity $\bar{I}$ of the emitting body along the line of propagation between the emitting body and the incident surface, and (2) the solid angle subtense of the surface upon which the emissions are incident. The effective surface area of both surfaces is the projection normal to the line of propagation. If the areas of the two surfaces are not small relative to the distance between the two surfaces, an integration over the surface areas must be used as discussed in Section 7-3.

$$E_i = \bar{I} \psi \quad [W] \tag{7-19}$$

Where

$\bar{I} =$ the intensity of the emitting surface $\quad [W/st]$

$\psi =$ the solid angle subtense of the incident surface $\quad [st]$

The magnitude of the radiation flux intensity of a surface can be determined in the following manner.

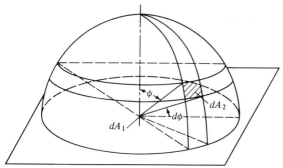

Figure 7-20   Nomenclature for intensity of radiation.

When a small surface area dA_1 is emitting radiations, some portion of the total emitted radiations will be incident upon a small surface dA_2 at a distance r from dA_1. To determine the portion of the radiations from dA_1 which are incident upon dA_2, a hemispherical surface with a radius r containing the area dA_2 can be positioned directly over the small flat radiating surface dA_1. This is illustrated in Figure 7-20.

For the purposes of this initial analysis, dA_2 will be assumed to lie on the surface of the hemisphere, which places it normal to the direction of the propagation of the radiant heat between the two surfaces. The effect of the surface dA_2 being at some angle other than normal is discussed later on. The radii of the hemisphere all emanate from the radiating surface dA_1, and all of the radiations emitted by the surface dA_1 will pass through the hemispherical surface. When the radiating area dA_1 is viewed from the hemispherical surface, its size (area) appears maximum when viewed normal to the surface of dA_1. When viewed from a point on the hemisphere displaced from the normal by the angle ϕ, the surface element dA_1 will appear smaller in size because the projected area is $dA_1 \cos \phi$.

To determine the radiant heat per unit time emitted by dA_1 which reaches an area dA_2 of the surface of a hemisphere or radius r, a parameter called intensity (I) is used. The intensity of radiation I from dA_1 is defined as the radiant energy propagated in a particular direction per unit solid angle and per unit of area dA_1 as projected on a plane perpendicular to the direction of propagation (which is the surface of the hemisphere in Figure 7-20). For a diffuse surface, the emission intensity does not vary with the emission angle ϕ. The energy flux (W) radiating per unit area of a diffuse surface A_1 in terms of the intensity of radiation (I) is

$$W = \pi I \quad \left[\text{W/m}^2 \right] \tag{7-20}$$

since

$$W_1 = \varepsilon_1 \sigma T_1^4 \quad \left[\text{W/m}^2 \right]$$

$$I = \frac{\varepsilon \sigma T_1^4}{\pi} \quad \left[\text{W/m}^2 \cdot \text{st} \right] \tag{7-21}$$

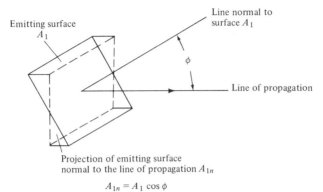

$$A_{1n} = A_1 \cos \phi$$

Figure 7-21 The projection of an emitting surface normal to the line of propagation.

The intensity $(\bar{I})$ of the emitting surface (A) as viewed normal to the surface is

$$\bar{I} = \frac{\varepsilon_1 \sigma T_1^4}{\pi}(A_{1n}) \quad [\mathrm{W/st}] \tag{7-22}$$

If the radiations are from a sphere, such as the sun, the radiating area is the silhouette of the sphere $(\pi d /4)$ and the angle ϕ between the normal to the emitting surface and the line of radiation propagation is zero (0).

If the emitting surface (A_1) is viewed at an angle away from normal to the surface (A_1), the projection of the surface normal to the viewing line (line of propagation) (A_{1n}) must be used to determine the intensity of the emitting surface as viewed by the incident surface (along the line of propagation). This is illustrated in Figure 7-21.

Then for any geometrical relative disposition of an emitting and incident surface, the intensity $(\bar{I}_1)$ of the emitting surface A_1 as viewed by the incident surface A_2 is

$$\bar{I}_1 = A_1 \cos \phi \frac{\varepsilon_1 \sigma T_1^4}{\pi} \quad [\mathrm{W/st}] \tag{7-23}$$

The solid angle subtended by the incident surface A_2 as viewed from the emitting surface A_1 is

$$\psi = \frac{A_{2n}}{r^2} = \frac{A_2 \cos \theta}{r^2} \quad [\mathrm{st}]$$

as the energy E_{1-2} incident upon the incident surface A_2 is

$$E_{1-2} = \frac{A_1 A_2 \cos \theta \cos \phi \varepsilon_1 \sigma T_1^4}{r^2 \pi} \quad [\mathrm{W}] \tag{7-24}$$

Equation (7-24) is valid when the subtended solid angles of the incident surface with respect to the emitting surface and the emitting surface with respect to the incident surface are small. Otherwise, an integration must be performed resulting in the introduction of configuration or shape factors, which is discussed in Section 7-3.

Example Problem 7-5

Calculate the solar constant (energy incident on a square meter of surface normal to the line of radiation propagation) at the earth's orbit.

The diameter of the sun is 1.39×10^9 m

The average distance from the sun to the earth's orbit is 1.49×10^{11} m.

The nominal temperature of the sun is 10,200 K.

The emissivity of the sun is 1.0.

SOLUTION

SKETCH

$\phi = 0$

Figure 7-22

EQUATIONS

$$E_{1-2} = A_1 A_2 \frac{\cos \phi}{r^2 \pi} \varepsilon \sigma T_1^4 \quad [\text{W}] \qquad [\text{Eq. (7-24x)}]$$

PARAMETERS

$$A_1 = \frac{d^2}{\pi 4}, \qquad [\text{m}^2]$$

$$d = 1{,}390{,}000{,}000 \text{ m} \qquad (\text{given}) \rightarrow 1.39 \times 10^9 \text{ m}$$

$$= \frac{(1.39 \times 10^9)^2}{\pi 4} \quad [\text{m}]^2$$

$$A_1 = 1.538 \times 10^{17} \quad [\text{m}^2]$$

$$A_2 = 1 \text{ m}^2 \qquad (\text{given})$$

$$\cos \phi = 1 \text{ as } \phi = 0° \quad [-]$$

$$r = 1.49 \times 10^{11} \text{ m} \quad \text{(given)}$$

$$\varepsilon_1 = 1.0 \quad [-] \quad \text{(given)}$$

$$\sigma = 5.67 \times 10^{-8} \quad [W/m^2 \cdot K^4]$$

$$T_1 = 10{,}200 \text{ K} \quad \text{(given)}$$

$$T_1^4 = 1.082 \times 10^{16} \quad [K^4]$$

SUBSTITUTION

$$E_{1-2} = (1.538 \times 10^{17}) \frac{(1 \times 1)}{\pi (1.49 \times 10^{11})^2} (1)(5.67 \times 10^{-8})(1.082 \times 10^{16})$$

$$[m^2] \frac{[m^2][-]}{[-][m^2]} [-][W/m^2 \cdot K^4][K^4] = [W]$$

ANSWER

$$E_{1-2} = 1353 \quad [W]$$

7-2.2 The Reaction Between a Body and Incident Radiations

The radiant energy incident upon a body is absorbed, reflected, and transmitted to various degrees in accordance with the absorptivity (α), reflectivity (ρ), and transmissivity (τ) characteristics of the body. Because the numerical value of the absorptivity, reflectivity, and transmissivity of a body represents the ratio of the amount of the incident radiant energy which is absorbed, reflected, or transmitted, the values are dimensionless and their sum always equals 1.0.

$$\alpha + \rho + \tau = 1 \quad [-] \tag{7-25}$$

where

α = absorptivity or the fraction of total energy absorbed $[-]$
ρ = reflectivity or the fraction of the total energy reflected $[-]$
τ = transmissivity or the fraction of the total energy transmitted through the body $[-]$

A sketch showing the distribution of incident radiant energy is shown in Figure 7-23. The absorptivity, reflectivity, and transmissivity of the body to the incident radiation energy is discussed in the following sections. The response of the body to the incident radiations is completely independent of and unaffected by the simultaneous emissions from the body.

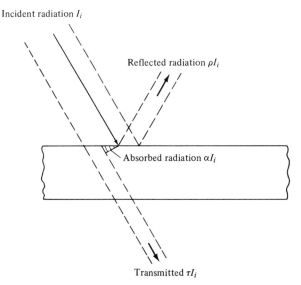

Incident radiation I_i

Reflected radiation ρI_i

Absorbed radiation αI_i

Transmitted τI_i

Figure 7-23   Sketch showing the distribution of incident radiant energy.

ABSORPTIVITY

In opaque materials the absorptivity is a surface phenomenon. In transparent materials the absorption occurs throughout the material. The absorptivity of a real material is usually a function of the wavelength of the incident radiations. Consequently, for example, the absorptivity of a surface to incident solar energy may be significantly different than the absorptivity to incident radiations from a 300 K surface. This is due to the wide difference in the spectral energy distribution in the radiations from the two different sources.

The absorptivity of a surface was established to be equal to the emissivity of the surface, at the same temperature, by Kirchhoff in 1859. The Kirchhoff law states:

> If a surface at any given temperature absorbs n times more of an arriving radiation than another surface of equal area at the same temperature, then at the same temperature the first surface emits n times more of the same kind of radiation than does the second surface.

Simply stated

$$\varepsilon = \alpha \quad [-] \tag{7-26}$$

where

ε = emissivity of material surface $[-]$

α = absorptivity of material surface $[-]$

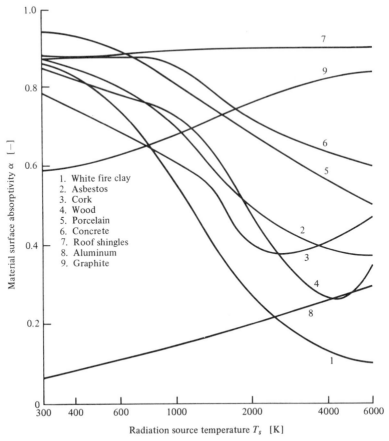

1. White fire clay
2. Asbestos
3. Cork
4. Wood
5. Porcelain
6. Concrete
7. Roof shingles
8. Aluminum
9. Graphite

Radiation source temperature T_s [K]

Figure 7-24 The absorptivity of various common materials for black- or gray-body radiation from a source at T_s. (SOURCE: W. Sieber, "Zusammenstezung der von Werk-und Baustoffen Zuruckge-worfenen Warmestrahlung," *Z. Tech. Physik* 22, 130–135 (1941).)

Values of absorptivity of various materials are shown as a function of the temperature of the black or gray body radiation source in Figure 7-24. This figure indicates that the absorptivity of some material surfaces to solar radiations (source temperature of 10,000 K) may be quite different from values for sources in the 300 to 600 K range.

To assist the student in differentiating between incident emissions from radiating bodies and solar radiations, tables of *emissivities* are given over the usable temperature range of the material, and tables of *absorptivity* are given for solar radiations (with a source temperature of 10,000 K). If material surface absorptivity values are needed, by Kirchhoff's law the value of the material absorptivity is equal to the material's emissivity (at the same temperature). Consequently the material emissivity values are used for absorptivity values in such cases.

Values for the absorptivity of typical surfaces to solar radiations are given in Table 7-7. Absorptivities of materials at other temperatures are the same as the emissivities listed for various materials and temperatures in Appendix B, Table B-9.

The amount of heat ($\dot{Q}_a$) absorbed by a body is a function of the radiant energy incident upon the surface (E_{1-2}) and the surface absorptivity (α).

$$\dot{Q}_a = \alpha E_{1-2} \quad [\text{W}]$$ (7-27)

where

$\dot{Q}_a$ = the heat absorbed by the body   [W]

α = the absorptivity of the body   [—]

E_{1-2} = the energy incident upon the surface from Eq. (7-24)   [W]

REFLECTIVITY

The reflectivity characteristics of the surface of a material follow conventional optics. A smooth surface will give a specular reflection, with the angle of reflection equal to the angle of incidence of the radiations. If the surface is rough, a diffuse reflection may occur. Depending upon the nature of the surface of the solid material, the reflectivity can vary from 0 to 1. If the material is transparent, a reflection will also occur. The magnitude of this reflection at an air material interface (for transparent surfaces without an antireflection coating) is a function of the index of refraction (n) of the transparent material. The equation for the reflectivity for normal incident radiation is:

$$\rho = \left(\frac{n-1}{n+1}\right)^2 \quad [—]$$ (7-28)

As the angle of incidence (measured from the normal to the surface) becomes larger, an increased reflectance will generally be experienced. At grazing incidence angles, the reflectivity approaches 1 (100%).

The reflectivity of a surface is a function of the incident radiation wavelength. As the spectral energy distribution of the incident radiations shifts, changes can be expected in the values of the surface reflectivity. Antireflection coatings generally reduce the reflectivity at one relatively narrow bandwidth of wavelengths and increase the reflectivity at other wavelengths.

The amount of heat ($\dot{Q}_r$) reflected by a body is a function of the radiant energy incident upon the surface (E_{1-2}) and the surface reflectivity (ρ).

$$\dot{Q}_r = \rho E_{1-2} \quad [\text{W}]$$ (7-29)

where

$\dot{Q}_r$ = the heat reflected by the body   [W]

ρ = the effective reflectivity of the body   [—]

E_{1-2} = the energy incident upon the surface from Eq. (7-24)   [W]

TRANSMISSIVITY

The majority of solid materials encountered in engineering, except for glasses, are opaque. Consequently, none of any incident radiant energy is transmitted through such materials by radiation. Many liquids and gases have relatively high transmissivities.

The monochromatic transmittance characteristics of materials is a function of the wavelength of the incident radiation. For example, common glass is highly transparent to visible light, but essentially opaque to long wavelength infrared radiations. The transmission characteristics of glasses can be controlled resulting in the availability of transparent materials with a wide variety of transmission characteristics throughout the wavelength spectrum of thermal radiations.

For opaque bodies (wherein the transmissivity is zero), Eq. (7-25) becomes

$$\alpha + \rho = 1 \quad [-] \tag{7-30}$$

or, by applying Kirchhoff's law

$$\varepsilon + \rho = 1 \quad [-]$$

7-3 HEAT EXCHANGE BY RADIATION

The general equation for heat exchange by radiation between two surfaces is

$$\dot{Q} = A_1 F_A F_\varepsilon \sigma \left(T_1^4 - T_2^4 \right) \quad [W] \tag{7-31}$$

where

A_1 = area of the radiating upon which F_A is based $[m^2]$

F_A = shape factor based upon the geometric configuration of the two radiating surfaces   [—]

F_ε = effective emissivity factor (for blackbodies $F_\varepsilon = 1.0$)   [—]

σ = Stefan-Boltzmann constant (5.670×10^{-8}) $[W/m^2 \cdot K^4]$

T_1 = absolute temperature of surface A_1 [K]

T_2 = absolute temperature of the second surface   [K]

The shape factor (F_A) and the emissivity factor (F_ε) are discussed in the following sections.

7-3.1 General Equations for Heat Exchange by Radiation Between Surfaces

The net radiant interchange between surfaces depends upon the geometrical arrangement of the surfaces. In many cases, the two radiating surfaces do not "see each other" completely, and thus, the view factor is important. Consider the two radiating areas shown in Figure 7-25. The radiant heat transferred from surface dA_1 to dA_2 is

$$d\dot{Q}_{12} = \left(\frac{\cos \phi_1 \cos \phi_2 \, dA_1 \, dA_2}{\pi r^2} \right) \varepsilon_1 \varepsilon_2 \sigma \left(T_1^4 - T_2^4 \right) \quad [\text{W}] \tag{7-32}$$

The term inside the parentheses in Eq. (7-31) can be replaced after the integration of the equation by two factors. One factor A represents one of the radiating surface areas, and the other factor F_A is a configuration factor (sometimes called a "shape" or "view" factor). Using these two factors, Eq. (7-32) can be rewritten as

$$\dot{Q}_{12} = A_1 F_{A_1} \varepsilon_1 \varepsilon_2 \sigma \left(T_1^4 - T_2^4 \right) \quad [\text{W}] \tag{7-33}$$

Likewise an effective emissivity factor F_ε can be substituted for $\varepsilon_1 \varepsilon_2$, resulting in

$$\dot{Q}_{12} = A_1 F_{A_1} F_\varepsilon \sigma \left(T_1^4 - T_2^4 \right) \quad [\text{W}] \qquad [\text{Eq. (7-31)}]$$

where

$\dot{Q}_{12} =$ heat transferred from surface 1 to 2 $[\text{W}]$

$\sigma =$ Stefan-Boltzmann constant $[\text{W}/\text{m}^2 \cdot \text{K}^4]$

$A_1 =$ area of surface 1 $[\text{m}^2]$

$F_{A_1} =$ form factor between surfaces 1 and 2 based on area A_1 $[-]$

$T =$ surface temperature $[\text{K}]$

A similar expression can be developed based on the second surface:

$$\dot{Q}_{21} = A_2 F_{A_2} F_\varepsilon \sigma \left(T_2^4 - T_1^4 \right) \quad [\text{W}] \tag{7-34}$$

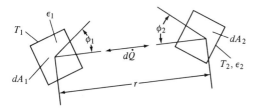

Figure 7-25   Geometric configuration of two radiating surfaces.

since

$$\dot{Q}_{21} = -\dot{Q}_{21} \quad [\text{W}]$$

then

$$A_1 F_{A_1} = A_2 F_{A_2} \quad [\text{m}^2] \tag{7-35}$$

This important relation indicates that the net radiant energy exchange between the two surfaces may be evaluated by calculating the configuration factor from either surface to the other.

For an enclosure consisting of several surfaces, the net energy transferred from one of the surfaces (A_1) at temperature T_1, to other surfaces can be expressed as follows:

$$\dot{Q}_{\text{net}} = A_1 F_{A_{12}} F_\varepsilon \sigma (T_1^4 - T_2^4) + A_1 F_{A_{13}} F_\varepsilon \sigma (T_1^4 - T_3^4) + \cdots \quad [\text{W}] \tag{7-36}$$

In the case of most radiation heat exchange configurations, the solid angle subtended by each surface with respect to the other surface is large. This results in a varying value of both angles ϕ_1 and ϕ_2 in Figure 7-25. Consequently, an integration of Eq. (7-32) must be performed to obtain the configuration factor F_A for Eq. (7-31).

Values of the configuration factor F have been calculated by Hottel and are given in Table 7-4 and Figures 7-26, 27, 28, and 29.

7-3.2 Heat Exchange by Radiation Between Large Parallel Planes

Effectively all of the energy radiated by one plane of two large parallel black planes will be incident upon the other plane. The factor F_A in Eq. (7-31) will be 1.0, as indicated on line 1 of Table 7-4. Because both surfaces are black, the receiving surface will absorb all of the incident radiation, and the radiating surface will radiate with an emissivity of 1. Consequently, the emissivity factor F_ε will be 1, as indicated in line 1 of Table 7-4. Thus the net energy exchange between the two surfaces will be

$$\dot{Q} = A F_A F_\varepsilon \sigma (T_1^4 - T_2^4) \quad [\text{W}] \quad [\text{Eq. (7-31)}]$$

$$F_A = 1.0 \quad [-]$$

$$F_\varepsilon = 1.0 \quad [-]$$

$$\dot{Q} = A\sigma (T_1^4 - T_2^4) \quad [\text{W}] \tag{7-37}$$

Example Problem 7-6

Calculate the net radiant energy exchange per square meter between two very large parallel black planes (infinite area) at temperatures of 50°C and 250°C, respectively.

Table 7-4 Radiation Between Solids, Factors For Use in Eq. (7-31)

SURFACES BETWEEN WHICH RADIATION IS BEING INTERCHANGED	AREA A	F_A	F_ϵ
1. Infinite parallel planes.	A_1 or A_2	1	$\dfrac{1}{(1/\varepsilon_1)+(1/\varepsilon_2)-1}$
2. Completely enclosed body, small compared with enclosing body. (Subscript 1 refers to enclosed body.)	A_1	1	ε_1
3. Completely enclosed body, large compared with enclosing body. (Subscript I refers to enclosed body.)	A_1	1	$\dfrac{1}{(1/\varepsilon_1)+(1/\varepsilon_2)-1}$
4. Intermediate case between 2 and 3. (Incapable of exact treatment except for special shapes.) (Subscript 1 refers to enclosed body.)	A_1	1	$\varepsilon_1 > F_\epsilon > \dfrac{1}{(1/\varepsilon_1)+(1/\varepsilon_2)-1}$
5. Concentric spheres or infinite cylinders, special case of 4. (Subscript 1 refers to enclosed body.)	A_1	1	$\dfrac{1}{(1/\varepsilon_1)+(A_1/A_2)[(1/\varepsilon_2)-1]}$*

Table 7-4 (*Continued*)

SURFACES BETWEEN WHICH RADIATION IS BEING INTERCHANGED	AREA A	F_A	F_ϵ
6. Surface element dA and area A_1. There are various special cases of 6 with results presentable in graphical form. They follow as Cases 7, 8, 9.	dA	See special Cases 7, 8, 9[†]	$\epsilon_1 \epsilon_2$
7. Element dA and rectangular surface above and parallel to it, with one corner of rectangle contained in normal to dA.	dA	Figure 7-26	$\epsilon_1 \epsilon_2$
8. Element dA and any rectangular surface above and parallel to it. Split rectangle into 4 having common corner at normal to dA and treat as in Case 7.	dA	Sum of F_A's determined for each rectangle as in Case 7	$\epsilon_1 \epsilon_2$
9. Element dA and circular disk in plane parallel to plane of dA.	dA	Formula below[‡]	$\epsilon_1 \epsilon_2$
10. Two parallel and equal squares or disks of width or diameter D and distance between of L.	A_1 or A_2	Figure 7-27 curves 1 and 2	$\epsilon_1 \epsilon_2$
11. Same as 10 except planes connected by nonconduct-	A_1 or A_2	Figure 7-27 curve 3	$\epsilon_1 \epsilon_2$

12. Two equal rectangles in parallel planes directly opposite each other and distance L between.	A_1 or A_2	$\sqrt{F_A' F_A''}$ §	$\varepsilon_1\varepsilon_2$ or $\dfrac{1}{(1/\varepsilon_1)+(1/\varepsilon_2)-1}$ §
13. Two rectangles with common sides, in perpendicular planes.	A_1 or A_2	Figure 7-2	$\varepsilon_1\varepsilon_2$
14. Radiation from a plane to a tube bank (1 or 2 rows) above and parallel to the plane.		Figure 7-2	$\varepsilon_1\varepsilon_2$

SOURCE: H. C. Hottel in *Mech. Eng.* 52 (7), 699 (July, 1930).

*This form results from assumption of completely diffuse reflection. If reflection is completely specular (mirrorlike), then $F_\varepsilon = 1/[(1/\varepsilon_1 + 1/\varepsilon_2) - 1]$.

†A complete treatment of this subject, including formulas for special complicated cases and the description of a mechanical device for solving problems in radiation, is given in source.

‡Case 9, $R =$ radius of disk ÷ distance between planes; $x =$ distance from dA to normal through center of disk ÷ distance between planes.

$$F_A = \frac{1}{2}\left[1 - \frac{x^2 + 1 - R_x^2}{\sqrt{x^4 + 2(1-R^2)x^2 + (1+R^2)^2}}\right]$$

§$F_A' = F_A$ for squares equivalent to short side of rectangle (Fig. 7-27 curve 2) and $F_A'' = F_A$ for squares equivalent to long side of rectangle (Fig. 7-27 curve 2).
$F_\varepsilon = \varepsilon_1\varepsilon_2$ if the areas are small compared with D.
$F_\varepsilon = 1/[(1/\varepsilon_1 + 1/\varepsilon_2) - 1]$ if the areas are large compared with D where D is the distance between the planes.

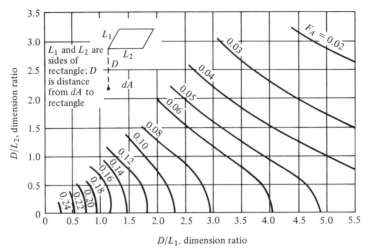

Figure 7-26 Radiation between surface element and rectangle above and parallel to it. (SOURCE: H. C. Hottel, "Radiant Heat Transmission, *Mech. Eng.* 52 (1930).)

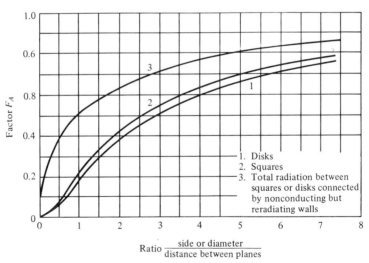

Figure 7-27 Direct radiation between equal disks or squares in parallel planes directly opposed. (SOURCE: H. C. Hottel, "Radiant Heat Transmission," *Mech. Eng.* 52 (1930).)

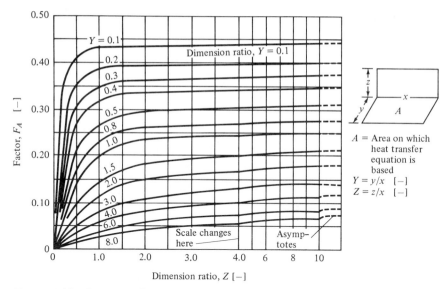

Figure 7-28   Radiation between adjacent rectangles in perpendicular planes. (SOURCE: H. C. Hottel, "Radiant Heat Transmission," *Mech. Eng.* 52 (1930).)

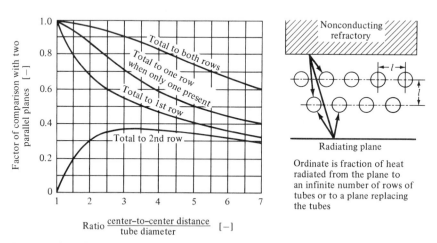

Figure 7-29   Radiation from a plane to one or two rows of tubes above and parallel to the plane. (SOURCE: H. C. Hottel, "Radiant Heat Transmission," *Mech. Eng.* 52 (1930).)

SOLUTION

SKETCH

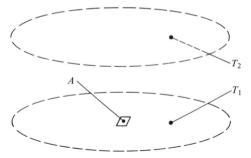

Figure 7-30

EQUATION

$$\dot{Q}_{12}/A = \sigma(T_1^4 - T_2^4) \quad [\text{W}/\text{m}^2] \qquad [\text{Eq. (7-36)}]$$

PARAMETERS

$$\sigma = 5.670 \times 10^{-8} \quad [\text{W}/\text{m}^2 \cdot \text{K}^4]$$
$$T = {}^\circ\text{C} + 273 \quad [\text{K}]$$
$$T_1 = 50 + 273 = 323 \ K \qquad \text{(given)}$$
$$T_2 = 250 + 273 = 523 \ K \qquad \text{(given)}$$

SUBSTITUTIONS

$$\dot{Q}_{12}/A = 5.670 \times 10^{-8}(323^4 - 523^4)$$
$$[\text{W}/\text{m}^2 \cdot \text{K}^4][\text{K}^4 - \text{K}^4] = [\text{W}/\text{m}^2]$$
$$= 5.670 \times 10^{-8}(108.8 \times 10^8 - 748.2 \times 10^8)$$
$$= 5.670 \times 10^{-8}(-639.4 \times 10^8)$$

ANSWER

$$\dot{Q}_{12}/A = -3625 \quad [\text{W}/\text{m}^2]$$

If we have two large parallel nonblack planes of different emissivities, effectively all of the energy radiated by one plane will be incident upon the other plane. The shape factor F_A in Eq. (7-31) will be 1.0 as indicated in line 1 of Table 7-4. Because of the emissivity of the "radiating" surface (A_1), the amount of the radiant energy incident upon the second surface will be less than that of a black surface at the same temperature. The amount of the energy incident upon the second surface will not be totally absorbed, as in the case of a black surface. The emissivity factor accounting for the reduced amount of radiant energy as well as the reduced absorption of the incident energy for the case of the two infinite parallel plates is given in line 1 of Table 7-4 as

$$\frac{\dot{Q}_{12}}{A} = F_A F_\varepsilon \sigma \left(T_1^4 - T_2^4 \right) \quad [\text{W}] \qquad [\text{Eq. (7-31)}]$$

$$F_A = 1.0 \quad [-] \qquad (\text{Table 7-4, line 1})$$

$$F_\varepsilon = \frac{1}{(1/\varepsilon_1) + (1/\varepsilon_2) - 1} \quad [-] \qquad (\text{Table 7-4, line 1})$$

Substituting in Eq. (7-31),

$$\frac{\dot{Q}_{12}}{A} = \left[\frac{1}{(1/\varepsilon_1) + (1/\varepsilon_2) - 1} \right] \sigma \left(T_1^4 - T_2^4 \right) \quad [\text{W/m}^2] \qquad (7\text{-}38)$$

Example Problem 7-7

Calculate the net radiant energy exchange per square meter between two very large parallel planes at temperatures of 600°C and 100°C if the emissivities of the hot and cold planes are 0.9 and 0.7, respectively.

SOLUTION

SKETCH

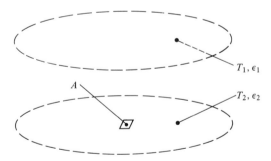

Figure 7-31

EQUATION

$$\frac{\dot{Q}_{12}}{A} = \left[\frac{1}{(1/\varepsilon_1) + (1/\varepsilon_2) - 1} \right] \sigma \left(T_1^4 - T_2^4 \right) \quad [\text{W/m}^2] \qquad [\text{Eq. (7-38)}]$$

PARAMETERS

$$\varepsilon_1 = 0.9 \quad [\,-\,] \qquad \text{(given)}$$

$$\varepsilon_2 = 0.7 \quad [\,-\,] \qquad \text{(given)}$$

$$\sigma = 5.670 \times 10^{-8} \quad [\text{W}/\text{m}^2 \cdot \text{K}^4]$$

$$T_1 = 600 + 273 = 873 \text{ K}$$

$$T_2 = 100 + 273 = 373 \text{ K}$$

SUBSTITUTIONS

$$= \frac{5.670 \times 10^{-8}}{(1/0.9) + (1/0.7) - 1} \left[(873)^4 - (373)^4 \right] \frac{[\text{W}/\text{m}^2 \cdot \text{K}^4] [[\text{K}^4] - [\text{K}^4]]}{[\,-\,] + [\,-\,]}$$

$$= \frac{5.670 \times 10^{-8}}{(1.111 + 1.429 - 1)} \left[(5808) - (194) \right] \times 10^8 \quad [\text{W}/\text{m}^2]$$

$$= \frac{5.670}{1.540} (5614)$$

ANSWER

$$\dot{Q}_{12}/A = 20{,}674 \; [\text{W}/\text{m}^2] \text{ or } 20.7 \; [\text{kW}/\text{m}^2]$$

7-3.3 Heat Exchange by Radiation Between Small Parallel Plane Segments

When two parallel plane segments exchange energy by radiation, only a portion of the total energy radiated by the first surface is incident upon the second surface. Consequently, the shape factor F_A in Eq. (7-31) will be less than 1. Figure 7-27 presents curves of the shape factors for equal size surfaces (circles or squares) that are centered around a normal axis between two parallel planes. The shape factor between an elemental area and a rectangle with one corner of the rectangle contained in the normal to the elemental area is given in Figure 7-26. The emissivity factor in this case is the product of the emissivities of the two surfaces as indicated in row 6, Table 7-4.

Example Problem 7-8

(a) A square surface 1 m² is parallel to a flat rectangular surface 10 by 12 m and is located opposite the center of the large rectangular surface with the two surfaces 20 m apart. What is the radiation heat transfer shape factor F_A?

(b) If the surface of the 1 m² has an emissivity of 0.3 and a temperature of T_s, and the large surface has an emissivity of 0.8

and a temperature of T_L, what is the net radiation heat transfer rate between the two surfaces expressed in terms of T_s and T_L?

SOLUTION

SKETCH

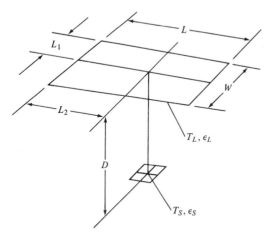

Figure 7-32

PLAN FOR (a)
Use Figure 7-26. Consider the small surface to be an "elemental" area.

EQUATIONS FOR (a)
We need to evaluate D/L_1 and D/L_2 of Figure 7-26.

D/L_1:

$$D = 20 \quad [\text{m}] \quad \text{(given)}$$
$$L_1 = W/2 \quad [\text{m}]$$
$$W = 10 \quad [\text{m}] \quad \text{(given)}$$
$$L_1 = 10/2 = 5 \quad [\text{m}]$$
$$D/L_1 = 20/5 \quad [\text{m}]/[\text{m}] = [-]$$
$$D/L_1 = 4.0 \quad [-]$$

D/L_2:

$$L_2 = L/2 \quad [\text{m}]$$
$$L = 12 \quad [\text{m}] \quad \text{(given)}$$
$$L_2 = 12/2 = 6 \quad [\text{m}]$$
$$D/L_2 = 20/6 \quad [\text{m}]/[\text{m}] = [-]$$
$$D/L_2 = 3.33 \quad [-]$$

From Figure 7-23 at $D/L_1 = 4.0$ and $D/L_2 = 3.33$,

$$F_A = 0.022 \quad [-]$$

ANSWER FOR (a)

$$F_A = 0.022 \quad [-]$$

PLAN FOR (b)
Use equation

$$\dot{Q} = 4 \left[dA \, F_A F_\varepsilon \sigma \left(T_s^4 - T_L^4 \right) \right] \quad [\text{W}] \qquad [\text{Eq. (7-31)}]$$

PARAMETERS FOR (b)

$$dA = \tfrac{1}{4} \quad [\text{m}^2] \qquad (\text{given})$$
$$F_A = 0.022 \quad [-] \qquad [\text{Example Problem 7-8}(a)]$$
$$F_\varepsilon = \varepsilon_L \varepsilon_s \quad [-] \qquad (\text{line 7, Table 7-4})$$
$$\varepsilon_L = 0.8 \quad [-] \qquad (\text{given})$$
$$\varepsilon_s = 0.3 \quad [-] \qquad (\text{given})$$
$$F_\varepsilon = (0.8)(0.3) \quad [-][-] = [-]$$
$$F_\varepsilon = 0.24 \quad [-]$$
$$\sigma = 5.67 \times 10^{-8} \quad [\text{W}/\text{m}^2 \cdot \text{K}^4]$$

SUBSTITUTION FOR (b)

$$\dot{Q} = (4)(\tfrac{1}{4})(0.022)(0.24)(5.67 \times 10^{-8})\left(T_s^4 - T_L^4 \right)$$
$$[\text{m}^2][-][-][\text{W}/\text{m}^2 \cdot \text{K}^4][\text{K}^4 - \text{K}^4] = [\text{W}]$$

ANSWER FOR (b)

$$\dot{Q} = 0.030 \times 10^{-8} \left(T_s^4 - T_L^4 \right) \quad [\text{W}]$$

7-3.4 Heat Exchange by Radiation Between an Enclosed Body and the Enclosure

A SMALL BODY IN A LARGE ENCLOSURE
A large cavity, or enclosure, acts as a blackbody regardless of the emissivity of its surface. Any radiation incident upon the surface of the cavity is partially absorbed and the balance reflected. The reflected radiation strikes the surface of the cavity again, where the absorption and reflection process repeats itself. After repeated reflections, for all practical purposes, all of the original incident radiant energy will be absorbed by the surface of the cavity, which effectively makes the cavity a blackbody.

When a body is positioned inside a cavity, and the dimensions of the body are small relative to the cavity, the shape factor F_A based upon the area of the small body is 1, and the emissivity factor F_e is the emissivity of the surface of the small body. The equation for the radiant energy transfer between the small body and the surrounding cavity is

$$\dot{Q}/A = \varepsilon_1 \sigma \left(T_1^4 - T_2^4 \right) \quad \left[W/m^2 \right] \tag{7-39}$$

where

$A =$ the surface area of the small body $\left[m^2 \right]$

$\varepsilon_1 =$ the emissivity of the surface of the small body $[-]$

$T_1 =$ the temperature of the surface of the small body $[K]$

$T_2 =$ the temperature of the surface of the cavity $[K]$

Example Problem 7-9

An iron pipe at $125°C$ has an o.d. of 5 cm. If the emissivity of the pipe surface is 0.8 and the surrounding room is at $27°C$, what is the net energy exchange per meter of pipe length?

SOLUTION

SKETCH

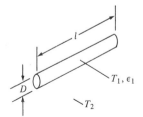

Figure 7-33

EQUATION

$$\frac{\dot{Q}_{12}}{l} = \frac{A_1}{l} \varepsilon_1 \sigma \left(T_1^4 - T_2^4 \right) \quad [W/m] \qquad [\text{Eq. (7-39)}]$$

PARAMETERS

$$\frac{A_1}{l} = \pi D$$

$D = 5 \text{ cm} \rightarrow 0.05 \text{ m} \qquad (\text{given})$

$\quad = \pi(0.05)$

$$\frac{A_1}{l} = 0.157 \text{ m}$$

$\varepsilon_1 = 0.8 \quad [-] \qquad \text{(given)}$

$\sigma = 5.670 \times 10^{-8} \quad [\text{W}/\text{m}^2\,\text{K}^4]$

$T_1 = 125 + 273 = 398 \text{ K}$

$T_2 = 27 + 273 = 300 \text{ K}$

SUBSTITUTION

$$\frac{\dot{Q}_{12}}{l} = \frac{A_1}{l}\varepsilon_1\sigma(T_1^4 - T_2^4) \quad [\text{W}/\text{m}]$$

$$= (0.157)(0.8)(5.670 \times 10^{-8})\left[(398)^4 - (300)^4\right]$$

$$\left[\text{m}^2/\text{m}\right][-]\left[\text{W}/\text{m}^2\cdot\text{K}^4\right]\left[[\text{K}^4] - [\text{K}^4]\right] = [\text{W}/\text{m}]$$

$$= 0.712 \times 10^{-8}(250.9 - 81) \times 10^8$$

$$= 0.712(169.9)$$

ANSWER

$$\frac{\dot{Q}_{12}}{l} = 121 \quad [\text{W}/\text{m}]$$

COMPLETELY ENCLOSED BODY, LARGE COMPARED WITH
ENCLOSURE

When a body is completely enclosed and is large compared with the enclosure, the condition approaches that of two parallel planes of infinite extent. Both radiating surfaces are equal in area, and all of the radiation from one surface is incident upon the other surface. For the solution of this type of a problem, the applicable equations in Section 7-3.2 are used.

COMPLETELY ENCLOSED BODY, INTERMEDIATE SIZE
COMPARED WITH THE ENCLOSURE

For configurations, such as concentric spheres or concentric cylinders of infinite length, the shape factor is 1 because all of the radiations emitted by the enclosed body are incident upon the surface of the enclosure. If the two surfaces are specular (mirrorlike), the emissivity factor is the same as for infinite parallel planes. If the surfaces give completely diffuse reflection, the emissivity factor is affected by the ratio of the areas of the two surfaces as indicated in line 5 of Table 7-4.

Example Problem 7-10

An insulated steam pipe 0.25 m in diameter passes along a brick-lined tunnel which has a passage area 10 times the cross-sectional area of the insulated steam pipe. What difference would there be in the net radiation heat transfer between the pipe and the tunnel surfaces (radiation heat loss from the pipe) when the surface of the tunnel is at 25°C and the surface of the steam pipe insulation is at 100°C between the two following cases:

Case (a)—the surfaces of the tunnel are brick and the rough surface of the steam pipe insulation has been painted with aluminum paint;

Case (b)—the tunnel has been lined with reflective aluminum sheet and the steam pipe insulation has been covered with reflective aluminum foil.

SOLUTION

SKETCH

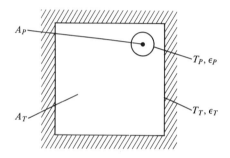

Figure 7-34

Treat the pipe as a completely enclosed body intermediate in size compared with the enclosure. Define the difference between the two cases as

$$\text{difference } \Delta = \frac{\dot{Q}_b}{\dot{Q}_a} \quad [-]$$

Equation for $\dot{Q}$ is

$$\dot{Q} = A \cdot F_A \cdot F_\epsilon \cdot \sigma (T^4 - T^4) \quad [\text{W}]$$

EQUATION

$$\Delta = \frac{\dot{Q}_b}{\dot{Q}_a} = \frac{F_{A_b} F_{\epsilon_b}}{F_{A_a} F_{\epsilon_a}} \quad [-]$$

PARAMETERS

$F_{A_b} = 1.0 \quad [-]$ (Table 7-4   Line 4)

$F_{A_a} = 1.0 \quad [-]$ (Table 7-4   Line 4)

$$F_{\varepsilon_b} = \frac{1}{(1/\varepsilon_P)+(1/\varepsilon_T)-1} \quad [-] \quad \text{(Table 7-4 Line 4)}$$

$\varepsilon_P = 0.040$ $[-]$ (Table B-9, Appendix B polished aluminum plate)

$\varepsilon_T = 0.055$ $[-]$ (Table B-9, Appendix B rough aluminum plate)

$$F_{\varepsilon_b} = \frac{1}{(1/0.040)+(1/0.055)-1} = \frac{1}{25.0+18.2-1.0} = \frac{1}{42.2}$$

$F_{\varepsilon_b} = 0.024$ $[-]$

$$F_{\varepsilon_a} = \frac{1}{(1/\varepsilon_P)+(A_P/A_T)(1/\varepsilon_T)-1} \quad [-] \quad \text{(page 333)}$$

$\varepsilon_P = 0.52$ $[-]$ (Table B-9, Appendix B 10% Al 22% lacquer)

$\varepsilon_T = 0.93$ $[-]$ (Table B-9, Appendix B red rough brick)

$$\frac{A_P}{A_T} = \frac{1}{10} \quad [-] \quad \text{(given)}$$

$$F_{\varepsilon_a} = \frac{1}{(1/0.52)+(1/10)(1/0.93-1)} \quad [-]$$

$$\frac{1}{1.923+0.1(1.075-1)} = \frac{1}{1.923+0.008} = \frac{1}{1.931}$$

$F_{\varepsilon_a} = 0.518$ $[-]$

SUBSTITUTION

$$\Delta = \frac{(1.0)(0.024)}{(1.0)(0.518)} \quad \frac{[-][-]}{[-][-]} = [-]$$

ANSWER

$\Delta = 0.046$ $[-]$

COMBUSTION CHAMBER WALLS TO TUBES

Radiation between the walls of a combustion chamber and tubes can be calculated using

$$\dot{Q} = A_1 F_A F_\varepsilon F_T \sigma (T_1^4 - T_2^4) \quad [\text{W}] \tag{7-40}$$

For a configuration, such as a refractory floor and four walls with a ceiling of parallel tubes, the radiant heat transfer would be calculated separately for the floor and walls and summed to obtain the total. The evaluation of $F_A F_\varepsilon F_T$ for these separate calculations will be discussed in the following sections.

• *Floor to Tubes.* The shape factor F_A for two parallel rectangles of equal area is calculated using Figure 7-27 as indicated by Table 7-4, line 12.

$$F_A = (F_A' F_A'')^{1/2} \quad [-]$$

where

F_A' = is for parallel squares of one dimension $\quad [-]$

F_A'' = is for parallel squares of the other dimension $\quad [-]$

The emissivity factor F_ε for two parallel rectangles of equal area is calculated as indicated by Table 7-4, line 12.

$F_\varepsilon = \varepsilon_1 \varepsilon_2$ if the areas are small compared with their separation distance

$F_\varepsilon = \dfrac{1}{(1/\varepsilon_1) + (1/\varepsilon_2) - 1}$ if the areas are large compared with their separation distance

The emissivity of the tubes is considered to be 0.9 by authors such as Brown and Marco[4] due to reflected radiation between the tubes, and the wall and the tubes. The tube configuration factor F_T is obtained from Figure 7-29.

• *Wall to Tubes.* The shape factor F_A for two adjacent rectangles in perpendicular planes is obtained from Figure 7-28. The emissivity factor F indicated by Table 7-4, line 13 is

$$F_\varepsilon = \varepsilon_1 \varepsilon_2 \quad [-]$$

As discussed above the emissivity of the tubes is considered to be 0.9. The tube configuration factor F_T is obtained from Figure 7-29.

COMBUSTION GASES TO BOILER TUBES FOR LUMINOUS FLAMES

The net heat transfer between a luminous flame and a tube array can be treated in terms of the general radiation heat transfer equation.

$$\dot{Q} = A F_A F_\varepsilon \sigma (T_F^4 - T_t^4) \quad [\text{W}]$$

The shape factor F_A can be based on radiation heat transfer between two parallel planes modified by a factor given in Figure 7-29, which can be applied to a basic configuration factor of 1.

The emissivity factor (F_ε) is the product of the effective emissivity of the luminous flame (ε_F) and the tubes (ε_t).

$$F_\varepsilon = \varepsilon_F \varepsilon_t \quad [-]$$

[4] A. I. Brown, and Marco, S. M., *Introduction to Heat Transfer*, 3rd ed., McGraw-Hill, New York, 1958.

The emissivity of the luminous flame (ε_F) can be established through the use of Figures 7-14 and 7-14a. The flame temperature can be established by Figure 7-14.

7-3.5 Adjacent Rectangles in Perpendicular Planes

The shape factor F_A for radiant heat exchange between two adjacent perpendicular planes is shown in Figure 7-28. The emissivity factor is the product of the emissivities of the two rectangles.

Example Problem 7-11

A 3-m high side of a house is adjacent to a 5-m wide black top driveway which runs the 10-m length of the house side. The side of the house is painted clapboard. What is the net radiant heat exchange between the driveway and the side of the house when the surface of the house side is 30°C and the driveway surface is 45°C?

SOLUTION

SKETCH

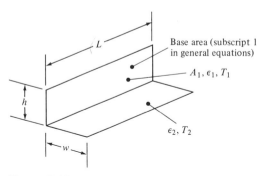

Figure 7-35

EQUATION

$$\dot{Q} = A_1 F_{A_1} F_\varepsilon \sigma \left(T_1^4 - T_2^4 \right) \quad [\text{W}] \qquad [\text{Eq. (7-31)}]$$

PLAN

Select the side of the house as the basic area—subscript 1 in the equations.

Evaluate F_{A_1} from Figure 7-28, F_ε from Table 7-4, line 13

PARAMETERS

$A_1 = (h)(L)$ $[m^2]$

$h = 3$ m   (given)

$L = 10$ m   (given)

$A_1 = (3)(10)$ $[m][m] = [m^2]$

$A_1 = 30$ m^3

Obtaining F_{A_1} from Figure 7-28, we need dimension ratios Y and Z.

$Y = h/L = 3/10 = 0.3$ $[-]$

$Z = w/L$ $[-]$

$w = 5$ m   (given)

$Z = 5/10 = 0.50$ $[-]$

From Figure 7-28 at $Z = 0.5$ and $Y = 0.3$,

$F_{A_1} = 0.315$ $[-]$

$F_\varepsilon = \varepsilon_1 \varepsilon_2$ $[-]$ (Table 7-4, line 13)

$\varepsilon_1 = 0.94$ $[-]$ (Table B-9 Appendix B oil paints, all colors)

$\varepsilon_2 = 0.91$ $[-]$ (Table B-9, Appendix B roofing paper)

$F_\varepsilon = (0.94)(0.91)$ $[-][-] = [-]$

$F_\varepsilon = 0.855$ $[-]$

$\sigma = 5.670 \times 10^{-8}$ $[W/m^2 \cdot K^4]$

$T_1 = 30 + 273 = 303$ K   (given)

$T_1^4 = 84.3 \times 10^8$ K^4

$T_2 = 45 + 273 = 318$ K   (given)

$T_2^4 = 102.3 \times 10^8$ K^4

SUBSTITUTION

$\dot{Q} = (30)(0.315)(0.855)(5.670 \times 10^{-8})(84.3 - 102.3) \times 10^8$

$\quad\quad [m^2][-][-][W/m^2 \cdot K^4][K^4 - K^4] = [W]$

$\quad = (45.8)(-18.0)$

ANSWER

$\dot{Q} = -824.4$ $[W]$

Note: the negative sign indicates a net heat gain by the base area with the subscript 1 in the equations, which is the side of the house.

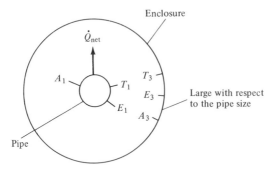

Figure 7-36 Radiation heat exchange between a pipe and a surrounding enclosure without a radiation shield.

7-3.6 Radiation Shields

Radiation heat transfer between two objects can be reduced by introducing opaque, highly reflective surface(s) between the two objects, as illustrated in Figures 7-36 and 7-37.

Figure 7-36 illustrates a pipe passing through an enclosure large in size relative to the pipe diameter such as an attic of a house. Considering radiation heat transfer only

$$\dot{Q}_n = A_1 F_A F_E \sigma (T_1^4 - T_2^4) \quad [\text{W}] \qquad [\text{Eq. (7-31)}]$$

$$F_A = 1.0 \qquad (\text{from Table 7-4})$$

$$F_E = E_1 \qquad (\text{from Table 7-4})$$

$$\dot{Q}_n = A_1 \varepsilon_1 \sigma (T_1^4 - T_2^4) \quad [\text{W}] \tag{7-41}$$

When a radiation shield is placed over the pipe spaced away from the pipe to avoid physical contact, a series radiation heat transfer path is established, as illustrated in Figure 7-37.

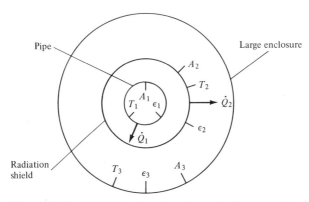

Figure 7-37 Radiation heat exchange between a pipe and a surrounding enclosure through a radiation shield.

Once more, considering radiation heat transfer only, the heat flow from (or to) the pipe to the radiation shield must equal the heat flow from (or to) the radiation shield and the enclosure.

$$\dot{Q}_1 = \dot{Q}_2 \quad [W]$$

$$\dot{Q}_1 = A_1\left(\frac{1}{(1/\varepsilon_1)+(1/\varepsilon_2)-1}\right)\sigma(T_1^4 - T_2^4) \quad [W] \qquad (7\text{-}42)$$

$$\dot{Q}_2 = A_2\varepsilon_2\sigma(T_2^4 - T_3^4) \quad [W] \qquad (7\text{-}43)$$

Solving for T_2, using $\dot{Q}_1 = \dot{Q}_2$

$$A_1\left(\frac{1}{(1/\varepsilon_1)+(1/\varepsilon_2)-1}\right)\sigma(T_1^4 - T_2^4) = A_2\varepsilon_2\sigma(T_2^4 - T_3^4)$$

$$T_2^4 = \frac{A_1\left(\dfrac{1}{(1/\varepsilon_1)+(1/\varepsilon_2)-1}\right)T_1^4 + A_2\varepsilon_2 T_3^4}{A_1\left(\dfrac{1}{(1/\varepsilon_1)+(1/\varepsilon_2)-1}\right) + A_2\varepsilon_2} \quad [K^4] \qquad (7\text{-}44)$$

To obtain the radiation heat flow with the radiation shield, once T_2^4 has been determined by Eq. (7-44), either Eq. (7-41) or (7-42) can be used.

Example Problem 7-12

A bare galvanized iron duct 25 cm in diameter extends through an attic carrying hot air. If the attic surfaces are at 5°C and the surface temperature of the bare duct is 55°C, what is the radiation loss per meter of duct length?

SOLUTION

SKETCH

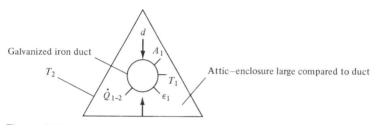

Galvanized iron duct

Attic–enclosure large compared to duct

Figure 7-38

EQUATION

$$\dot{Q}_{1-2} = A_1 F_A F_e \sigma \left(T_1^4 - T_2^4 \right) \quad [\text{W/m}] \qquad [\text{Eq. (7-31)}]$$

$$A_1 = \pi d = (\pi)(0.25) = 0.785 \text{ m}^2/\text{m}$$

PARAMETERS

$F_A = 1 \quad [-] \qquad$ (Table 7-4)

$F_e = \varepsilon_1 \quad [-] \qquad$ (Table 7-4)

$\varepsilon_1 = 0.276 \quad [-] \qquad$ (Appendix B, Table B-9)

$\sigma = 5.67 \times 10^{-8} \quad \left[\text{W/m}^2 \cdot \text{K}^4 \right]$

$T_1 = 55 + 273 = 328 \text{ K} \qquad T_1^4 = 115.7 \times 10^8 \text{ K}^4$

$T_2 = 5 + 273 = 278 \text{ K} \qquad T_1^4 = 59.7 \times 10^8 \text{ K}^4$

SUBSTITUTION

$$\dot{Q}_{1-2} = 0.785 \times 1 \times 0.276 \times 5.67 \times 10^{-8} (115.7 \times 10^8 - 59.7 \times 10^8)$$

$$\left[\text{m}^2/\text{m} \right][-][-]\left[\text{W/m}^2 \cdot \text{K}^4 \right]\left[\text{K}^4 - \text{K}^4 \right] = [\text{W/m}]$$

ANSWER

$$\dot{Q}_{1-2} = 68.8 \; [\text{W/m}]$$

Example Problem 7-13

If the duct is loosely wrapped with one layer of aluminum foil and the temperature of the galvanized iron duct surface remains at 55°C, what is the radiation loss per meter of duct length?

SKETCH

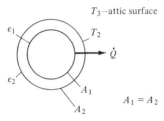

Figure 7-39

EQUATION

$$\dot{Q}_{1-2} = \dot{Q}_{2-3} \quad [\text{W/m}]$$

$$\dot{Q}_{1-2} = A_1 F_A F_\varepsilon \sigma (T_1^4 - T_2^4) \quad [\text{W/m}] \qquad [\text{Eq. (7-31)}]$$

PARAMETERS

$$A_1 = \pi d = (\pi)(0.25) = 0.785 \text{ m}^2/\text{m}$$

$$F_A = 1 \quad [-] \qquad (\text{Table 7-4})$$

$$F_\varepsilon = \frac{1}{(1/\varepsilon_1) + (1/\varepsilon_2) - 1} \quad [-] \qquad (\text{Table 7-4})$$

$$\varepsilon_1 = 0.276 \quad [-] \qquad (\text{Appendix B, Table B-9})$$

$$\varepsilon_2 = 0.04 \quad [-] \qquad (\text{Appendix B, Table B-9})$$

$$F_\varepsilon = \frac{1}{(1/0.276) + (1/0.04) - 1} = \frac{1}{3.62 + 25 - 1} = 0.036 \quad [-]$$

$$\sigma = 5.67 \times 10^{-8} \quad [\text{W/m}^2 \cdot \text{K}^4]$$

$$T_1 = 55 + 273 = 328 \text{ K} \qquad T_1^4 = 115.7 \times 10^8 \text{ K}^4$$

$$T_2 = \text{unknown} \quad [\text{K}]$$

SUBSTITUTION

$$\dot{Q}_{1-2} = 0.785 \times 1 \times 0.036 \times 5.67 \times 10^{-8} (115.7 \times 10^8 - T_2^4)$$

$$[\text{m}^2/\text{m}][-][\text{W/m}^2 \cdot \text{K}^4][\text{K}^4 - \text{K}^4] = [\text{W/m}]$$

$$= 0.1602 \times 10^{-8} (115.7 \times 10^8 - T_2^4) = 18.535 - 0.1602 \times 10^{-8} T_2^4$$

$$\dot{Q}_{2-3} = A_2 F_A F_\varepsilon \sigma (T_2^4 - T_3^4) \quad [\text{W/m}]$$

$$A_2 = A_1 = 0.785 \text{ m}^2/\text{m} \qquad (\text{assumed})$$

$$F_A = 1 \quad [-] \qquad (\text{Table 7-4})$$

$$F_\varepsilon = \varepsilon_2 \quad [-] \qquad (\text{Table 7-4})$$

$$\varepsilon_2 = 0.04 \quad [-]$$

$$\sigma = 5.67 \times 10^{-8} \quad [\text{W/m}^2 \cdot \text{K}^4]$$

$$T_2 = \text{unknown}$$

$$T_3 = 5 + 273 = 278 \text{ K} \qquad T_3^4 = 59.7 \times 10^8 \text{ K}^4$$

$$\dot{Q}_{2-3} = 0.785 \times 1 \times 0.04 \times 5.67 \times 10^{-8} (T_2^4 - 59.7 \times 10^8)$$

$$[\text{m}^2/\text{m}][-][-][\text{W/m}^2 \cdot \text{K}^4][\text{K}^4 - \text{K}^4] = [\text{W/m}]$$

$$= 0.178 \times 10^{-8} (T_2^4 - 59.7 \times 10^8) = 0.178 \times 10^{-8} T_2^4 - 10.627$$

$$[\text{W/m}]$$

Equating $\dot{Q}_{1-2} = \dot{Q}_{2-3}$ [W/m],

$$18.535 - 0.1602 \times 10^{-8} T_2^4 = 0.178 \times 10^{-8} T_2^4 - 10.627$$

$$29.162 = 0.338 \times 10^{-8} T_2^4$$

$$86.28 \times 10^8 = T_2^4$$

$$T_2 = 305 \text{ K}$$

Substituting T_2 into the equation for $\dot{Q}_{2-3}$,

$$\dot{Q}_{2-3} = 0.785 \times 1 \times 0.04 \times 5.67 \times 10^{-8} (86.28 \times 10^8 - 59.7 \times 10^8)$$

$$[m^2/m][-][-][W/m^2 \cdot K^4][K^4 - K^4] = [W/m]$$

ANSWER

$$\dot{Q}_{2-3} = 4.7 \text{ [W/m]}$$

If more than one radiation shield is involved, the same basic method is used. Each heat shield transfers the same amount of heat by net absorption on one side and net emission on the other side.

In cases where the $F_A F_\varepsilon$ factor product is approximately the same for all radiation heat transfer steps, the radiation heat transfer through the shields will be approximately

$$\dot{Q} = \frac{A_A (F_A F_\varepsilon) \sigma (T_1^4 - T_F^4)}{n + 1} \quad \text{[W]} \qquad \text{[Eq. (7-42)]}$$

where

$$A_A = \text{average area} \quad [m^2]$$

$(F_A F_\varepsilon) = $ shape factor/emissivity factor product

(typical for all radiation heat transfer steps) $[-]$

$T_1 = $ pipe surface temperature $[K]$

$T_F = $ surrounding enclosure surface temperature $[K]$

$n = $ number of radiation shields $[-]$

7-4 FUNDAMENTALS OF SOLAR RADIATION

In houses with a south aspect, the sun's rays penetrate into the porticos in winter, but in summer the path of the sun is right over our heads and above the roofs so that there is shade.

Socrates, ca.400 B.C.

Solar energy has been used to various degrees by people throughout the ages. Solar energy is characterized by continual variations at a point (any location) on the earth's surface in both the angle of the line of propagation and the intensity of the solar energy at the earth's surface. These variations in the line of propagation are a function of both the hour (earth's rotation) and the season (tilt of the earth's axis with respect to the sun). The latitude of the point on the earth's surface also has an effect.

The following introduction discusses:

- the general characteristics of the source (the sun);
- the general characteristics of the solar radiations at the surface of the earth (after passage through the earth's atmosphere);
- the geometrical considerations between an incident surface and the line of propagation of the sun's rays.

The discussion in this text is primarily directed toward

- thermal input into buildings from solar radiations incident on the walls and roofs;
- the energy input to solar collectors.

7-4.1 The Solar Constant (I_0) (1.353 [kW/m^2])

The magnitude of the incoming (incident) solar radiation flux density (normal to the beam) at the orbit of the earth is given by the symbol I_0. As discussed in Section 7-2.1, the magnitude of I_0 [from Eq. (7-24)] is

- proportional to the effective radiating area of the sun;
- proportional to the emissivity of the sun;
- proportional to the fourth power of the temperature of the sun;
- inversely proportional to the square of the distance of the earth's orbit from the sun.

Because humankind's existence is dependent upon a certain consistency in the emitted radiant energy from the sun, people, since they were first able, have continuously made measurements of the intensity of solar radiations. Throughout the period since accurate measurements have been taken, the emitted solar energy has remained constant. The magnitude of the incident solar radiation flux density just outside the earth's atmosphere has been termed the "solar constant." However, the incident solar radiation flux density at the earth's orbit will vary throughout the year because the earth's orbit is elliptical rather than circular. The major axis of the earth's elliptical orbit is 3.4% greater than the minor axis. This causes a variation in the incident solar flux density of $\pm 3\frac{1}{2}\%$ from an average value. This average value of the solar radiation flux density at the earth's orbit (often referred to as near-earth space) is called the solar constant I_0. Its measured value by international agreement is 1.353 kW/m^2.

Because the solar radiations are an incoming radiation flux from essentially a point source, the primary principles are those discussed in

Section 7-2.1. These involve the factors of (1) the incident angle of the sun's rays with the surface involved and (2) the incident radiation flux density at the surface. With solar radiations both of these factors change continually with time.

The following discussion of solar radiations has been organized to

- define the geometric angles involved with the motion of the sun with respect to a surface of interest;
- discuss and evaluate the solar radiation flux density incident upon a surface;
- discuss the net thermal input into buildings from incident solar radiations;
- discuss factors pertaining to solar collectors.

7-4.2 Solar Radiation Incident upon the Earth's Surface

The incoming solar radiation interacts with the earth's atmosphere which is illustrated in Figure 7-40. The earth's atmosphere reflects, scatters, and absorbs the solar radiations passing through to various degrees depending upon the clarity of the air and the length of the path length through the atmosphere. This results in a combination of direct (or beam) (I_b) and diffuse (I_d) incident radiation flux densities at the earth's surface, with a total (often called global) flux density (I_g) that is less than the value of the incoming solar radiant flux density (I_0) at the outer edge of the atmosphere.

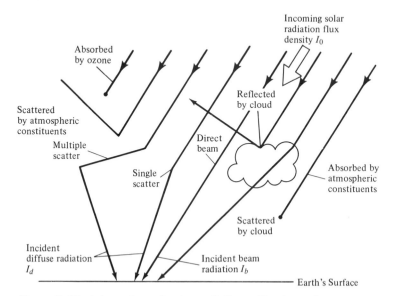

Figure 7-40 Interaction of solar radiation with atmosphere.

The resultant solar radiation that is incident upon the earth's surface has two components, the sum of which is the total incident radiation. These components are:

- *Direct radiation*: Direct radiation is defined as the solar radiation intercepted by the earth's surface with negligible direction change and scattering in the atmosphere. Direct radiation is also referred to as beam radiation, and is subject to optical concentration by the use of focusing lenses or mirrors.

- *Diffuse radiation*: Diffuse radiation is defined as the solar radiation scattered by constituents of the atmosphere such as molecules of water vapor, carbon dioxide, dust, and aerosols and redirected onto the earth's surface as seen in Figure 7-40.

- *Total radiation*: Total radiation is the sum of the direct and diffuse, and is sometimes referred to as global radiation.

INCIDENT DIRECT (BEAM) SOLAR RADIATION FLUX DENSITY (I_b)

The direct solar radiation flux (I_0) is subject to attenuation as it passes through the earth's atmosphere due to

- absorption by the constituents of the atmosphere;
- reflection by particulate matter and refraction gradients;
- scattering caused by aerosol and particulate matter.

The incident direct (beam) radiation flux density (I_b) normal to the beam at the earth's surface can be described by the conventional transmissivity equation

$$I_b = I_0 \bar{\tau} \text{atm} \quad [\text{W}/\text{m}^2] \tag{7-45}$$

where

I_0 = the solar constant $[\text{W}/\text{m}^2]$

$\bar{\tau}$atm = the atmospheric transmittance of the beam radiation $[-]$

The magnitude of the atmospheric transmittance of the beam radiation ($\bar{\tau}$atm) can be related to the

- mass of the atmosphere through which the beam passes;
- the clarity of the atmosphere.

The mass of the atmosphere through which the beam passes is primarily a function of the path length of the beam through the atmosphere. The relative geometric path length is in turn a function of the solar altitude angle:

- the solar altitude angle (α_s);
- the altitude (Z) of the surface upon which I_b is incident.

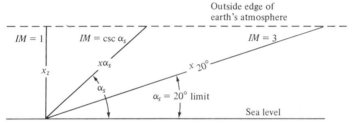

Figure 7-41 Beam radiation path length air mass (*IM*).

The effects of the solar altitude angle are illustrated in Figure 7-41. The effects of altitude are illustrated in Figure 7-42.

To relate the different geometric path lengths to the mass of the atmosphere through which the beam travels, the concept of a parameter called "air mass" has been introduced. The air mass is described by the length of the path of the incoming beam radiation through the earth's atmosphere. The reference (or basic) path length is the zenith path length through the atmosphere from sea level. The air mass (*IM*) is defined as the ratio of the geometric path length through the atmosphere at the solar altitude angle (α_s) to the zenith path length, as illustrated in Figure 7-41. The mathematical equation for *IM* disregarding the curvature of the earth is then

$$IM = \frac{x_{\alpha_s}}{x_z} = \frac{1}{\sin \alpha_s} = \csc \alpha_s \quad [-] \tag{7-46}$$

This equation is usually limited to solar altitude angles between 90° and 20° as illustrated in Figure 7-41.

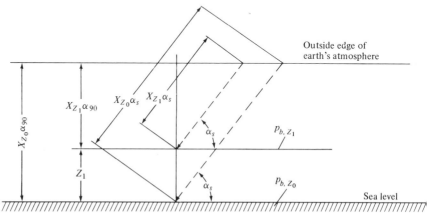

Figure 7-42 Relative geometric path lengths for incoming beam radiation for various altitudes (Z_i) and solar altitude angles (α_s).

Example Problem 7-14

For solar altitudes of 90°, 30°, and 20°, the air masses, IM, is given by

$$IM = \frac{1}{\sin 90°} = \frac{1}{1} = 1$$

and

$$IM = \frac{1}{\sin 30°} = \frac{1}{0.5} = 2$$

and

$$IM = \frac{1}{\sin 20°} = \frac{1}{0.3420} \simeq 3$$

Based on a standard atmosphere in which air pollution is absent, Kreith and Kreider[5] developed the following equation for the sea level air mass ratio as a function of the solar altitude angle (IM_{0, α_s}):

$$IM_{0, \alpha_s} = \left[1299 + (614 \sin \alpha_s)^2 \right]^{1/2} - 614 \sin \alpha_s \quad [-] \tag{7-47}$$

When the solar radiations are incident at a significant elevation above sea level, the incoming beam path is shorter as illustrated in Figure 7-42. This shows that the incoming path length at altitude $Z_1(X_{Z_1}\alpha_s)$ is shorter than the sea level path length ($X_{Z_0}\alpha_s$). As the total mass of air through which the incoming radiations pass is the important factor, not the physical length of the path without regard to the atmospheric density, the barometric pressure (p_b) is a suitable measurement of the effective zenith distance to the outside edge of the earth's atmosphere. At sea level the barometric pressure p_{b, Z_0} would be a measurement of $X_{Z_0}\alpha_{90}$, and at elevation Z_1 the barometric pressure p_{b, Z_1} would be a measurement of $X_{Z_1}\alpha_{90}$. Equation (7-46) then becomes

$$IM = \frac{p_{b, Z_1}}{p_{b, Z_0}} \csc \alpha \quad [-] \tag{7-48}$$

• *Evaluating the Direct Solar Radiation (I_b) at the Earth's Surface.* The direct (beam) radiation energy flux (I_b) incident upon a surface on the earth's surface placed normal to the sun's rays can be evaluated by a conventional transmissivity equation as

$$I_b = I_0 \bar{\tau} \text{atm} \quad \left[W/m^2 \right] \tag{7-49}$$

where I_b is the direct (beam) radiation energy flux, or the terrestrial intensity of the direct radiation. I_0 is the solar constant or the extrater-

[5] Kreider, J. F., and Kreith, F., *Solar Heating and Cooling, Engineering Practical Design, and Economics*, Scripta, Washington D.C., 1975.

Table 7-5 COEFFICIENTS a_0, a_1, AND k CALCULATED
FOR THE 1962 STANDARD ATMOSPHERE FOR USE
IN DETERMINING SOLAR TRANSMITTANCE $\bar{\tau}$atm

	ALTITUDE ABOVE SEA LEVEL [km]					
	0	0.5	1.0	1.5	2.0	(2.5)*
	23-km HAZE MODEL					
a_0	0.1283	0.1742	0.2195	0.2582	0.2915	(0.320)
a_1	0.7559	0.7214	0.6848	0.6532	0.6265	(0.602)
k	0.3876	0.3436	0.3139	0.2910	0.2745	(0.268)
	5-km HAZE MODEL					
a_0	0.0270	(0.063)	0.0964	(0.126)	(0.153)	(0.177)
a_1	0.8101	(0.804)	0.7978	(0.793)	(0.788)	(0.784)
k	0.7552	(0.573)	0.4313	(0.330)	(0.269)	(0.249)

SOURCE: Adapted from Hottel.[6]
*Values in parentheses indicate interpolated or extrapolated values.

restrial intensity of direct radiation, and $\bar{\tau}$atm is the atmospheric transmittance of the extraterrestrial intensity of direct radiation.

Kreith and Kreider[5] suggest that in general, atmospheric transmittance $\bar{\tau}$atm ($\equiv I_b/I_0$) is given to within 3% accuracy for clear skies by the relation

$$\bar{\tau}\text{atm} = 0.5\left[e^{-0.65IM}(Z,\alpha_s) + e^{-0.095IM}(Z,\alpha_s)\right] \quad [-] \qquad (7\text{-}50)$$

where

$$\bar{\tau}\text{atm} = \text{atmospheric transmittance} \quad [-]$$
$$IM = \text{air mass ratio} \quad [-]$$
$$\alpha_s = \text{solar altitude angle} \quad [\text{rad}]$$
$$Z = \text{altitude} \quad [\text{m}]$$

Hottel[6] has made modifications to the conventional transmittance equation to account for particulates and water vapor in the air. Hottel's equation is

$$\bar{\tau}\text{atm} = a_0 + a_1 e^{-k\csc\alpha_s} \quad [-] \qquad (7\text{-}51)$$

where a_0, a_1, and k are functions only of altitude and visibility as shown in Table 7-5. The coefficient values in the table were calculated for the 1962 standard atmosphere. Other climate types including tropical, subarctic, and temperate show an effect on the values of a_0, a_1, and k of less than 4% in most cases.

[6] H. C. Hottel, and Broughton, F. P., *Ind. Eng. Chem. Anal. Ed.* 4, 166 (1932).

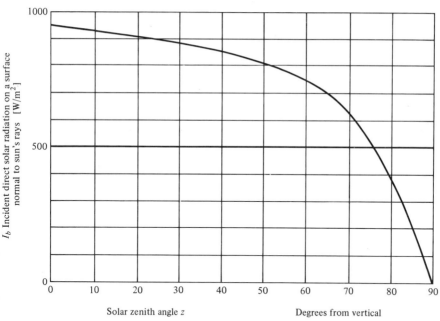

Figure 7-43 Direct solar radiation received by a surface normal to the incident radiation beam for the period from May to August in northern latitudes. (SOURCE: A. I. Brown and S. M. Marco, *Introduction to Heat Transfer*, 3rd ed., McGraw-Hill, New York, 1958.)

The value of I_b (I_0 attenuated by the length of the path through the atmosphere) suggested by Brown and Marco[7] for sea level condition on a clear summer day are plotted as a function of the zenith angle z in Figure 7-43. The value of I_b is obtained form the intersection of the angle z with the I_b curve.

DIFFUSE SOLAR RADIATION

Diffuse or scattered radiation is not associated with a specific direction as is direct (beam) radiation. Consequently, diffuse radiation comprises components coming from all directions. Because of this the diffuse solar radiation is not subject to optical manipulation, such as concentrating, focusing, and the like. The magnitude of the diffuse energy incident upon a surface is primarily affected by the incident angle (i) of the direct (beam) solar radiation. The larger the incidence angle, the longer the path through the earth's atmosphere, and the greater the amount of scattered energy. For convenience the diffuse radiation is often evaluated as a ratio (f) of

[7]A. I. Brown, and Marco, S. M., *Introduction to Heat Transfer*, 3rd ed., McGraw-Hill, New York, 1958.

Table 7-6 APPROXIMATE RATIO OF SKY RADIATION TO DIRECT SOLAR RADIATION RECEIVED BY A HORIZONTAL SURFACE (EARTH'S SURFACE) ON A CLEAR DAY

INCIDENCE ANGLE (i)	RATIO (f)
0°	0.16
10	0.17
20	0.18
30	0.19
40	0.22
50	0.26
60	0.32
70	0.44
80	0.71
85	1.33

the direct (beam) component, and is given by

$$I_d = I_b(f) \quad [\text{W/m}^2] \tag{7-52}$$

where I_b is the direct (beam) solar radiation, and f is the ratio of the sky radiation to direct solar radiation. Brown and Marco[7a] tabulated the suggested values of f given in Table 7-6.

Diffuse radiation amounts to approximately 10 to 15% of the total radiation on a clear day. In partly cloudy weather diffuse radiation can readily become 50% of the total incident solar radiation. On a completely overcast day all of the incident solar radiation is diffuse due to the fact that the direct solar radiation has been either absorbed or scattered.

The night sky is often treated as a blackbody with an effective temperature of 227 K.[8] This will of course vary upon the clearness of the night atmosphere and being lower on clear nights and higher on cloudy nights. These nocturnal radiations are diffuse radiations.

TOTAL INCIDENT RADIATIONS

The total incident radiations are the sum of the direct beam (I_b) and the diffuse (I_d). The direct beam radiations act upon the projected area of the incident surface normal to the sun's rays. The diffuse beam acts upon the actual area of the incident surface with an attenuation factor based upon the disposition of the surface from horizontal.

An example of typical diurnal measurement, of the direct, diffuse, and total incident solar radiations on a horizontal surface, is shown in Figure 7-44.

[7a] Ibid.
[8] F. Kreith, *Principles of Heat Transfer*, 1962.

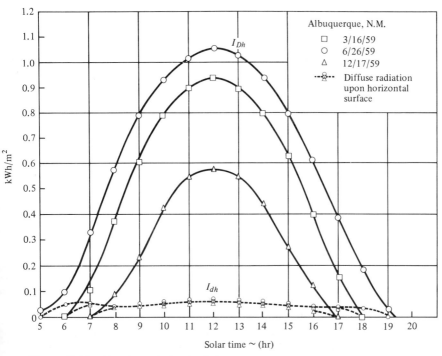

Figure 7-44   Hourly variations of direct and diffuse solar radiation on a horizontal surface for the three typical days. (SOURCE: From Report NSF/RANN/SE/G1-37815/PR/73/3.)

7-4.3 Geometrical Considerations Between the Incident Surface and the Sun's Rays

The geometrical considerations between the incident surface and the sun's rays are defined by the following primary angles.

- solar declination angle (δ_s)
- latitude of the surface's position on the earth (L)
- solar hour angle (h_s)
- solar altitude angle (α_s)
- solar azimuth angle (a_s)
- solar incidence angle on the surface (i)

These angles are discussed in the following sections.

THE SOLAR DECLINATION ANGLE (δ_s)

The solar declination angle (δ_s) is defined as the angle at a point on the equator of the earth at "high noon" between the ray from the center of the sun and the vertical line from the point. "High noon" (often referred to as

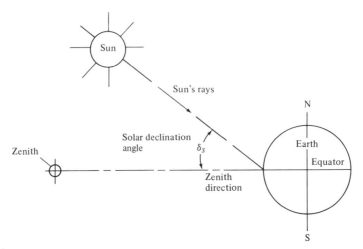

Figure 7-45 The solar declination angle and the zenith.

the solar noon) is the instant in time when a shadow (from the sun) falls in exactly a north–south direction. Because the zenith is defined as the point directly overhead, the vertical line from a point on the earth's surface is often called the zenith direction. These are illustrated in Figure 7-45.

The solar declination angle continuously changes, being a function of the day of the year. This is because the earth's rotational axis is tilted 23.5° with respect to the normal to the plane of the earth's orbit around the sun. The two opposing points in the orbit of the solar declination angle are (a)

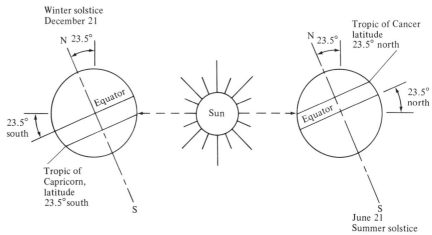

Figure 7-46 Solar declination angles at the winter and summer solstices of the northern hemisphere.

23.5° north (at the summer solstice) and (b) 23.5° south (at the winter solstice). This is illustrated in Figure 7-46. Likewise, at the spring (vernal) and the fall (autumnal) equinox, the solar declination angle is 0°.

The solar declination angle can be calculated for any day of the year using Eq. (7-53), which is given below:

$$\delta_s = 23.45 \sin\left(360\frac{284+N}{365}\right) \quad [-] \tag{7-53}$$

where

$N = (\text{Julian})$ date or the day of the year starting with January 1 as $N = 1$ $\quad [-]$

THE LATITUDE (L) ON THE EARTH'S SURFACE

The latitude of the location on the earth's surface is measured in degrees north or south of the earth's equator. The equator is 0° latitude and the two poles are at 90° north and 90° south. In latitudes between the Tropic of Cancer and the Tropic of Capricorn the sun will progress during the seasons over the zenith between north and south of the vertical. At latitudes north of the Tropic of Cancer and south of the Tropic of Capricorn, the sun will not reach the zenith at any time in the year.

THE SOLAR ANGLE (h_s)

The time of the solar day is based upon the "hour angle" (h_s), and is expressed either in degrees or in hours and minutes where one (1) hour is equal to 15°. The hour angle is a measurement of the earth's rotational position about the earth's rotational axis. The solar hour angle is measured east and west from the local solar noon (often called high noon) which is the datum (0°), each hour being 15° longitude with mornings positive and afternoons negative. For example, $h_s = +15°$ for 11 A.M. and $-37.5°$ for 2:30 P.M. The hour angle of sunrise and sunset varies with the season (solar declination) and the latitude. These values can be obtained for any day of the year directly from an ephemeris.*

THE SOLAR ALTITUDE ANGLE (α_s)

The solar altitude angle (α_s) is measured from the local horizontal plane (at the point of interest on the earth's surface) upward to the ray from the center of the sun. The solar altitude angle is a factor in the evaluation of atmospheric absorption and scattering. The complementary angle (measured from a vertical line) is called the zenith angle (z). These angles are illustrated in Figure 7-47. The solar altitude angle (α_s) may be calculated and is given by

$$\sin \alpha_s = \sin L \sin \delta_s + \cos L \cos \delta_s \cos h_s \quad [-] \tag{7-54}$$

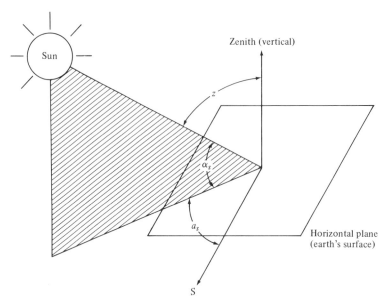

Figure 7-47 Diagram showing solar-altitude angle α, solar-azimuth angle a_s, and zenith angle z.

where

L = the latitude [degrees]

δ_s = the declination angle [degrees]

h_s = the solar hour angle [degrees]

Since the solar altitude angle is continually changing, for many purposes nominal values are adequate. Brown and Marco[7] have prepared graphs to give such nominal values. Their graph shown in Figure 7-48 plots the average relationship during the period from May to August between the zenith angle (z) and the time of the solar day for latitudes between 25° and 50° north latitude.

Example Problem 7-15

Determine the solar altitude angle at a position on the earth's surface at 30°N latitude on March 1 at 2:00 P.M. local solar time.

The American Ephemeris and Nautical Almanac is published yearly and may be obtained from the Superintendent of Documents, Washington, D.C., or from the regional office.

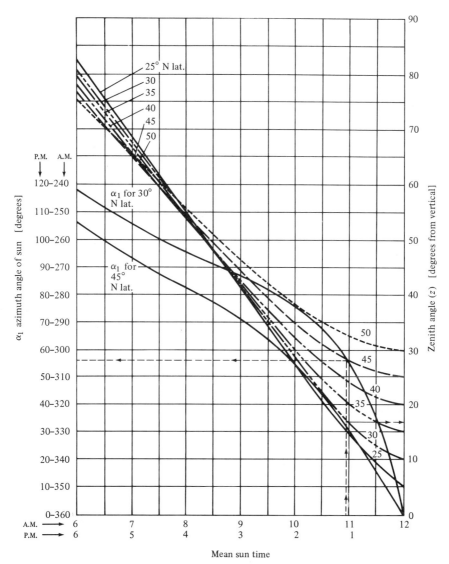

Figure 7-48   Solar angles for the period from May to August in northern latitudes. (SOURCE: A. I. Brown and S. M. Marco, *Introduction to Heat Transfer*, 3rd ed., McGraw-Hill, New York, 1958.)

SOLUTION

SKETCH

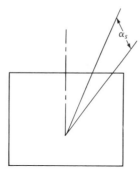

Figure 7-49

EQUATION

$$\sin \alpha_s = \sin L \cdot \sin \delta_s + \cos L \cdot \cos \delta_s \cdot \cos h_s \quad [-]$$

PLAN
Evaluate $\sin \alpha_s$, then from trigonometric tables evaluate α_s.

PARAMETERS

$$L = 30° \quad \text{(given)}$$
$$\sin L = 0.500$$
$$\delta_s = 23.45 \sin\left(360 \frac{284 + N}{365}\right) \quad [\text{Eq. (7-53)}]$$
$$N = 31 + 28 + 1 = 60$$
$$= 23.45 \sin\left(360 \frac{284 + 60}{365}\right)$$
$$= 23.45 \sin\left(360 \frac{344}{365}\right) = 23.45 \sin 339.29°$$
$$= 23.45 \sin -20.71° = (23.45)(-0.354)$$
$$\delta_s = -8.30°$$
$$\sin \delta_s = -0.1444$$
$$\cos L = 0.866$$
$$\cos \delta_s = 0.9895$$
$$h_s = -30°$$
$$\cos h_s = 0.866$$

SUBSTITUTION

$$\sin \alpha_s = (0.500)(-0.1444) + (0.866)(0.9895)(0.866)$$
$$= -0.0722 + 0.7421 = 0.6699$$

ANSWER

$$\alpha = 42.05°$$

THE SOLAR AZIMUTH ANGLE (a_s)

The solar azimuth angle (a_s) is measured in the horizontal plane between a due south line and the projection of the site-to-sun line on the horizontal plane as illustrated in Figures 7-47 and 7-50. The solar azimuth angle (a_s) may be computed from

$$\sin a_s = \frac{\cos \delta_s \sin h_s}{\cos \alpha_s} \quad [-] \tag{7-55}$$

where

δ_s = the solar declination angle [degrees]

h_s = the solar hour angle [degrees]

α_s = the solar altitude angle [degrees]

THE SOLAR INCIDENCE ANGLE (i)

The solar incidence angle (i) is measured between the ray to the center of the sun and the normal to the surface of interest, as illustrated in Figure 7-51. The solar incidence angle (i) is the same as the angle in Figure 7-18 for receiving surfaces not normal to the line of radiation propagation, and the associated Eq. (7-16).

For a fixed surface on the surface of the earth, the solar incidence angle (i) continually changes as it is a function of

- the solar declination angle (δ_s)
- the solar hour angle (h_s)

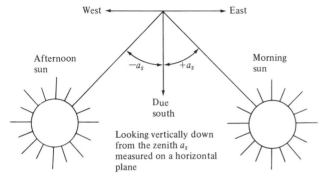

Figure 7-50 The solar azimuth angle (a_s).

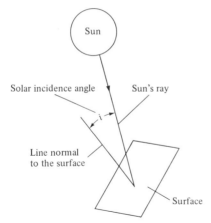

Figure 7-51   The solar incidence angle (i).

Fixed angles that are also involved are

- the latitude of the surface of interest
- the orientation of the surface of interest with respect to the surface of the earth

The orientation of the surface of interest is characterized with respect to the earth's surface by the following two angles, which are illustrated in Figure 7-52. These two angles are called

- the wall azimuth (a_w)
- the tilt angle (β_w)

and are illustrated in Figures 7-52 and 7-53.

The wall azimuth angle (a_w) is measured in a horizontal plane between a line pointing due south and the intersection with the horizontal plane of a plane through the "fall line" of the surface and also normal to the plane of the surface. The fall line can be further described as a line on the surface normal to a horizontal line on the surface.

The tilt angle (β_w) is measured, in a north–south vertical plane, between a horizontal line and the line of intersection of the surface plane with the vertical north–south vertical plane. This is illustrated in Figure 7-53.

From this geometry, equations for evaluating the solar incidence angle (i) have been developed. The generalized equation for calculating the incidence angle (i) for a fixed surface is given below, followed by simplifications for

- horizontal surface
- south facing vertical surfaces

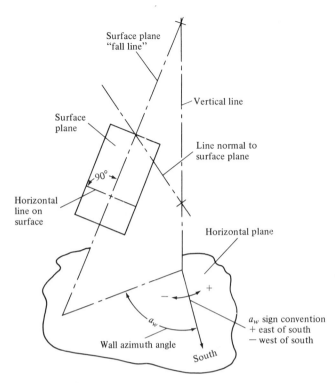

Figure 7-52   Wall azimuth angle geometry.

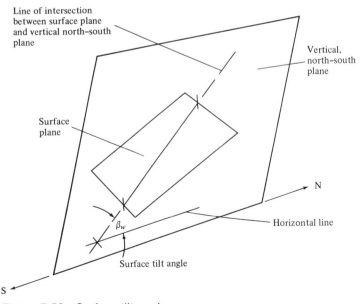

Figure 7-53   Surface tilt angle.

- south facing tilted surfaces
- non-south facing tilted surfaces

- *Generalized Equation.* The generalized equation [Eq. (7-56)] is given below. In cases where the simplified equations are applicable, they will be found easier to use.

$$\cos i = \sin \delta_s (\sin L \cos \beta_w - \cos L \sin \beta_w \cos a_w)$$
$$+ \cos \delta_s \cos h_s (\cos L \cos \beta_w + \sin L \sin \beta_w \cos a_w)$$
$$+ \cos \delta_s \sin \beta_w \sin a_w \sin h_s \quad [-] \tag{7-56}$$

- *Equation for Horizontal Surfaces.* The incident angle for a horizontal surface is given by

$$\cos i = \sin \alpha_s$$

Also

$$\cos i = \sin L \sin \delta_s + \cos L \cos \delta_s \cos h_s \quad [-] \tag{7-57}$$

- *Equation for South Facing Vertical Surfaces.* The incident angle for a south facing vertical surface is given by

$$\cos i = \cos L \sin \delta_s + \sin L \cos \delta_s \cos h_s \quad [-] \tag{7-58}$$

- *Equation for South Facing Tilted Surfaces.* The incident angle for a south facing tilted surface is given by

$$\cos i = \sin(L - \beta_w) \sin \delta_s = \cos(L - \beta_w) \cos \delta_s \cos h_s \quad [-] \tag{7-59}$$

- *Equation for Non-South Facing Tilted Surfaces.* If a tilted surface faces a direction other than due south, the following equation is used to calculate the incident angle (i):

$$\cos i = \cos(a_s - a_w) \cos \alpha_s \sin \beta_w + \sin \alpha_s \cos \beta_w \quad [-] \tag{7-60}$$

Example Problem 7-16

A tilted surface generally looking to the south and west has

- a tilt angle β_w of 45°;
- a wall azimuth angle a_w of 15° to the west.

When the solar altitude angle (α_s) is 60° and the solar azimuth angle (a_s) is 30° to the west, determine the solar incidence angle (i) on the tilted surface.

SOLUTION

SKETCH

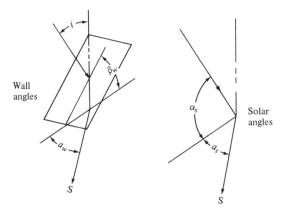

Figure 7-54

EQUATION

$$\cos i = \cos(a_s - a_w)\cos \alpha_s \cdot \sin \beta_w + \sin \alpha_s \cdot \cos \beta_w \qquad [\text{Eq. (7-60)}]$$

PLAN
Determine $\cos i$ and then evaluate i using trigonometric tables.

PARAMETERS

$$a_s = -30° \qquad \text{(given)}$$
$$a_w = -15° \qquad \text{(given)}$$
$$(a_s - a_w) = -30 - (-15) = -15°$$
$$\cos(a_s - a_w) = 0.9659$$
$$\alpha_s = 60° \qquad \text{(given)}$$
$$\cos \alpha_s = 0.500$$
$$\sin \alpha_s = 0.866$$
$$\beta_w = 45° \qquad \text{(given)}$$
$$\sin \beta_w = 0.7071$$
$$\cos \beta_w = 0.7071$$

SUBSTITUTION

$$\cos i = (0.9659)(0.500)(0.7071) + (0.866)(0.7071)$$
$$= 0.3415 + 0.6123 = 0.9538$$

ANSWER

$i = 17.5°$

7-4.4 Radiant Energy Exchange for Structural Surfaces

In the general case, the surface of a body in the open will absorb incident radiant energy from the sun, sky, and surroundings, and it will emit radiant energy. The net energy received by a surface may be expressed as follows:

$$\dot{Q} = \dot{Q}_{sun} + \dot{Q}_{sky} + \dot{Q}_s - \dot{Q}_\varepsilon \quad [W] \tag{7-61}$$

where

$\dot{Q}_{sun} =$ absorbed direct solar energy $\quad [W]$

$\dot{Q}_{sky} =$ absorbed energy from the sky $\quad [W]$

$\dot{Q}_s =$ absorbed energy from the surroundings $\quad [W]$

$\dot{Q}_\varepsilon =$ energy emitted from the structural surface $\quad [W]$

In many cases not all of these elements of the total radiant energy exchange need to be considered, either because of the relatively low magnitude, or because certain incident and emitted energies can be assumed to be essentially equal, and hence can be neglected. Each of the four elements will be discussed in the following sections.

ABSORBED DIRECT SOLAR ENERGY (Q_{sun})
The portion of the incident direct solar energy (Q_{sun}) that is absorbed by a structural surface is given by

$$\dot{Q}_{sun} = \alpha_{sun} I_b A \cos i \quad [W] \tag{7-62}$$

where

$\alpha_{sun} =$ absorptivity of the surface for solar radiation $\quad [-]$

$I_b =$ direct (beam) solar radiation flux normal to the sun's rays at the earth's surface $\quad [W/m^2]$

$A =$ surface area $\quad [m^2]$

$i =$ angle between the sun's rays and the normal to the surface [degrees]

The absorptivity (α_{sun}) of a surface for solar radiation (radiation from a 10,000 K blackbody) is usually significantly different from the absorptivity of the same surface for radiation from blackbodies at lower temperatures, such as 200–400 K. Some absorptivities of various surfaces for solar radiations are given in Table 7-7. The absorptivities of the surfaces for radiation from lower temperature blackbodies are given in Appendix B, Table B-9.

Table 7-7   ABSORPTIVITIES OF VARIOUS SURFACES FOR
SOLAR RADIATION

SUBSTANCE	α_{sun}
Building materials:	
Brick, red	0.70–0.77
Clay tiles, red and red-brown	0.65–0.74
Slate	0.79–0.93
Other roofing materials:	
Galvanized iron, new	0.66
Galvanized iron, dirty	0.89
Roofing paper	0.88
Asphalt	0.89
Paints:	
Black flat	0.97–0.99
White flat	0.12–0.26
Metals:	
Aluminum polished	0.26
Copper polished	0.26
Iron polished	0.45
Iron oxide (red)	0.74
Duralumin	0.53
Monel metal	0.43
Miscellaneous:	
White paper	0.27
Asphalt pavement	0.85

SOURCE: From M. Fishenden and O. A. Saunders, "Calculations of Heat Transmission," H. M. Stationery Office, London, 1932.

• *Variation of Surface Absorptivity with Incident Angle.* For incident angles up to 45° from normal to the surface, there is usually no significant change in the absorptivity. In general, surfaces with a high emissivity (absorptivity) tend to have little change from the value at normal incidence until the angle of incidence approaches 60° from normal. As the angle of incidence increases further from the normal, the value of the emissivity decreases rapidly. An example of the variation of emissivity with the angle of incidence for several high emissivity surfaces is illustrated by the measurements taken by Schmidt and Eckert[9] presented by the polar plot in Figure 7-55.

Similar measurements of low emissivity metallic surfaces show a constant emissivity for a range of incident angles from the normal. Beyond a certain angle the values of the emissivities increase with an increase in the deviation of the incident angle from normal as illustrated in Figure 7-56.

[9]E. Schmidt and E. Echert, "Uber die Richtung Verteilung der Warmestrahlung," Forsch, Gebiete Ingenieur wesen 6 (1935).

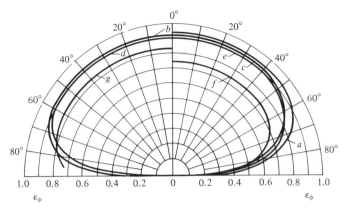

Figure 7-55 Directional variation of emittance for several electrical nonconductors. (SOURCE: E. Schmidt and E. Eckert, "Uber die Richtungsverteilung der Warmestrahlung," *Forsch. Gebeite Ingenieurwesen* 6 (1935).)

ABSORBED ENERGY FROM THE SKY ($\dot{Q}_{sky}$)

The energy received from the sky which is absorbed by the surface is different between day and night. During the daytime this energy is the absorbed diffuse solar radiation ($\dot{Q}_{sky,\,d}$), which is evaluated by

$$\dot{Q}_{sky,\,d} = \alpha_{sun} I_d A \quad [\text{W}] \tag{7-63}$$

or, using Eq. (7-49) to evaluate I_d

$$\dot{Q}_{sky,\,d} = \alpha_{sun} f I_b A \quad [\text{W}] \tag{7-64}$$

where

α_{sun} = absorptivity of the surface for solar radiation $\quad [-]$

I_d = diffuse solar radiation $\quad [\text{W}/\text{m}^2]$

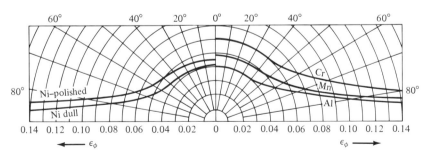

Figure 7-56 Directional variation of emissivity for several metals. (SOURCE: E. Schmidt and E. Eckert, "Uber die Richtungsverteilung der Warmestrahlung," *Forsch. Gebeite Ingenieurwesen* 6 (1935).)

f = ratio of sky radiation to direct solar radiation
(from Table 7-7) $[-]$

I_b = direct (beam) solar radiation at the earth's surface $[W/m^2]$

A = surface area $[m^2]$

During the nighttime the surface is receiving radiation from the sky which is characterized as a blackbody with a certain effective temperature. For a clear night the effective blackbody temperature of the sky is considered to be 227 K. For hazy or cloudy nights the effective blackbody temperature of the sky will be higher. The absorbed energy from the sky at night $(\dot{Q}_{sky,\,n})$ is evaluated by

$$\dot{Q}_{sky,\,n} = A\varepsilon\sigma T_{sky}^4 \quad [W] \tag{7-65}$$

where

A = surface area $[m^2]$

ε = emissivity of the surface at the effective sky temperature* $[-]$

σ = Stefan-Boltzmann constant 5.670×10^{-8} $[W/m^2\cdot K^4]$

T_{sky} = effective blackbody temperature of the sky $[K]$

ABSORBED ENERGY FROM THE SURROUNDINGS ($\dot{Q}_S$)

The surroundings from which a surface may receive energy include the surface of the earth, walls of other structures, foliage, and so forth. The absorbed portion of the incident radiation from such sources would be the sum of the absorbed portion of the incident radiation from each source involved. The equation for the absorbed incident radiation ($\dot{Q}_s$) for each source is

$$\dot{Q}_s = AF_\varepsilon F_A\sigma T^4 \quad [W] \tag{7-66}$$

where

A = surface area $[m^2]$

F_ε = emissivity factor (see Table 7-4) $[-]$

F_A = configuration factor (see Table 7-4) $[-]$

σ = Stefan-Boltzmann constant 5.670×10^{-8} $[W/m^2\cdot K^4]$

T_s = temperature of the source surface $[K]$

Usually $\dot{Q}_s$ is considered to be O[W] as F_A and T^4 are usually small and the resultant value of $\dot{Q}_s$ is negligible relative to the values of $\dot{Q}_{sun}$, $\dot{Q}_{sky}$, and $\dot{Q}_\varepsilon$.

---

*Values of emissivities of various surfaces are given in Appendix B, Table B-9.

ENERGY EMITTED BY THE STRUCTURAL SURFACE ($\dot{Q}_\varepsilon$)

The energy emitted by a surface ($\dot{Q}_\varepsilon$) can be calculated by the following equation

$$\dot{Q}_\varepsilon = A\varepsilon\sigma T_s^4 \quad [\text{W}] \tag{7-67}$$

where

$A =$ surface area $[\text{m}^2]$

$\varepsilon =$ emissivity of the surface at the surface temperature $[-]$

$\sigma =$ Stefan-Boltzmann constant 5.670×10^{-8} $[\text{W}/\text{m}^2 \cdot \text{K}^4]$

$T_s =$ temperature of the surface $[\text{K}]$

Example Problem 7-17

Determine the radiation energy exchange on March 15 at 1:00 P.M. local time between a 5 by 8 m new galvanized iron roof and its surroundings. The roof is located at 30°N latitude and at a longitude where the local time is one-half hour ahead of the solar time. It makes an angle with the horizontal of 30° and faces directly west, and has a surface temperature of 37°C. The sky is clear.

SOLUTION

SKETCH

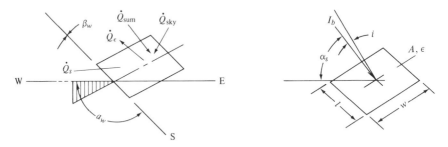

Figure 7-57

EQUATION
The net energy received by the roof surface can be calculated from Eq. (7-61).

$$\dot{Q} = \dot{Q}_{\text{sun}} + \dot{Q}_{\text{sky}} + \dot{Q}_s - \dot{Q}_\varepsilon \quad [\text{W}] \qquad [\text{Eq. (7-61)}]$$

PROBLEM EQUATION PARAMETERS

$\dot{Q}_{sun}$—the absorbed direct (beam) radiation is obtained from the following equation:

$$\dot{Q}_{sun} = \alpha_{sun} I_b A \cos i \quad [\text{W}] \quad [\text{Eq. (7-62)}]$$

$\dot{Q}_{sky}$—the absorbed diffuse radiation is determined from the following equation:

$$\dot{Q}_{sky} = \alpha_{sun} f I_b A \quad [\text{W}] \quad [\text{Eq. (7-64)}]$$

$\dot{Q}_s$—the radiation emissions from the earth's surroundings absorbed by the roof can be evaluated by

$$\dot{Q}_s = A F_e F_A \sigma T^4 \quad [\text{W}] \quad [\text{Eq. (7-66)}]$$

$\dot{Q}_\varepsilon$—the emitted radiation from the surface of the roof can be obtained from the following equation:

$$\dot{Q}_\varepsilon = A \varepsilon \sigma T_s^4 \quad [\text{W}] \quad [\text{Eq. (7-6)}]$$

The sequence of steps required to evaluate these equations follows:
The direct (beam) solar radiation ($\dot{Q}_{sun}$) is obtained from the following equation:

$$\dot{Q}_{sun} = \alpha_{sun} I_b A \cos i \quad [\text{W}] \quad [\text{Eq. (7-62)}] \tag{1}$$

PARAMETERS (1)

$$\alpha_{sun} = 0.66 \quad [-] \quad (\text{Table 7-7})$$

$$I_b = I_0 \bar{\tau} \text{ atm} \quad [\text{W}/\text{m}^2] \quad [\text{Eq. (7-49)}] \tag{1a}$$

$$I_0 = \text{solar constant} = 1353 \quad [\text{W}/\text{m}^2]$$

$$\bar{\tau} \text{atm} = a_0 + a_1 e^{-k \csc \alpha_s} \quad [-] \tag{1b}$$

From Table 7-5 the coefficients a_0, a_1, and k on a clear day (assume 23-km visibility) and zero altitude (sea level) are as follows:

$$a_0 = 0.1283$$
$$a_1 = 0.7559$$
$$k = 0.3876$$

$$\csc \alpha_s = \frac{1}{\sin \alpha_s} \tag{1c}$$

$$\sin \alpha_s = \sin L \sin \delta_s + \cos L \cos \delta_s \cos h_s \tag{1d}$$

PARAMETERS FOR $\sin \alpha_s$ (1d)

$$L = \text{latitude } 30°\text{N} \quad [\text{degrees}]$$
$$h_s = 15° \, (1 \text{ hr}) = -15° \, (\text{for 1 P.M.}) \quad [\text{degrees}]$$

$$\delta_s = 23.45 \sin\left(360 \frac{284+N}{365}\right) \quad [\text{degrees}] \qquad [\text{Eq. (7-53)}]$$

$$N = 74 \text{ (for March 15)}$$

$$\delta_s = -2.82°$$

SUBSTITUTION IN EQUATION FOR $\sin \alpha_s$ *(1d)*

$$\sin \alpha_s = (\sin 30°)(\sin -2.82°) + (\cos 30°)(\cos -2.82°)(\cos -15°)$$

$$= (0.5)(-0.0492) + (0.8660)(0.99879)(0.96593)$$

$$= -0.0246 + 0.8355$$

$$\sin \alpha_s = 0.8109$$

SUBSTITUTION IN EQUATION FOR $\csc \alpha_s$ *(1c)*

$$\csc \alpha_s = \frac{1}{0.8109} = 1.23$$

SUBSTITUTION IN EQUATION FOR $\bar{\tau}$ atm *(1b)*

$$\bar{\tau}\text{atm} = 0.1283 + (0.7559)(e^{-0.3876\,(1.23)})$$

$$= 0.1283 + (0.7559)(e^{-0.4763})$$

$$\bar{\tau}\text{atm} = 0.59756 \quad [-]$$

SUBSTITUTION IN EQUATION FOR I_b *(1a)*

$$I_b = (1353)(0.59746) = [\text{W/m}^2][-] = [\text{W/m}^2]$$

$$I_b = 808 \text{ W/m}^2$$

$$A = l \cdot w \quad [\text{m}^2]$$

$$l = 5 \text{ m} \qquad (\text{given})$$

$$w = 8 \text{ m} \qquad (\text{given})$$

$$A = 5 \text{ m} \times 8 \text{ m} = 40 \text{ m}^2 \tag{1e}$$

$$\cos i = \cos(a_s - a_w)\cos \alpha_s \sin \beta_w + \sin \alpha_s \cos \beta_w \quad [\text{Eq. (7-60)}] \tag{1f}$$

where

a_s = the solar azimuth angle [degrees]

a_w = the azimuth of the roof with respect to due south [degrees]

a_s = the solar altitude angle [degrees]

β_w = the tilt angle of the roof [degrees]

To evaluate a_s

$$\sin a_s = \frac{\cos \delta_s \sin h_s}{\cos \alpha_s} \quad [-] \qquad [\text{Eq. (7-55)}] \tag{1g}$$

PARAMETERS FOR $\sin a_s$ *(1g)*

$$\delta_s = -2.82° \quad (1d)$$

Then $\cos \delta_s = 0.99879$ [—].

$$h_s = -15° \quad (given)$$

Then $\sin h_s = -0.25882$ [—].

$$\alpha_s = 54.2° \quad (1d) \quad [\sin \alpha_s = 0.8109]$$

Then $\cos \alpha_s = 0.5852$ [—].

SUBSTITUTION IN EQUATION FOR $\sin a_s$ *(1g)*

$$\sin a_s = \frac{(0.99879)(-0.25882)}{(0.5852)} \quad \frac{[-][-]}{[-]}$$

$$= -0.44174 \quad [-]$$

$$a_s = -26.22°$$

To evaluate a_w, since the roof is facing directly west, the angle

$$a_w = -90°$$

$$\cos \alpha_s = 0.5852 \quad (1d)$$

To evaluate β_w, since the roof faces directly west $\beta_w = 0$ (see Figure 7-53)

$$\sin \beta_w = 0$$

$$\cos \beta_w = 1.0$$

SUBSTITUTION IN EQUATION FOR $\cos i$ *(1f)*

$$\cos i = \{\cos[-26.22° - (-90°)]\}$$

$$(\cos 54.2°)(\sin 0°) + (\sin 54.2°)(\cos 0°)$$

$$= (\cos 63.78°)(0.5852)(0) + (0.8109)(1)$$

$$= 0.0 + 0.8109$$

$$\cos i = 0.8109$$

SUBSTITUTION IN EQUATION FOR $\dot{Q}_{sun}$ *(1)*

$$\dot{Q}_{sun} = (0.66)(808)(40)(0.8109) \quad [-][W/m^2][m^2][-] = [W]$$

ANSWER (1)

$$\dot{Q}_{sun} = 17{,}300 \quad [W] \quad (direct\ solar\ radiation) \tag{1}$$

The radiation received by the roof from the sky ($\dot{Q}_{sky}$) can be determined from Eq. (7-64) as follows:

$$\dot{Q}_{sky} = \alpha_{sun} f I_b A \quad [W] \tag{2}$$

PARAMETERS (2)

$$\alpha_{\text{sun}} = 0.66 \qquad \text{(Table 7-7)}$$

f = ratio of sky radiation to direct (beam)

radiation = 0.20   (Table 7-6)   [for $i = 36°$ from (1f)]

$$I_b = 808 \quad [\text{W/m}^2] \qquad \text{(1a)}$$

$$A = 40 \text{ m}^2 \qquad \text{(1e)}$$

SUBSTITUTION (2)

$$\dot{Q}_{\text{sky}} = (0.66)(0.20)(808)(40) \quad [-][-][\text{W/m}^2][\text{m}^2] = [\text{W}]$$

ANSWER (2)

$$\dot{Q}_{\text{sky}} = 4266 \quad [\text{W}]$$

The radiation received from the earth is negligible due to the geometry of the problem.

$$\dot{Q}_s = 0 \quad [\text{W}] \tag{3}$$

The radiation emitted by the surface of the roof can be determined from Eq. (7-67) as follows:

$$\dot{Q}_\varepsilon = A\varepsilon\sigma T_s^4 \quad [\text{W}] \tag{4}$$

PARAMETERS (4)

$$A = 40 \text{ m}^2 \qquad \text{(1e)}$$

$$\varepsilon = 0.23 \quad [-] \qquad \text{(Appendix B, Table B-8)}$$

$$\sigma = 5.670 \times 10^{-8} \quad [\text{W/m}^2 \cdot \text{K}^4]$$

$$T_s = 37 + 273 = 310 \text{ K} \qquad \text{(given)}$$

SUBSTITUTION (4)

$$\dot{Q}_\varepsilon = (40)(0.23)(5.670 \times 10^{-8})(310)^4$$

$$= [\text{m}^2][-][\text{W/m}^2 \cdot \text{K}^4][\text{K}^4] = [\text{W}]$$

$$= (52.16)(92.35)$$

ANSWER (4)

$$\dot{Q}_\varepsilon = 4817 \quad [\text{W}]$$

PROBLEM EQUATION

$$\dot{Q} = \dot{Q}_{\text{sun}} + \dot{Q}_{\text{sky}} + \dot{Q}_s - \dot{Q}_\varepsilon \quad [-]$$

PARAMETERS

$\dot{Q}_{sun} = 17{,}300 \quad [\text{W}] \qquad (1)$

$\dot{Q}_{sky} = 4266 \quad [\text{W}] \qquad (2)$

$\dot{Q}_s = 0 \quad [\text{W}] \qquad (3)$

$\dot{Q}_\varepsilon = 4817 \quad [\text{W}] \qquad (4)$

SUBSTITUTION

$\dot{Q} = 17{,}300 + 4266 + 0 - 4817 \quad [\text{W}] + [\text{W}] + [\text{W}] - [\text{W}] = [\text{W}]$

ANSWER

$\dot{Q} = 16{,}750 \quad [\text{W}]$

7-5 SOLAR ENERGY SYSTEMS

Solar energy systems for heating, cooling, or power generation are largely composed of components in which heat transfer plays a dominant role in the design. The net output of the system for a given collector area is the solar insolation for the subtended collector area normal to the sun's rays less the thermal losses of the system. The thermal losses for a solar system comprising

- a collector
- ducts for the circulating heat transfer fluid
- thermal storage

are identified in more detail below.

The thermal losses from a collector include:

- solar radiation transmission losses from absorption while passing through any transparent coverings, lenses, or windows;
- solar radiation reflection losses from transparent surfaces or mirror reflecting surfaces for concentration;
- incomplete absorption and reflection of the solar radiations at the absorbing surface ($\alpha \neq 1$; $\rho \neq 0$);
- radiation losses:
  - emissions by the absorbing surface passing through any transparent covering;
  - emissions by the transparent covering;
- convection heat transfer to the surrounding air:
  - from the transparent covering surface or the uncovered absorbing surface;
  - from the external surfaces of the back and sides of the structure covering the absorbing surface.

The thermal losses from the ducts for the circulating heat transfer fluid include convection heat transfer from the duct to the surrounding air.

The thermal losses from the thermal storage include conduction and/or convection heat transfer from the thermal storage unit to the surroundings.

The determination of relative magnitudes and effects of these losses on the solar energy system performance is a complex analysis. Trends for guidance in the design of a solar energy system, as well as suggestions for nominal collector areas and thermal storage capacities for typical applications are given in references such as ERDA's *Pacific Regional Solar Heating Handbook*[10] and *The Solar Home Book*.[11] More detailed component design information can be found in references such as Kreith and Kreider, *Solar Heating and Cooling*.[5]

PROBLEMS

7-1. Over what spectral band width in the electromagnetic spectrum of wavelengths are thermal radiations? What limits the band width on the short wavelength side? What limits the band width on the long wavelength side?

7-2. (a) What is the emitted radiant energy flux density from a blackbody surface at 400°C?

(b) If the emitted radiant energy flux density were to be doubled, to what temperature would the surface of the blackbody need to be raised?

7-3. (a) Define the term emissivity (ε).

(b) What is a blackbody? Does its emissivity (ε_b) vary with temperature? Explain your answer.

(c) What is a gray body? Does its emissivity (ε) vary with temperature? Explain your answer.

(d) How does a real body differ from a black or a gray body? Explain.

7-4. (a) Describe how the monochromatic emissive power (W_λ) (spectral emitted radiant energy flux density) varies with the wavelength (λ) for emissions from a blackbody.

(b) At what wavelength is the blackbody monochromatic emissive power (W_λ) the maximum? What factors control the magnitude and wavelength?

(c) Describe how the monochromatic emissive power (W_λ) varies with the wavelength (λ) for emissions from a real body.

[10] *Pacific Regional Heating Handbook*, ERDA San Francisco Operations Office, by Los Alamos Scientific Laboratory, 2nd ed., Superintendent of Documents, Washington D.C., 1976.

[11] Bruce Anderson, and Riordan, Michael, *The Solar Home Book*, Cheshire Books, Harrisville, N.H.,1976.

7-5. A radiant wall heater has a sheet steel radiating surface 25 cm wide by 1 m high. When the radiation surface of the heater is at 200°C, how much radiant heat is emitted?

7-6. How does the character of the emitted radiant energy from a gas mass differ from the emitted radiant energy from a solid surface?

7-7. (a) What gases tend to emit very little radiant energy?
(b) What common gas products from combustion emit significant amounts of radiant energy?

7-8. What factors affect the magnitude of the radiant energy flux density (I) emitted by a gas mass?

7-9. If a water tube boiler furnace were a cube 5 m on a side with flue gases composed of a composition of 0.8 mol H_2O/mol CO_2 with a temperature of 1500°C, what would be the radiant energy flux density (I) at the furnace wall with a combustion pressure of 1 atm?

7-10. If the combustion chamber in a fire tube boiler is 1 m in diameter and 5 m long with the same flue gas composition and temperature as in Problem 7-9, what would be the radiant energy flux density (I) at the surface of the cylindrical combustion chamber wall?

7-11. If the boiler furnace in Problem 7-9 were fired by oil producing a luminous flame with a "red brightness temperature" of 1600 K and a "green brightness temperature" of 1640 K, what would be the radiant energy flux density at the furnace wall?

7-12. If the boiler furnace in Problem 7-9 were fired by coal (equivalent to coal No. 1 powdered to 75% through 200 mesh) producing 14 kg of combustion products per kilogram of coal, with the same flame temperature as in Problem 7-11, what would be the radiant energy flux density at the furnace wall?

7-13. Describe or define the intensity ($\bar{I}$) of an emitting body.

7-14. A surface 1 m^2 is located 100 m from a radiating source with an intensity ($\bar{I}$) of 1000 kW/st. How much radiant energy is incident on the surface?

7-15. What is the intensity of the sun? Its diameter is 1.39×10^6 km and its effective surface temperature is 10,200 K.

7-16. Would the sun's intensity change as one traveled in space toward or away from the sun? Explain your answer.

7-17. What would be the sun's incident radiant energy flux density just outside the atmosphere of the planet Mars (mean distance from the sun is 228×10^6 km); what would it be at the planet Venus (mean distance from the sun is 108×10^6 km)? How do these compare with the Earth's "solar constant?"

7-18. What happens to radiant energy when it is incident upon a surface? Give the general equation that relates reflectivity (ρ), absorptivity (α), and transmissivity (τ).

7-19. What happens to incident radiant energy that is reflected? Describe the two general types of reflection from a surface.

7-20. What causes a reflection from the surface of a transparent material if no antireflection coating has been applied?

7-21. What happens to incident radiant energy that is absorbed?

7-22. What happens to incident radiant energy that is transmitted?

7-23. A surface, upon which radiant energy in the amount of 380 W/m² is incident, absorbs 240 W/m², transmits 30 W/m², and reflects the remainder. Calculate the values for absorptivity, reflectivity, and transmissivity.

7-24. Give the general equation for the radiant energy heat exchange between two surfaces. What portion of the total hemispheric emitted radiant energy (intensity $\bar{I}$) of each surface is involved in this radiant energy heat exchange?

7-25. Calculate the radiant energy exchange per square meter between two infinite parallel black planes, the temperatures of which are 260°C and 95°C.

7-26. Two equal parallel black disks 1 m in diameter are spaced 0.5 m apart and directly opposite each other. If the temperatures of the disks are 150°C and 375°C, determine the net radiant energy exchange between the two surfaces.

7-27. Two parallel black surfaces of rectangular shape, 1.5 by 2 m, are located 3 m apart. If the temperatures of the two surfaces are 400°C and 325°C, determine the net radiant energy exchange.

7-28. If the distance between the two surfaces in Problem 7-27 is increased to 6 m, what will be the change in the radiant energy exchanged?

7-29. A 10 by 10 m room, 3 m high, is heated by hot water pipes laid beneath an oak floor. The walls and ceiling of the room are coated with an oil paint. The surface temperature of the floor is 27°C, and the surface temperature of the ceiling is 15°C. What is the rate of radiant heat exchange between the floor and the ceiling? Assume the sidewalls to be nonconducting, but reradiating.

7-30. If a bare steel steam pipe 20 cm in diameter carrying steam at 150°C were wrapped with four layers of aluminum foil to provide insulation by radiation shields, how much reduction in the radiant energy loss would be realized (neglect any incident thermal radiations from the surroundings)? How much reduction would be realized if 10 layers were applied?

7-31. What happens to the beam of solar radiations when it passes through the earth's atmosphere?

7-32. How can the magnitude of the direct solar radiation at the earth's surface be evaluated? What two equations were given in the text to evaluate the atmospheric transmittance for incoming extraterrestrial radiant energy?

7-33. Calculate I_b at sea level for the solar beam with a solar altitude angle (α_s) of 60° using (a) the Kreith and Kreider Eq. (7-50) and (b) the

Hottel Eq. (7-51) evaluated for the 23-km haze model. How do these values compare with the curve suggested by Brown and Marco (Figure 7-43)?

7-34. How did Brown and Marco evaluate the magnitude of the diffuse solar radiant energy at the earth's surface? How can the ratio (f) exceed 1.0?

7-35. List and define six solar angles used in evaluating the solar radiant beam energy incident on surfaces.

7-36. Why are the values of absorptivity (α) of a surface for solar radiations different from the emissivity valves (ε) for the surface?

7-37. An 8 by 14 m flat roof is covered with new galvanized iron. The net radiant energy exchange between the roof and the surroundings is 52 kW gained by the roof. If this occurs at noon (sun time) at a 40° north latitude on a clear summer day, what is the temperature of the roof surface?

7-38. A 5 by 7 m asphalt roof absorbs solar radiation at 1:30 P.M. local time on a clear summer day. The roof is situated at 45° north latitude at a longitude 7.5° east of the standard meridian for its time zone, and slopes directly south at an angle of 20° with the horizontal. Determine the absorbed portion of the radiation received by the roof from the sky, including both the direct solar beam and diffuse radiations.

Chapter 8
Overall Heat Transfer

The overall heat transfer coefficient (U) is the thermal conductance of a heat transfer path from one end to the other. A common heat transfer path is from one fluid to another through some sort of a wall. For example:

- in a building—from the inside air, through the wall, and into the outside air;
- in a boiler—from the hot combustion gases, through the boiler tube wall, through a layer of scale (if any), and into the boiler water;
- in a boiler—from hot gas or the external surface of the boiler, through insulation, and into the surrounding air;
- in smokestacks—from the flue gas, through the wall of the smoke stack, and into the surrounding air;
- in heat exchangers—from one fluid, through the heat exchanger heat transfer surface, and into the other fluid.

In general the majority of heat transfer paths comprise a series of thermal resistances, which can be separated into three groups:

- hot end fluid/surface convective film resistance;

- total thermal resistance of the conductive path between the two surfaces in contact with the fluids;
- cold end surface/fluid convective film resistance.

The nature and methods of evaluating fluid/surface convective film coefficients have been discussed in Chapters 3, 4, 5, and 6.

The nature and methods of evaluating the total thermal resistance of a conductive path have been discussed in Section 2-3.

The usual units for U are $[W/m^2 \cdot \Delta_1 {}^\circ C]$. The basic heat flow rate equation using U is

$$\dot{Q} = UA \, \Delta T_m \quad [W] \tag{8-1}$$

where

$\dot{Q} =$ heat flow $[W]$

$U =$ overall heat transfer coefficient $[W/m^2 \cdot \Delta_1 {}^\circ C]$

$A =$ area $[m^2]$

$\Delta T_m =$ effective mean temperature difference $[\Delta {}^\circ C]$

When one is interested in the temperature of a surface in the heat transfer path, such as boiler tubes when scale is present (to avoid tube overheating and resultant mechanical failure), or the temperature of the inside of a smoke stack (to avoid condensation and resultant corrosion), the temperature of the surface of interest can be found, as discussed in Section 2-2.1.3 using $1/U = \mathcal{R}_t$.

The following discussion will begin with heat transfer through the walls of buildings, and then heat exchangers. The purpose is to introduce the student to the two common major applications of heat transfer design, namely:

- buildings
- heat exchangers

Based upon these applications, extensions to other heat transfer configurations can be made.

Any particular application of heat transfer equipment or system design will have requirements which will indicate relative values of economic considerations, such as size, weight, pumping power for forced convective fluid flows, and so forth. In practice these factors must be analyzed and evaluated to achieve an optimum design for the application. However, in this introductory text, the scope of the discussion of overall heat transfer will be limited to a technical discussion of the methods of predicting overall heat transfer performance and the size and type of heat exchangers to accomplish typical heat exchange tasks.

8-1 BUILDING HEAT TRANSFER

In the design of buildings or process structures, such as furnaces, kilns, and so forth, the magnitude of the heat flow through the walls can be of importance. The amount of heat that flows through the wall is a function of the temperature drop across the wall and the thermal resistance of the wall. The factors affecting the design of a building or process structure for minimum cost of the required heating/cooling are relatively complex. The initial cost of the walls, insulation, and heating/cooling equipment, heating/cooling operating costs, and maintenance must all be considered.

The scope of the following discussion includes:

- the total amount of heat flow into or out of a structure
- the thermal conductance of typical walls
- the factors involved in reducing the heat loss or cooling load

8-1.1 The Total Amount of Heat Flow into or out of a Structure

The total amount of heat flowing into (cooling load) or out of (heating load) a structure is the summation of the heat flow over the entire external surface of the structure. Usually the external surfaces can be categorized according to areas with similar thermal conductances of the wall. The external surface of a house, for example, could be categorized into the following sections:

- walls
- windows
- doors
- ceiling
- floor

The amount of heat flowing through each section is

$$\dot{Q} = AU\Delta T \qquad [\text{Eq. (2-10a)}]$$

where

$\dot{Q} =$ heat flow $\quad [\text{W}]$

$A =$ area of the heat flow path (wall surface area) $\quad [\text{m}^2]$

$\Delta T =$ temperature drop along the heat flow path $\quad [\Delta°C]$

      (difference between inside and outside temperatures)

$U =$ overall thermal conductance of the heat flow path

$$[\text{W/m}^2 \cdot \Delta_1 °C]$$

If the temperature drop for all of the sections is the same, the total heat loss can be calculated as a parallel path system. This is discussed on page 48. If the temperature drop is different for some of the sections, for example, such as for a house with some unheated rooms or an apartment in a complex with common walls with adjacent apartments, the heat loss through each section can be calculated separately, and then summed to obtain the total heat loss.

Example Problem 8-1

A house can be characterized by the following:

SECTION	AREA A [m^2]	OVERALL THERMAL CONDUCTANCE (U) [W/m$^2 \cdot \Delta_1$°C]
Solid wall	150	1.5
Ceiling	130	1.1
Floor	130	1.9
Windows	20	6.5
Doors	6	3.0

What is the conductive heat loss from the house in watts if the inside and outside temperatures are 25°C and 5°C, respectively, for all of the sections?

SOLUTION
Treat as a parallel conduction path system (page 48) Basic equation (from page 49):

$$\dot{Q}_t = \mathcal{K}_t \Delta T \quad [\text{Eq. (2-10)}]$$
$$\mathcal{K}_t = \mathcal{K}_1 + \mathcal{K}_2 + \mathcal{K}_3 + \mathcal{K}_4 + \mathcal{K}_5 \quad [\text{Eq. (2-15)}]$$
$$\mathcal{K} = AU \quad [\text{W}/\Delta°\text{C}]$$
$$\mathcal{K}_1 = (150)(1.5) = 225 \quad [\text{W}/\Delta°\text{C}] \quad \text{(solid wall)}$$
$$\mathcal{K}_2 = (130)(1.1) = 143 \quad [\text{W}/\Delta°\text{C}] \quad \text{(ceiling)}$$
$$\mathcal{K}_3 = (130)(1.9) = 247 \quad [\text{W}/\Delta°\text{C}] \quad \text{(floor)}$$
$$\mathcal{K}_4 = (20)(6.5) = 130 \quad [\text{W}/\Delta°\text{C}] \quad \text{(windows)}$$
$$\mathcal{K}_5 = (6)(3.0) = 18 \quad [\text{W}/\Delta°\text{C}] \quad \text{(doors)}$$

SUBSTITUTING IN EQ. (2-15)

$$\mathcal{K}_t = 225 + 143 + 247 + 130 + 18 = 763 \quad [\text{W}/\Delta°\text{C}]$$
$$\Delta T = T_i - T_o \quad [\Delta°\text{C}]$$
$$T_i = 25°\text{C} \quad \text{(given)}$$
$$T_o = 5°\text{C} \quad \text{(given)}$$
$$\Delta T = 25 - 5 = 20 \quad [\Delta°\text{C}]$$

SUBSTITUTING IN EQ. (2-10)

$$\dot{Q} = (763)(20) \quad [\text{W}/\Delta°\text{C}][\Delta°\text{C}] = [\text{W}]$$

ANSWER

$$\dot{Q} = 15{,}260 \quad [\text{W}]$$

8-1.2 Thermal Conductance of Walls

The temperature profile along the heat flow path through a wall from the inside air to the outside air is illustrated in Figure 8-1. The path is comprised of a series of thermal resistances, namely

- inside surface air boundary layer resistance
- wall resistance
- outside surface air boundary layer

When heat transfer by radiation is negligible, the total thermal resistance of such a heat flow path is the sum of the thermal resistances of each section of the heat flow path, as discussed on page 54.

$$\mathscr{R}_t = \mathscr{R}_1 + \mathscr{R}_2 + \mathscr{R}_3 \qquad [\text{Eq. (2-18a)}]$$

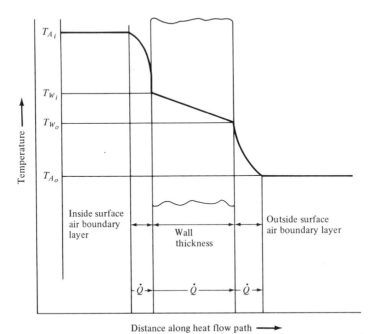

Figure 8-1   Temperature profile along the heat flow path.

The thermal resistance ($\Re_1$) of the heat transfer from the air to the inside surface of the wall is controlled by the air boundary layer. Usually the boundary layer is caused by free convection, as contrasted to forced convection. Free convection is discussed in Chapter 3, and simplified equations for calculating h_c for air are given in Section 3-4. The thermal resistance is the reciprocal of the convective film coefficient h_c.

$$\Re_1 = \frac{1}{h_c} \quad [\text{m}^2 \cdot \Delta°\text{C/W}]$$

The thermal resistance ($\Re_2$) of a homogenous wall is a function of the thermal conductivity of the wall material and the wall thickness, as discussed on page 45.

$$\Re_2 = \frac{x}{k} \quad [\text{m}^2 \cdot \Delta°\text{C/W}]$$

where

x = wall thickness (length of heat flow path)   [m]

k = thermal conductivity of the wall material $[\text{W/m} \cdot \Delta°\text{C}]$

In the case of a composite wall constructed of layers of different materials, such as face brick (b) plus concrete (c) plus plaster (p), the combined thermal resistance ($\Re_2$) is the sum of the thermal resistance of the three layers.

$$\Re_2 = \Re_b + \Re_c + \Re_p$$

$$\Re_b = \frac{x_b}{k_b}, \qquad \Re_c = \frac{x_c}{k_c}, \qquad \Re_p = \frac{x_p}{k_p}$$

In the case of a more complex wall structure a combination of parallel path and series path heat flow systems may be encountered. The total thermal resistance can be calculated by the methods discussed in Section 2-2.1.

The thermal resistance ($\Re_3$) from the outside surface of the wall is also controlled by the air boundary layer. Usually the boundary layer is caused by forced convection due to the wind velocity. The design wind velocity is usually specified in the building heating (or cooling) design requirements or specifications. Forced convection is discussed in Chapter 4 and equations for calculating h_c are given for smooth surfaces. Substantial differences may be encountered between values calculated by these formulae and those actually experienced with typical external surfaces of buildings.

THE OVERALL HEAT TRANSFER COEFFICIENT (U)

In the calculation of the heat flow in a composite system, use is made of the overall heat transfer coefficient (U), as illustrated in the following formula for calculating the heat flow:

$$\dot{Q} = UA\,\Delta T \qquad [\text{Eq. (8-1)}]$$

where

$\dot{Q}$ = heat flow $\;[\text{W}]$

U = overall heat transfer coefficient $\;[\text{W}/\text{m}^2 \cdot \Delta_1 {}^\circ\text{C}]$

A = heat flow path area $\;[\text{m}^2]$

ΔT = temperature drop across the heat flow path $\;[\Delta{}^\circ\text{C}]$

The overall heat transfer coefficient is then the reciprocal of the overall thermal resistance of the heat flow path comprising the following thermal resistances in series:

- the inside convective film $(1/h_{c_i})$
- the wall $(\mathcal{R}_w)$
- the outside convective film $(1/h_{c_o})$

$$U = \frac{1}{\mathcal{R}_t} = \left[\frac{1}{\dfrac{1}{h_{c_i}} + \mathcal{R}_w + \dfrac{1}{h_{c_o}}} \right] \quad [\text{W}/\text{m}^2 \cdot \Delta_1 {}^\circ\text{C}] \qquad (8\text{-}2)$$

The overall heat transfer coefficient (U) can be evaluated in several ways. One way is to use published experimental test measurements and computed values for typical wall configurations. The other way is to compute the overall heat transfer coefficient for the conditions of the wall design and the inside and outside convection involved in the problem. These computations are usually relatively cumbersome, often requiring that values be assumed and the final solution sought by an iterative process. Consequently most experimental data applicable to building walls have been directed toward determining the overall heat transfer coefficient (U) directly.

Values of U have been experimentally measured and calculated for ordinary types of building construction, comprising a variety of materials and thickness. Tabulations of these values have been published by numerous authors, such as the Carrier Air Conditioning Company's *System Design Manual, Part 1*. A few representative values are listed in Appendix B, Table B-6.

Usually this type of data is presented for conditions of a stated wind velocity, such as 24 km/h (15 mph), and often the temperature at which the tests were made is not indicated. This immediately raises two questions:

1. What is the effect of changes in the temperature level and should temperature compensations be made?

2. What is the effect of higher or lower wind velocities and how could compensations be made?

The value of U for a heat flow path through a wall structure and the convective air films on both sides is relatively insensitive to temperature and usually no correction for the temperature level is necessary. The temperature coefficients of the convective films and the wall structure compensate each other to a degree as some increase with increasing temperature and some decrease. For example, illustrative approximate temperature coefficients for increasing temperature are given below:

Free convection film (inside wall), $-0.2\%/\Delta_1°C$
Forced convection film (outside wall), $+0.3\%/\Delta_1°C$
Wall air spaces, $+0.3\%/\Delta_1°C$
. Crystalline solid materials, $-0.3\%/\Delta_1°C$

The magnitude of the effect of the temperature coefficient of each resistance will be dependent upon the relative magnitude of that thermal resistance to the total thermal resistance of the heat flow path. In most cases the temperature coefficient of U for walls will be less than $0.2\%/\Delta_1°C$. For example, a 25°C variance between the actual temperature level and the test temperature level for the data used would result in a 5% error in the calculated heat flow. For most building applications this magnitude of error is negligible.

The effect of the wind can be significant, and is dependent upon the relationship between the thermal resistance of the outside convection film to the thermal resistance of the rest of the heat flow path. The higher the resistance of the rest of the heat flow path, the less the effect of the wind. The effect of wind upon the overall heat transfer coefficient (U) is discussed on page 399.

When published values of U cannot be found for a wall configuration of interest, modifications can readily be made for the value of U of the closest wall configuration by subtracting and adding thermal resistances as appropriate for the changes in the wall configuration. By way of an example the following illustrative problem is given.

If it is not practical to modify a published value of U for a particular configuration, a composite calculation of U can be made. This involves calculating the thermal resistance of each section of the total heat flow path through the wall, which includes:

1. External surface conductance as a function of the wind velocity—Appendix B, Table B-11.
2. Conductivity of the wall material: dry conductivity—Appendix B, Table B-8; effect of density and moisture content of wood—page 49.
3. Conductance of air spaces—Section 3-5.3.
4. Inside wall surface convection—Section 3-4.1.

Example Problem 8-2

The wall of a building is comprised of

> external wood (white pine) sheathing 2.00 cm thick;
> a vertical air space 10.0 cm wide;
> metal lath and plaster 1 cm thick inside surface finish

When the wind is blowing 24 km/h, what is the overall heat transfer coefficient (U) between the inside and outside air when the average temperature between inside and outside air is 16°C?

SOLUTION

SKETCH

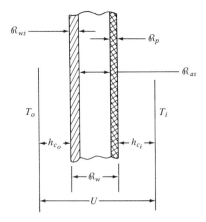

Figure 8-2

PLAN
Use Eq. (8-2) to determine U.

EQUATION

$$U = \cfrac{1}{\cfrac{1}{h_{c_i}} + \mathfrak{R}_w + \cfrac{1}{h_{c_o}}} \quad \left[\text{W/m}^2 \cdot \Delta_1 °\text{C}\right] \quad \left[\text{Eq. (8-2)}\right]$$

PARAMETERS

$$h_{c_i} = 10.8 \quad \left[\text{W/m}^2 \cdot \Delta_1 °\text{C}\right] \quad \text{(Appendix B, Table B-11 for}$$

$$= \text{air velocity and plaster} = 0.0 \text{ km/h)}$$

$$\mathcal{R}_w = \mathcal{R}_p + \mathcal{R}_{as} + \mathcal{R}_{ws} \quad [\text{m}^2 \cdot \Delta°\text{C/W}] \quad [\text{Eq. (2-18a)}]$$

$$\mathcal{R}_p = \frac{x_p}{k_p} [\text{m}^2 \cdot \Delta°\text{C/W}] \quad [\text{Eq. (2-15)}]$$

$$x_p = 1 \text{ cm} \rightarrow 0.010 \quad [\text{m}] \quad (\text{given})$$

$$k_p = 0.47 \quad [\text{W/m} \cdot \Delta_1°\text{C}]$$

(Appendix B, Table B-8 for plaster, gypsum–metal lath)

$$\mathcal{R}_p = \frac{0.010}{0.47} \frac{[\text{m}]}{[\text{W/m} \cdot \Delta_1°\text{C}]} = [\text{m}^2 \cdot \Delta°\text{C/W}]$$

$$\mathcal{R}_p = 0.0213 \quad [\text{m}^2 \cdot \Delta°\text{C/W}]$$

$$\mathcal{R}_{as} = \frac{1}{h_{as}} \quad [\text{m}^2 \cdot \Delta°\text{C/W}]$$

$$\frac{1}{h_{as}} = 6.5 \quad [\text{W/m}^2 \cdot \Delta_1°\text{C}]$$

[Figure 3-16 at 16°C (extrapolate)]

$$\mathcal{R}_{as} = \frac{1}{6.5} \frac{1}{[\text{W/m}^2 \cdot \Delta_1°\text{C}]} = [\text{m}^2 \cdot \Delta°\text{C/W}]$$

$$\mathcal{R}_{as} = 0.154 \quad [\text{m}^2 \cdot \Delta°\text{C/W}]$$

$$\mathcal{R}_{ws} = \frac{x_{ws}}{k_{ws}} \quad [\text{m}^2 \cdot \Delta°\text{C/W}]$$

$$x_{ws} = 2 \text{ cm} \rightarrow 0.020 \quad [\text{m}] \quad (\text{given})$$

$$k_{ws} = 0.112 [\text{W/m} \cdot \Delta_1°\text{C}]$$

(Appendix B, Table B-8 (white pine))

$$\mathcal{R}_{ws} = \frac{0.020}{0.112} \frac{[\text{m}]}{[\text{W/m} \cdot \Delta_1°\text{C}]} = [\text{m}^2 \cdot \Delta°\text{C/W}]$$

$$\mathcal{R}_{ws} = 0.179 \quad [\text{m}^2 \cdot \Delta°\text{C/W}]$$

$$\mathcal{R}_w = 0.0213 + 0.154 + 0.179$$

$$[\text{m}^2 \cdot \Delta°\text{C/W}] + [\text{m}^2 \cdot \Delta°\text{C/W}] + [\text{m}^2 \cdot \Delta°\text{C/W}]$$

$$\mathcal{R}_w = 0.354 \quad [\text{m}^2 \cdot \Delta°\text{C/W}]$$

$$h_{c_o} = 32.4 \quad [\text{W/m}^2 \cdot \Delta_1°\text{C}] \quad (\text{Appendix B, Table B-11 for}$$

$$\text{air velocity} = 24 \text{ km/h and smooth pine})$$

SUBSTITUTION

$$U = \cfrac{1}{\cfrac{1}{10.8} + 0.354 + \cfrac{1}{32.4}}$$

$$\cfrac{1}{\cfrac{1}{\left[W/m^2 \cdot \Delta_1 {}^\circ C\right]} + \left[\cfrac{m^2 \cdot \Delta {}^\circ C}{W}\right] + \cfrac{1}{\left[W/m^2 \cdot \Delta_1 {}^\circ C\right]}}$$

$$= \cfrac{1}{0.093 + 0.354 + 0.031}$$

$$\cfrac{1}{\cfrac{1}{\left[W/m^2 \cdot \Delta_1 {}^\circ C\right]} + \cfrac{1}{\left[W/m^2 \cdot \Delta_1 {}^\circ C\right]} + \cfrac{1}{\left[W/m^2 \cdot \Delta_1 {}^\circ C\right]}}$$

$$= \cfrac{1}{0.478} \quad \cfrac{1}{\cfrac{1}{\left[W/m^2 \cdot \Delta_1 {}^\circ C\right]}} = \left[W/m^2 \cdot \Delta_1 {}^\circ C\right]$$

ANSWER

$$U = 2.09 \quad \left[W/m^2 \cdot \Delta_1 {}^\circ C\right]$$

Often the values of U from tables of published data may vary by more than a factor of 2 from values calculated using free and forced convection formulae (for smooth surfaces) in combination with published values of thermal conductivities of wall materials. The effect of such a variance (possible error) should be considered when values of U are calculated.

In many building heat load problems this degree of accuracy is acceptable because uncontrollable outside air leakage and heat distribution may be major factors in the heating/cooling system design. The difference in the calculated (estimated) heat flow between the two methods is largely due to variations in the convective film coefficient on the inside wall surface, as well as the use of different values of thermal conductivities for the wall materials. Most tables of material thermal conductivities give generic values, and significant variations from these values can be encountered for specific materials of the general classification. Usually tables of overall heat transfer coefficients do not give a breakdown of specific data relative to the actual material thermal conductivities or the convective film coefficient of the two surfaces.

When a high degree of accuracy is required in the heat flow evaluation, care must be exercised to:

1. Use accurate values of thermal conductivities for the wall materials involved.

2. Make accurate evaluations of the convective film coefficients for the combination of surface texture and air current flow pattern over the surfaces.

3. Establish the effective temperature differences between the inside and outside air temperatures for the various wall sections of the building.

EFFECT OF WIND

The effect of wind on the heat transfer rate through a building wall is to reduce the thermal resistance of the convective film on the outside of the wall. The magnitude of the effect will depend upon the magnitude of this film resistance to the total thermal resistance of the heat transfer path from the outside air, through the wall, and into the inside air.

The basic heat flow equation is

$$\dot{Q} = A\,\Delta T / \mathcal{R}_t \quad [\text{W}] \qquad [\text{Eq. (2-12)}]$$

$$\mathcal{R}_t = \mathcal{R}_o + \mathcal{R}_w + \mathcal{R}_i \qquad [\text{Eq. (2-16)}]$$

The effect (change) on the outside wall convective film resistance due to various wind velocities can be determined from data such as presented in Appendix B, Table B-11. This table gives surface conductances (h_o) as a function of wind velocity for various typical building wall surfaces. The outside wall thermal resistance ($\mathcal{R}_o$) is the inverse of the surface conductance (h_o).

$$\mathcal{R}_o = \frac{1}{h_o} \quad [\text{m}^2 \cdot \Delta°\text{C}/\text{W}]$$

Often data are available listing the overall heat transfer coefficient (U) of a wall for an "average" wind velocity. To determine the (U) value for some other wind velocity, the surface conductance as a function of wind velocity data can be used as follows:

1. Separate the total thermal resistance into the outside film and the balance of the wall:

$$\mathcal{R}_t = \mathcal{R}_o + (\mathcal{R}_w + \mathcal{R}_i) \quad [\text{m}^2 \cdot \Delta°\text{C}/\text{W}]$$

2. Compute $\mathcal{R}_{t_1}$ from the available value of U for the average wind velocity

$$\mathcal{R}_{t_1} = \frac{1}{U} \quad [\text{m}^2 \cdot \Delta°\text{C}/\text{W}]$$

3. Determine $\mathcal{R}_{o_1}$ for the external surface of the wall in question and for the wind velocity associated with the (U_1) value used. The available surface conductance data may give the thermal resistance directly, or may give thermal conductances (h_o), which must be converted into the thermal resistance.

4. Calculate $(\mathcal{R}_w + \mathcal{R}_i)$ by using the values of $\mathcal{R}_{t_1}$ and $\mathcal{R}_{o_1}$ obtained from the tables or other data.
5. Determine the $\mathcal{R}_{o_2}$ for the external surface of the wall in question for the wind velocity of interest.
6. Calculate the total thermal resistance $(\mathcal{R}_{t_2})$ for the wind velocity of interest by

$$\mathcal{R}_{t_2} = \mathcal{R}_{o_2} = (\mathcal{R}_w + \mathcal{R}_i)$$

7. Then calculate the heat transfer $(\dot{Q}_2)$,

$$\dot{Q}_2 = \frac{A \, \Delta T}{\mathcal{R}_{t_2}}$$

Example Problem 8-3

What would be the change in $\mathcal{R}_t$ of an ordinary window when the wind changed from 24 km/h to 48 km/h?

SOLUTION

SKETCH

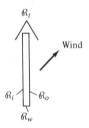

Figure 8-3

EQUATION

$$\text{change in } \mathcal{R}_t = \frac{\mathcal{R}_{t_{48}}}{\mathcal{R}_{t_{24}}} \quad [-]$$

PARAMETERS

$$\mathcal{R}_{t_{24}} = \frac{1}{U_{24}} \quad [\text{m}^2 \cdot \Delta °C/W]$$

$$U_{24} = 6.42 [W/\text{m}^2 \cdot \Delta_1 °C] \qquad (\text{Table B-10, Appendix B})$$

$$\mathcal{R}_{t_{24}} = \frac{1}{(6.42)} \left[\frac{1}{(W/\text{m}^2 \cdot \Delta_1 °C)} \right] = ([\text{m}^2 \cdot \Delta °C/W])$$

$$\mathcal{R}_{t_{24}} = 0.156 \quad \left[m^2 \cdot \Delta°C/W \right]$$

$$\mathcal{R}_{t_{48}} = \left(\mathcal{R}_w + \mathcal{R}_i \right) + \mathcal{R}_{o_{48}} \quad \left[m^2 \cdot \Delta°C/W \right]$$

$$\left(\mathcal{R}_w + \mathcal{R}_1 \right) = \mathcal{R}_{t_{24}} - \mathcal{R}_{o_{24}} \quad \left[m^2 \cdot \Delta°C/W \right]$$

$$\mathcal{R}_{o_{24}} = \frac{1}{h_{o_{24}}} \quad \left[m^2 \cdot \Delta°C/W \right]$$

$$h_{o_{24}} = 29.5 \quad \left[W/m^2 \cdot \Delta_1°C \right] \qquad (\text{Table B-11, Appendix B})$$

$$\mathcal{R}_{o_{24}} = \frac{1}{29.5} = 0.0339 \quad \left[m^2 \cdot \Delta°C/W \right]$$

$$\left(\mathcal{R}_w + \mathcal{R}_1 \right) = 0.156 - 0.0339$$

$$\left[m^2 \cdot \Delta°C/W - m^2 \cdot \Delta°C/W \right] = \left[m^2 \cdot \Delta°C/W \right]$$

$$\left(\mathcal{R}_w + \mathcal{R}_1 \right) = 0.122 \quad \left[m^2 \cdot \Delta°C/W \right]$$

$$\mathcal{R}_{o_{48}} = \frac{1}{h_{o_{48}}} \quad \left[m^2 \cdot \Delta°C/W \right]$$

$$h_{o_{48}} = 47.14 \quad \left[W/m^2 \cdot \Delta_1°C \right]$$

$$(\text{Table B-11, Appendix B})$$

$$\mathcal{R}_{o_{48}} = \frac{1}{47.14} \left[\frac{1}{W/m^2 \cdot \Delta_1°C} \right] = \left[m^2 \cdot \Delta°C/W \right]$$

$$\mathcal{R}_{o_{48}} = 0.021 \quad \left[m^2 \cdot \Delta°C/W \right]$$

$$\mathcal{R}_{t_{48}} = \left(\mathcal{R}_w + \mathcal{R}_i \right) + \mathcal{R}_{o_{48}} \quad \left[m^2 \cdot \Delta°C/W \right]$$

$$= (0.122) + (0.021)$$

$$\left[m^2 \cdot \Delta°C/W + m^2 \cdot \Delta°C/W \right] = \left[m^2 \cdot \Delta°C/W \right]$$

$$\mathcal{R}_{t_{48}} = 0.143 \quad \left[m^2 \cdot \Delta°C/W \right]$$

$$\text{change in } \mathcal{R}_t = \frac{\mathcal{R}_{t_{48}}}{\mathcal{R}_{t_{24}}} = \frac{0.143}{0.156} \frac{\left[m^2 \cdot \Delta°C/W \right]}{\left[m^2 \cdot \Delta°C/W \right]} = [-]$$

ANSWER
change in $\mathcal{R}_t = 0.917 \quad [-]$

THE APPLICATION OF INSULATION
In a heat flow path system one component of the system can exercise a dominant control over the amount of heat flowing in the system. In a parallel flow path system a high thermal conductance path (low thermal

resistance) can be the dominant controlling path. This is discussed on page 50. Consequently, applying insulation to this area would be the most effective. By applying insulation (increasing the thermal resistance) to the dominant path will have the greatest effect in reducing the total amount of heat flow. In a series path system, the high thermal resistance path section can be the dominant controlling resistance. This is discussed on page 59. The application of these concepts to the controlling of the heat flow in or out of a building is discussed in more detail in the following sections.

In order to effectively make an application of insulation to a building to reduce the amount of heat flow in or out of the building, an analysis of the building wall section heat flows should be made to determine what areas transfer the most heat. Table 8-1 gives a characterization of a house and the thermal conductance of the heat transfer path of the various sections, and the percent of the total thermal path conductance for each path.

From this table it can be seen that a major portion of the heat is being transferred through the ceiling. Consequently applying insulation to this area would be most effective. By applying a 15-cm thick layer of rock wool to the ceiling the ceiling conductance is calculated to be 0.246 $[W/m^2 \cdot \Delta_1 °C]$ resulting in a path conductance of 41.4 $[W/\Delta_1 °C]$.

The next significant reduction in the heat flow would be obtained by insulating the floor. By applying a 10-cm thick layer of rock wool, or equivalent insulation under the floor, the thermal conductance is calculated to be 0.332 $[W/m^2 \cdot \Delta_1 °C]$, resulting in a path conductance of 55.7 $[W/\Delta_1 °C]$.

The next significant reduction in the heat flow would be obtained by replacing the ordinary windows with double windows. This would reduce the window conductance to 2.56 $[W/m^2 \cdot \Delta_1 °C]$ resulting in a path conductance of 51.2 $[W/\Delta_1 °C]$.

Table 8-1 CHARACTERIZATION OF HEAT FLOW PATHS IN OR OUT OF A HOUSE

			THERMAL CONDUCTANCE		
SECTION	AREA $[m^2]$	DESCRIPTION	WALL $[W/m^2 \cdot \Delta_1 °C]$	SECTION PATH $[W/\Delta_1 °C]$	TOTAL [%]
Floor	168	wood subfloor and floor on joists	2.00	336.0	28.4
Ceiling	168	lath and plaster, no floor above	3.52	591.4	50.0
Wall	80	frame wall, clapboard 1-in. wood sheathing, lath and plaster	1.42	113.6	9.6
Door	6	3.8-cm wood ($1\frac{1}{2}$-in.)	2.20	13.2	1.1
Windows	20	ordinary glass	6.42	128.4	10.9
				1182.6	100.0

Table 8-2 EFFECT OF INSULATION IN HEAT FLOW PATHS IN OR OUT OF A HOUSE

| | | | THERMAL CONDUCTANCE | | |
SECTION	AREA [m^2]	DESCRIPTION	WALL [W/m^2·Δ_1°C]	SECTION PATH [W/Δ_1°C]	TOTAL [%]
Floor	168	wood subfloor and floor on joists, 10 cm rock wool underneath	0.332	55.7	20.3
Ceiling	168	lath and plaster with 15 cm rock wool above	0.246	41.4	15.0
Wall	80	framewall, clapboard 2.5 cm (1-in.) wood sheathing, lath and plaster	1.42	113.6	41.3
Door	6	3.8-cm wood (1½-in.)	2.20	13.2	4.8
Windows	20	double glass	2.56	51.2	18.6
				275.1	100.0

As existing walls are relatively difficult to insulate and the heat flow through the doors is a small portion of the total, insulation of these sections is not recommended. The resulting heat flow with the above insulation is summarized in Table 8-2.

The insulation applied in this case resulted in a heat flow reduction to 23.3% of the original heat flow. The uninsulated wall now accounts for 41.3% of the heat flow. If this were used in the planning for a new building, insulating the wall with glass wool would reduce the wall section path conductance to 31.8 [W/Δ_1°C]. This would reduce the total heat flow to 16.3% of the original heat flow. The heat flow through the insulated wall would be reduced to 16.5% of the total heat flow.

In order to effectively make an application of insulation to a wall to reduce the amount of heat flow in or out of the building, an analysis of the thermal resistances of the elements of the heat flow path through the wall should be made to determine the magnitudes of the thermal resistances. From this analysis the controlling resistance can be identified and means of increasing the total heat flow path thermal resistance can be selected. Increasing minor thermal resistances will have relatively little effect. Increasing the major or controlling resistance will have a significant effect. Adding thermal resistances of similar or greater magnitude than the controlling resistance will have a significant effect.

This is illustrated in the following discussion, which uses as an example a 4-mm thick glass window exposed to a 24 km/h wind. The heat flow path has three resistances:

- outside surface film resistance
- conductive resistance through the glass
- inside surface film resistance

Three thermal resistances can be evaluated as follows.

• *Outside Surface Film Resistance.* Table B-11 in Appendix B presents test data indicating a thermal resistance of the surface film on glass for a 24 km/h wind as 0.0352 [m²·Δ°C/W]

• *Conductive Resistance Through the Glass.*

$$\mathcal{R} = \frac{x}{k} \quad \left[\text{m}^2 \cdot \Delta°\text{C}/\text{W}\right]$$

where

$x =$ length of heat flow path $\quad [\text{m}]$

$k =$ thermal conductivity $\quad \left[\text{W}/\text{m} \cdot \Delta_1°\text{C}\right]$

For the window

$x = 0.004 \quad [\text{m}]$

$k = 0.78 \quad \left[\text{W}/\text{m} \cdot \Delta_1°\text{C}\right] \quad$ (from Appendix B, Table B-8)

$$\mathcal{R} = \frac{0.004}{0.78} \quad \frac{[\text{m}]}{\left[\text{W}/\text{m} \cdot \Delta_1°\text{C}\right]} = \left[\text{m}^2 \cdot \Delta_1°\text{C}/\text{W}\right]$$

$$\mathcal{R} = 0.0051 \quad \left[\text{m}^2 \cdot \Delta_1°\text{C}/\text{W}\right]$$

• *Inside Surface Film Resistance.* Table B-11 in Appendix B presents test data indicating a thermal resistance of the surface film on glass for zero velocity wind as 0.1174 [m²·Δ°C/W].

The total thermal resistance of the heat flow path and the relative magnitudes of the three components are listed in Table 8-3.

From these data it is readily seen that the inside surface film resistance is controlling. Obviously attempting to reduce the heat flow by increasing the glass thickness will have very little effect upon significantly increasing the total resistance of the heat flow path.

An effective way of increasing the thermal resistance is to add an air space by introducing a second pane of glass spaced a suitable distance from the original pane. From Figure 3-12 (in Chapter 3), the resistance of an air space with 4°C average air temperature reaches a value of $1/6.5 =$

Table 8-3 ANALYSIS OF HEAT FLOW PATH THROUGH A WINDOW

| | THERMAL RESISTANCE | |
| | MAGNITUDE | TOTAL |
PATH SECTION	$[\text{m}^2 \cdot \Delta_1°\text{C}/\text{W}]$	[%]
Outside surface film	0.0352	22.3
Glass	0.0051	3.2
Inside surface film	0.1174	74.5
	0.1577	100.0

Table 8-4 ANALYSIS OF HEAT FLOW PATH THROUGH
A DOUBLE WINDOW

| | THERMAL RESISTANCE | |
| | MAGNITUDE | TOTAL |
PATH SECTION	$[m^2 \cdot \Delta_1 °C/W]$	[%]
Outside surface film	0.0352	9.7
Glass	0.0051	1.4
Air gap (12 mm)	0.1538	42.5
Glass	0.0051	1.4
Inside surface film	0.1174	32.4
	0.3625	100.0

$0.1538 \ [m^2 \cdot \Delta_1 °C/W]$. The analysis of this double glass pane configuration is given in Table 8-4.

The addition of the air gap has increased the total thermal resistance of the heat flow path by a factor of 2.3. Attempting to insulate by increasing the thickness of the single glass pane by a factor, such as 4, would have resulted in increasing the total thermal resistance only by a factor of 1.1.

WHERE RADIATION HEAT TRANSFER IS A FACTOR

When solar radiations or radiation interchange between the sky or areas adjacent to building walls can affect the net heat transfer into or out of a building, the two major considerations are

- transmitted radiations through windows
- modification of the convective/conduction heat transfer through the walls by the radiations absorbed or emitted by the outside surface of the wall

These are discussed in the following sections.

An ordinary window glass transmits visible light, but not long wavelength IR. Incoming solar radiations are largely transmitted subject to the reflection and absorption characteristics of the glass. The long wavelength (IR) radiations emitted by the interior surfaces of the building are not transmitted out through the window, because the transmissivity of ordinary window glass is very low for these long wavelengths. During the day any sunshine passing through the windows is absorbed, becoming a heat input into the building. At night the window radiates to the night sky and surroundings as a gray body, but does not transmit the IR radiations.

The magnitude of the solar beam radiant energy transmission through a window is illustrated by the following sample problem which considers (a) sunshine passing through a window, and (b) the effect of blocking the sunshine by a sheet of aluminum foil placed on the inside of the windows.

Example Problem 8-4 (Radiant Energy Transmission)

A window in a house is exposed to direct sunlight. The characteristics of the glass are

$$\text{total reflection loss from both surfaces} = 5\%$$
$$\text{absorption of solar radiations} = 10\%$$

Considering the incident rays of the sun to be normal to the window, and the sun's rays have a zenith angle (z) of 60°.

(a) How much solar radiant energy [W/m²] enters the house through the window (transmitted by the glass)?
(b) If a piece of aluminum foil is placed over the inside of the window of the previous problem, how much of the radiant heat is reflected back out through the window?

SOLUTION

SKETCH FOR (a)

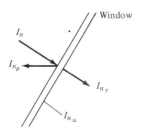

Figure 8-4

EQUATION FOR (a)

$$I_{n_\tau} = (I_n)(\tau) \quad [\text{W/m}^2]$$

PARAMETERS FOR (a)
I_n from Figure 7-43 for $z = 60° = 746$ [W/m²]

$$
\begin{aligned}
\tau &= 1 - \alpha - \rho \quad [-] \qquad [\text{Eq. (7-2)}] \\
&= \alpha = 0.10 \quad [-] \qquad \text{(given)} \\
&= \rho = 0.05 \quad [-] \qquad \text{(given)} \\
\tau &= 1 - 0.10 - 0.05 = 0.85 \quad [-]
\end{aligned}
$$

SUBSTITUTION FOR (a)

$$I_{n_\tau} = (746) \times (0.85) \quad [\text{W/m}^2][-] = [\text{W/m}^2]$$

ANSWER FOR (a)

$$I_{n_\tau} = 634 \quad [\text{W}/\text{m}^2]$$

SKETCH FOR (b)

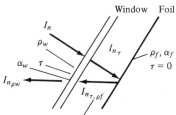

Solar energy reflected from the foil

Figure 8-5

EQUATION FOR (b)

$$I_{n_R} = I_{n_{\tau,R}} \cdot \tau_w [\text{W}/\text{m}^2] \qquad \text{(reflected energy passing back} \\ \qquad\qquad\qquad\qquad\qquad\qquad\qquad \text{through the glass)}$$

$$I_{n_{\tau,R}} = I_{n_\tau} \cdot \rho_f \quad [\text{W}/\text{m}^2] \qquad \text{(energy reflected by the foil)}$$

$$I_{n_\tau} = I_n \cdot \tau_w \quad [\text{W}/\text{m}^2] \qquad \text{(transmitted energy incident upon the foil)}$$

Substituting in the first equation

$$I_{n_R} = I_n (\tau_w \cdot \tau_w \cdot \rho_f) \quad [\text{W}/\text{m}^2]$$

Therefore,

$$I_{n_R} = I_n (\tau_w^2 \rho_f) \quad [\text{W}/\text{m}^2]$$

PARAMETERS FOR (b)

$$I_n = 746 \quad [\text{W}/\text{m}^2] \qquad \text{(from the previous problem)}$$

$$\tau_w = 1 - \alpha_w - \rho_w \quad [-] \qquad [\text{Eq. (8-9)}]$$

$$\alpha_w = 0.10 \quad [-] \qquad \text{(given)}$$

$$\rho_w = 0.05 \quad [-] \qquad \text{(given)}$$

$$\tau_w = 1 - 0.10 - 0.05 \quad ([-] - [-] - [-]) = [-]$$

$$\tau_w = 0.85 \quad [-]$$

$$\rho_f = 1 - \alpha_f \quad [-] \, (\text{as } \tau = 0) \qquad [\text{Eq. (8-9)}]$$

$$\alpha_f = 0.26 \quad [-] \qquad \text{(Table 7-7)}$$

$$\rho_f = 1 - 0.26 \quad ([-] - [-]) = [-]$$

$$\rho_f = 0.74 \quad [-]$$

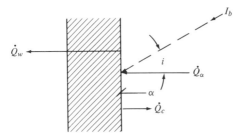

Figure 8-6 Heat flow through a wall with incident radiation. $\dot{Q}_{\hat{a}} = \alpha I_{\hat{b}} \cos i.$
See Sections 7.5.1 and 7.5.2.

SUBSTITUTION FOR (b)

$$I_{n_R} = (746)(0.85^2)(0.74) \quad [W/m^2][-][-] = [W/m^2]$$

ANSWER FOR (b)

$$I_{n_R} = 399.1 \quad [W/m^2]$$

When radiations are incident upon a building wall, the absorbed radiation energy results in an increased temperature of the outside surface of the wall, resulting in the absorbed heat being transferred through the wall into the building and/or transferred to the outside air. This is illustrated by Figure 8-6.

The heat flowing through the wall ($\dot{Q}_w$) is equal to the absorbed solar radiations ($\dot{Q}_\alpha$) less the heat flowing from the outside wall surface into the outside air ($\dot{Q}_c$).

$$\dot{Q}_w = (\dot{Q}_\alpha - \dot{Q}_c) \quad [W] \tag{8-3}$$

A method of calculating the heat flowing through a wall with incident solar radiation is illustrated by the following example problem.

Example Problem 8-5

The wall of a house is frame with clapboard, 2.5-cm (1-in.) wood sheathing, lath, and plaster. The temperature inside the room is 20°C. The outside air has a wind velocity of 24 km/h and is 20°C. The outside of the wall is painted brown with an absorptivity for solar radiation of 0.5. When the sun shines on the wall with a β angle of 60° and a ϕ angle of 45°, what will be the heat flow through the wall, considering convection on the outside wall and incident solar radiation?

SOLUTION

SKETCH

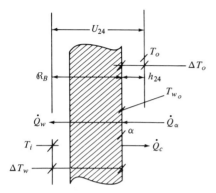

Figure 8-7

EQUATION

$$\dot{Q}_w/A = \dot{Q}_r/A - \dot{Q}_c/A \quad \left[\text{W}/\text{m}^2\right]$$

PARAMETERS

$$\dot{Q}_r/A = \alpha(I_b\cos\phi) \quad \left[\text{W}/\text{m}^2\right] \quad \left[\text{Eq. (7-62)}\right]$$

$$\alpha = 0.5 \quad (-) \quad (\text{given})$$

$$I_b = 235 \times 3.15 = 740 \quad \left[\text{W}/\text{m}^2\right] \quad (\text{Figure 7-26 for } \beta = 60°)$$

$$\phi = 45° \quad (\text{given})$$

$$\cos\phi = 0.707 \quad [-]$$

$$\dot{Q}_r/A = (0.5)(740)(0.707) \quad [-]\left[\text{W}/\text{m}^2\right][-] = \left[\text{W}/\text{m}^2\right]$$

$$\dot{Q}_r/A = 262 \quad \left[\text{W}/\text{m}^2\right]$$

$$\dot{Q}_c/A = (h_{24})(\Delta T_o) \quad \left[\text{W}/\text{m}^2\right]$$

$$h_{24} = 31.8 \quad \left[\text{W}/\text{m}^2 \cdot \Delta_1°\text{C}\right] \quad (\text{Appendix B, Table B-11})$$

$$\Delta T_o = (T_{w_o} - T_o) \quad [\Delta°\text{C}]$$

$$T_o = 20°\text{C} \quad (\text{given})$$

$$\Delta T_o = (T_{w_o} - 20) \quad [\Delta°\text{C}]$$

$$\dot{Q}_c/A = (31.8)(T_{w_o} - 20) \quad \left[\text{W}/\text{m}^2 \cdot \Delta_1°\text{C}\right][\Delta°\text{C}] = \left[\text{W}/\text{m}^2\right]$$

$$\dot{Q}_c/A = (31.8T_{w_o} - 636) \quad \left[\text{W}/\text{m}^2\right]$$

$$\dot{Q}_w/A = (262) - (31.8T_{w_o} - 636) \quad [W/m^2] - [W/m^2] = [W/m^2]$$

$$\dot{Q}_w/A = (898 - 31.8T_{w_o}) \quad [W/m^2]$$

$\dot{Q}_w/A$ also is

$$\dot{Q}_w/A = \frac{\Delta T_w}{\mathfrak{R}_B} \quad [W/m^2]$$

$$\Delta T_w = (T_{w_o} - T_i) \quad [\Delta°C]$$

$$T_i = 20°C \quad \text{(given)}$$

$$\Delta T_w = (T_{w_o} - 20) \quad [\Delta°C]$$

$$\mathfrak{R}_B = \frac{1}{U_{24}} - \frac{1}{h_{24}} \quad [m^2 \cdot \Delta_1°C/W]$$

$$U = 1.42 \quad [W/m^2 \cdot \Delta_1°C] \quad \text{(Appendix B, Table B-10)}$$

$$h_{24} = 31.8 \quad [W/m^2 \cdot \Delta_1°C] \quad \text{(Appendix B, Table B-11)}$$

$$\mathfrak{R}_B = \frac{1}{1.42} - \frac{1}{31.8} = 0.705 - 0.031 \quad [m^2 \cdot \Delta_1°C/W] - [m^2 \cdot \Delta_1°C/W]$$

$$= [m^2 \cdot \Delta_1°C/W]$$

$$\mathfrak{R}_B = 0.674 \quad [m^2 \cdot \Delta_1°C/W]$$

SUBSTITUTION

$$\dot{Q}_w/A = \frac{(T_{w_o} - 20)}{0.674} \frac{[\Delta°C]}{[m^2 \cdot \Delta_1°C/W]} = [W/m^2]$$

$$\dot{Q}_w/A = (1.484T_{w_o} - 29.67) \quad [W/m^2]$$

Combining the two equations for $\dot{Q}_w/A$ to solve for T_{w_o},

$$\dot{Q}_w/A = (1.485T_{w_o} - 29.67) = (898 - 31.8T_{w_o})$$

$$+ 33.29T_{w_o} = 927.7$$

$$T_{w_o} = 27.9°C$$

Using any of the equations for $\dot{Q}_w/A$,

$$\dot{Q}_w/A = 1.484T_{w_o} - 29.67 \quad [W/m^2]$$

$$= (1.484)(27.9) - 29.67$$

$$= 41.40 - 29.67$$

ANSWER

$$\dot{Q}_w/A = 11.7 \quad [W/m^2]$$

8-2 HEAT EXCHANGERS

Heat exchangers comprise the bulk of heat transfer applications. Heat exchangers transfer heat between two fluids. The heat transferred may be primarily in the form of latent heat, as is the case in condensers and evaporators; or the heat transferred may be primarily sensible heat, as is the case in heaters and coolers.

A heat exchanger has a purpose, either to remove heat from a fluid or to add heat to a fluid. Considering this fluid to be the dominant, or primary, fluid flow stream, the size of the heat exchanger (in terms of the heat transfer rate) is

$$\dot{Q} = \dot{M}c_p \Delta T \quad [\text{W}] \qquad \text{(for sensible heat transfer)}$$

or

$$\dot{Q} = \dot{M}r \quad [\text{W}] \qquad \text{(for latent heat transfer)}$$

The other, or secondary fluid flow stream, then has two imposed requirements:

- the heat transfer rate must be the same magnitude (but opposite sign) to that of the primary fluid flow stream;
- the initial and final temperature differences between the primary and secondary fluid flow streams must provide the required mean temperature difference to effect the heat transfer rate through the heat exchange from one fluid to the other.

A heat exchanger can be represented by the block diagram shown in Figure 8-8.

The indicated parameters are:

Hot Fluid

$$\dot{M}_h = \text{mass flow} \quad [\text{kg/s}]$$

$$c_{p_h} = \text{specific heat} \quad [\text{J/kg} \cdot \Delta_1 °\text{C}]$$

$$T_{h_{in}} = \text{fluid temperature entering the heat exchanger} \quad [°\text{C}]$$

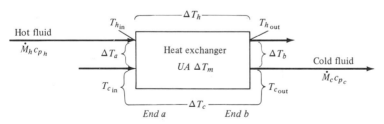

Figure 8-8 Block diagram of a heat exchanger.

$T_{h_{out}}$ = fluid temperature leaving the heat exchanger [°C]

ΔT_h = temperature drop of the fluid across the heat
    exchanger [Δ°C]

Coolant

$\dot{M}_c$ = mass flow [kg/s]

c_{p_c} = specific heat [J/kg·Δ_1°C]

$T_{c_{in}}$ = coolant temperature entering the heat exchanger [°C]

$T_{c_{out}}$ = coolant temperature leaving the heat exchanger [°C]

ΔT_c = temperature rise of the coolant across the heat
    exchanger [Δ°C]

Heat Exchanger Structure

ΔT_a = temperature difference between the hot fluid and coolant at the
    end of the heat exchanger where the hot fluids enters [Δ°C]

ΔT_b = temperature difference between the hot fluid and coolant at the
    end of the heat exchanger where the hot fluid exits [Δ°C]

U = overall heat transfer coefficient between the two fluids (through
    the respective films and the heat transfer wall of the heat
    exchanger) [W/m²·Δ_1°C]

A = effective direct heat transfer surface area [m²]

ΔT_m = effective mean temperature difference across the heat ex-
    changer structure for transferring heat through the heat ex-
    changer structure [Δ°C]

In the design and analysis of heat exchangers, the aspects of

- the hot fluid (being cooled)
- the coolant (being heated)
- the heat exchanger structure

must be considered individually.
The *hot fluid* gives up heat:

$$\dot{Q}_h = \dot{M}_h c_{p_h} (T_{h_{in}} - T_{h_{out}}) \quad [W] \tag{8-4}$$

The *coolant* picks up heat:

$$\dot{Q}_c = \dot{M}_c c_{p_c} (T_{c_{out}} - T_{c_{in}}) \quad [W] \tag{8-5}$$

The *heat exchanger structure* transfers the heat from the hot fluid to the coolant:

$$\dot{Q}_e = UA\,\Delta T_m \quad [\text{W}] \qquad [\text{Eq. (8-1)}]$$

As the amount of heat given up by the hot fluid goes into the coolant by being transferred through the heat exchanger structure

$$\dot{Q}_h = \dot{Q}_c = \dot{Q}_e \quad [\text{W}] \tag{8-6}$$

The ΔT_m, which is determined from the initial and final temperature differences between the two fluid stream temperatures, must be sufficient to effect the required heat transfer through the heat exchanger structure.

For purposes of heat transfer performance design, heat exchangers can be grouped according to the characteristic geometric arrangement between the direction of the flow paths of the two fluids passing through the heat exchanger. The usual four groupings are

- parallel flow
- counter flow
- cross flow
- multipass (with countless permutations)

In a performance and design analysis based on Eq. (8-1), each of these four groups of heat exchangers has substantially different performance characteristics. The values of U and A are determined in essentially the same manner for all of the four heat exchanger groupings. The differences in the performance characteristics of each of the four groups is considered in the evaluation of the effective mean temperature difference (ΔT_m). This is discussed in detail in Section 8-2.3.

8-2.1 The Overall Heat Transfer Coefficient

The heat flow path in a heat exchanger is a simple linear series conduction path, as illustrated in Figure 8-9. The resistances comprise

- convection resistance on tube inside wall, $(1/h_i)$
- wall conduction resistance, (x/k)
- convection resistance on tube outside wall, $(1/h_o)$

For most heat transfer surface geometries of thin walled tubes or flat plates, a linear series conduction heat transfer path with a constant cross-sectional area normal to the heat flow can be assumed. The area A in Eq. (8-1) would be that of the flat plate surface or that of a cylinder with the average diameter between the inside and outside diameters of the actual tube. In this case the three thermal resistances listed at the beginning of this section can be added in accordance with Eq. (2-16) and the resultant total resistance ($\mathcal{R}_t$) converted into U for use directly in Eq. (8-1)

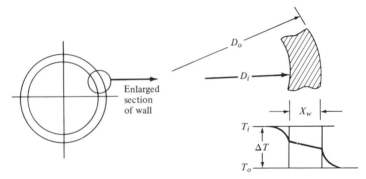

Figure 8-9 A typical heat transfer path through a heat exchanger tube wall.

by

$$U = \frac{1}{\mathfrak{R}_t} = \frac{1}{1/h_i + x/k + 1/h_o} \quad \left[\mathrm{W/m^2 \cdot \Delta_1 ^\circ C}\right] \tag{8-7}$$

In the case of a thick walled pipe (discussed on page 71) a constant cross-sectional area of the heat flow path cannot be assumed. Equation (8-1) must be modified to

$$\dot{Q}/L = \Delta T_m / \mathfrak{R}_u \quad [\mathrm{W/m}] \tag{8-8}$$

$$\mathfrak{R}_u = \mathfrak{R}_{c_i} + \mathfrak{R}_w + \mathfrak{R}_{c_o} \quad [\mathrm{m \cdot \Delta ^\circ C/W}] \tag{8-9}$$

Per unit length of tubing:

$$\mathfrak{R}_{c_i} = \frac{1}{h_i \pi D_i} \quad [\mathrm{m \cdot \Delta ^\circ C/W}]$$

$$\mathfrak{R}_w = \frac{\ln(D_o / D_i)}{2\pi k_w} \quad [\mathrm{m \cdot \Delta ^\circ C/W}]$$

$$\mathfrak{R}_{c_o} = \frac{1}{h_o \pi D_o} \quad [\mathrm{m \cdot \Delta ^\circ C/W}]$$

By substituting the values of these resistances into Eq. (8-9), the overall resistance $\mathfrak{R}_u$ is obtained. This value is used directly in Eq. (8-3).

As discussed on page 59, the large thermal resistance is the "controlling" resistance. For an efficient heat exchanger design, there should be no high thermal resistance in the heat flow path. Often when the heat exchanger is transferring heat between a gas and a liquid, there is a significant unbalance in the thermal resistances of the two convective films. The convection resistance in the gas is often high. In such a case, extended surfaces can be added to reduce the effective convection resistance in the gas. When extended surfaces are applied, the direct surface

Table 8-5 OVERALL COEFFICIENTS FOR METAL TUBES IN
POWER PLANT AND REFRIGERATION APPARATUS

	U $[\text{W}/\text{m}^2 \cdot \Delta_1 {}^\circ\text{C}]$
Steam boiler tubes	20–80
Economizers	10–60
Steam superheaters, close to furnace	45–85
Surface condenser (steam)	1100–5700
Feed-water heaters	1100–8500
Air preheaters	5–20
Ammonia condensers	
Double pipe	850–1400
Shell and tube	850–1700
Atmospheric	350–1100
Brine coolers:	
Double pipe	850–1700
Shell and tube	500–600
Water cooler, shell and coil	85–140
Cooling coils:	
Water to air in unit coolers	30–50
Boiling refrigerant to air in unit coolers	25–45
Brine to unagitated air	10–15

SOURCE: A. I. Brown and S. M. Marco, *Introduction to Heat Transfer*, 3rd
ed., New York, McGraw-Hill, 1958.

convection heat transfer coefficient would be increased by the extended
surface heat transfer improvement ratio $\mathcal{R}_{es}$ discussed on page 86. The
convective resistance with the extended surface is:

    Flat plate or thin walled tubing per unit area of original direct surface:

$$\mathcal{R}_c = \frac{1}{h_c \mathcal{R}_{es}} \quad [\text{m}^2 \cdot \Delta {}^\circ\text{C}/\text{W}] \tag{8-10}$$

Thick walled tubing per unit length of pipe:

$$\mathcal{R}_c = \frac{1}{h_c \mathcal{R}_{es} \pi D_o} \quad [\text{m} \cdot \Delta {}^\circ\text{C}/\text{W}] \tag{8-11}$$

    Typical overall coefficients of heat transfer for metal tubes as applied
in a variety of forms of heat exchangers are listed in Table 8-5.

8-2.2 Heat Transfer Area

The heat transfer area (A) for use in Eq. (8-1) is the area of the surface
that is between the two fluids and is in direct contact with both fluids
passing through the heat exchanger. This direct heat transfer area is the
wetted surface in contact with each fluid, and is the same for each of the
two fluids. This is obviously true for a flat plate surface. In a thin walled

tubular heat exchanger, the area is based upon the average tube diameter (between the tube outside and inside diameters). This surface is called the "direct" surface area.

In configurations involving thick walled tubes and the use of Eqs. (8-8) and (8-9), the direct surface area as outlined above is not involved because these equations do not contain the surface area parameter A.

In configurations using extended surface, the direct heat transfer surface area common to both fluids is used and the convection heat transfer coefficient is increased by the heat transfer improvement ratio.

8-2.3 Effective Mean Temperature Difference

The characteristics of the effective mean temperature difference vary between heat exchangers exchanging latent heat and those exchanging sensible heat. Consequently, the effective mean temperature difference for these two types of heat exchangers is discussed separately in the following sections.

HEAT EXCHANGERS TRANSFERRING LATENT HEAT
In heat exchangers transferring latent heat, the temperature of the fluid undergoing the phase change is constant. The latent heat is transferred as sensible heat into the other fluid stream, increasing or decreasing its temperature depending upon whether heat is being released (condensation) or absorbed (evaporation). The temperature profile along the flow path of the fluid transferring sensible heat is illustrated in Figures 8-10(a) and (b). The effective mean temperature difference is the same whether the fluid transferring sensible heat has a single or multiple pass configuration. The effective mean temperature difference is based upon the initial and final temperature difference and is defined by Eq. (8-12).

$$\Delta T_{\mathrm{lm}} = \frac{\Delta T_a - \Delta T_b}{\ln(\Delta T_a / \Delta T_b)} \quad [\Delta \degree C] \tag{8-12}$$

where

ΔT_{lm} = log mean temperature difference (LMTD) $\quad [\Delta \degree C]$

ΔT_a = larger temperature difference at the start of the heat exchange

ΔT_b = smaller temperature difference at the end of the heat exchange

$\ln$ = natural logarithm

In calculations of the mean temperature difference between the vapor and the cooling fluid in a condenser, the vapor is usually considered to be at the saturation temperature corresponding to the pressure, regardless of the level of superheat.

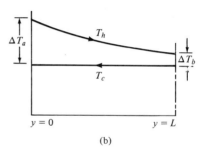

Figure 8-10 Fluid temperature profiles for condensing and boiling heat exchangers.

In calculations of the mean temperature difference between the heating fluid and the evaporating liquid in a boiler, the evaporating liquid and the generated vapor are usually considered to be at the saturation temperature corresponding to the pressure. In cases where superheated vapor is generated, the superheat is almost always added in a heat exchanger separated from the vaporizing heat exchanger.

Example Problem 8-6

A steam condenser is transferring 240 kW of thermal energy at a condensing temperature of 60°C. The cooling water enters the condenser at 20°C with a flow rate of 8000 kg/h.

(a) What is the LMTD (log mean temperature difference), (ΔT_{lm})?
(b) If the U for the condenser surface is 1000 W/m²·Δ_1°C, what is the required heat transfer surface area?

SOLUTION

SKETCH

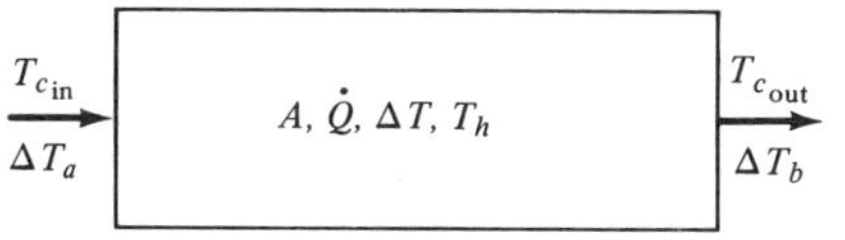

$T_{c_{in}}$ → | $A, \dot{Q}, \Delta T, T_h$ | → $T_{c_{out}}$
ΔT_a | | ΔT_b

Figure 8-11

EQUATION FOR (a)

$$\Delta T_{lm} = \frac{\Delta T_a - \Delta T_b}{\ln(\Delta T_a / \Delta T_b)} \quad [\Delta°C] \quad [\text{Eq. (8-12)}]$$

PARAMETERS FOR (a)

$$\Delta T_a = T_h - T_{c_{in}} \quad [\Delta°C]$$

$$T_h = 60°C \qquad \text{(given)}$$

$$T_{c_{in}} = 20°C \qquad \text{(given)}$$

$$\Delta T_a = 60 - 20 \quad [°C - °C] = [\Delta°C]$$

$$\Delta T_a = 40\Delta°C$$

$$\Delta T_b = T_h - T_{c_{out}} \quad [\Delta°C]$$

$$T_h = 60°C \quad \text{(given)}$$

To obtain the cooling water out temperature ($T_{c_{out}}$) calculate the temperature rise (ΔT) resulting from the transferred heat and add to the cooling water inlet temperature ($T_{c_{in}}$).

$$T_{c_{out}} = T_{c_{in}} + \Delta T \quad [°C]$$

$$\dot{Q} = \dot{M} c_p \Delta T \quad [W]$$

$$\Delta T = \frac{\dot{Q}}{\dot{M} c_p} \quad [\Delta°C]$$

$$\dot{Q} = 240 \quad [kW] \qquad \text{(given)}$$

$$c_p = 4.2 \quad [kJ/kg \cdot \Delta_1 K] \qquad \text{(Appendix B, Table B-11)}$$

$$\dot{M} = 8000 \ [kg/h] \rightarrow 2.22 \ [kg/s] \qquad \text{(given)}$$

$$\Delta T = \frac{240}{(2.22)(4.2)} \quad \frac{[kW]}{[kg/s][kJ/kg \cdot \Delta_1 K]}$$

$$= \frac{[kW][\Delta K]}{[kJ/s]} = [\Delta°C]$$

$$\Delta T = 25.7 \quad [\Delta°C]$$

$$T_{c_{out}} = T_{c_{in}} + \Delta T \quad [°C]$$

$$T_{c_{in}} = 20°C \qquad \text{(given)}$$

$$T_{c_{out}} = 20 + 25.7 = 45.7 \quad [°C] + [\Delta°C] = [°C]$$

$$\Delta T_b = T_h - T_{c_{out}}$$

$$T_h = 60°C \quad \text{(given)}$$

$$60 - 45.7 \quad [°C - °C] = [\Delta°C]$$

$$\Delta T_b = 14.3 [\Delta°C]$$

$$\ln(\Delta T_a/\Delta T_b)=\ln\frac{40}{14.3}=\ln 2.80=1.0296$$

$$\ln(\Delta T_a/\Delta T_b)=1.0296 \quad [-]$$

SUBSTITUTION FOR (a)

$$\Delta T_{lm}=\frac{40-14.3}{1.0296}\quad\frac{[\Delta°C-\Delta°C]}{[-]}=[\Delta°C]$$

$$=\frac{25.7}{1.0296}\quad[\Delta°C]$$

ANSWER FOR (a)

$$\Delta T_{ln}=24.9\quad[\Delta°C]$$

EQUATION FOR (b)

To obtain the surface area use the basic equation

$$\dot Q=AU\Delta T_{lm}\quad[W]$$

PARAMETERS FOR (b)

$$A=\frac{\dot Q}{U\Delta T_{lm}}\quad[m^2]$$

$$\dot Q=240,000\quad[W]\quad\text{(given)}$$

$$U=1000\quad[W/m^2\cdot\Delta_1°C]\quad\text{(given)}$$

$$\Delta T_{lm}=24.9\ \Delta°C\quad[\text{from part (a)}]$$

SUBSTITUTION FOR (b)

$$A=\frac{240,000}{1000\times24.9}\quad\frac{[W]}{[W/m^2\cdot\Delta_1°C][\Delta°C]}=[m^2]$$

ANSWER FOR (b)

$$A=9.64\quad[m^2]$$

HEAT EXCHANGERS TRANSFERRING SENSIBLE HEAT

Heat exchangers that are used to transfer sensible heat do not have one fluid at a constant temperature. Thus, the fluid temperatures vary with distance along the tube or heat exchanger element. The relative magnitude of the temperature change in each fluid is a function of the fluid weight flow and specific heat. Because the heat removed from one fluid must be transferred to the other fluid, the relationships of the ΔT of the two fluids flow are

$$\dot{Q} = \dot{M}_1 c_{p_1} \Delta T_1 = \dot{M}_2 c_{p_2} \Delta T_2 \quad [\text{W}] \tag{8-13}$$

$$\frac{\Delta T_1}{\Delta T_2} = \frac{\dot{M}_2 c_{p_2}}{\dot{M}_1 c_{p_1}} \quad [-] \tag{8-14}$$

where

$\dot{Q}$ = heat transfer rate $\quad [\text{W}]$

$\dot{M}$ = mass flow rate of fluid $\quad [\text{kg/s}]$

c_p = specific heat of fluid $\quad [\text{J/kg} \cdot \Delta_1 \text{K}]$

ΔT = temperature change in fluid during passage through

the heat exchanger $\quad [\Delta °\text{C}]$

Subscripts 1 and 2 refer to each of the two fluids.

For the same fluid inlet and outlet temperatures, the effective mean temperature difference (ΔT_m) will be significantly different for the configurations of

- parallel flow
- counter flow
- cross flow
- multipass flow

These configurations are discussed in the following sections. The choice of which geometric configuration to use for a given performance requirement will affect such aspects as size, weight, pressure drop, cost, and so forth.

• *Parallel Flow Configuration.* A parallel flow heat exchanger is one in which the two fluids flow in the same direction. In this configuration, the final temperature of the cold fluid does not reach the final temperature of the hot fluid as indicated in Figure 8-12.

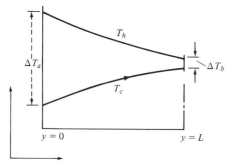

Figure 8-12 Fluid temperature profiles in a parallel flow heat exchanger.

The effective mean temperature difference between the two fluids is the log mean temperature difference (LMTD) calculated using Eq. (8-11). The effective mean temperature difference is characteristically relatively low. Hence this heat exchanger configuration is relatively inefficient.

Example Problem 8-7

Calculate the mean temperature difference in a double-pipe parallel flow heat exchanger where 500 kg/h of water enters at 80°C, flows through the outer annulus, and exists at 50°C. Inside the inner pipe water at 1000 kg/h enters at 20°C and exits at 35°C.

SOLUTION

SKETCH

Figure 8-13

EQUATION

$$\Delta T_{lm} = \frac{\Delta T_a - \Delta T_b}{\ln(\Delta T_a / \Delta T_b)} \quad [\Delta °C] \quad [Eq. (8-12)]$$

PARAMETERS

$$\Delta T_a = T_{h_{in}} - T_{c_{in}} \quad [\Delta °C]$$
$$T_{h_{in}} = 80°C \quad \text{(given)}$$
$$T_{c_{in}} = 20°C \quad \text{(given)}$$
$$\Delta T_a = 80 - 20 \quad [°C] - [°C] = [\Delta °C]$$
$$\Delta T_a = 60 \ \Delta °C$$
$$\Delta T_b = T_{h_{out}} - T_{c_{out}} \quad [\Delta °C]$$
$$T_{h_{out}} = 50°C \quad \text{(given)}$$
$$T_{c_{out}} = 35°C \quad \text{(given)}$$
$$\Delta T_b = 50 - 35 \quad [°C] - [°C] = [\Delta °C]$$
$$\Delta T_b = 15 \Delta °C$$

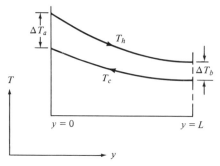

Figure 8-14 Fluid temperature profiles in a counterflow heat exchanger.

SUBSTITUTION

$$\Delta T_{lm} = \frac{(60-15)}{\ln(60/15)} \quad \frac{[\Delta°C - \Delta°C]}{[-]} = [\Delta°C]$$

$$= \frac{45}{\ln 4.00} = \frac{45}{1.3862}$$

ANSWER

$$\Delta T_{lm} = 32.5 \quad [\Delta°C]$$

• *Counterflow Configuration.* A counterflow heat exchanger is one in which the two fluids flow in opposite directions. In this configuration, the final temperature of the cold fluid may be higher than the final temperature of the hot fluid, as indicated in Figure 8-14.

The effective mean temperature difference between the two fluids is the log mean temperature difference (LMTD) calculated using Eq. (8-11). The effective mean temperature difference is characteristically relatively high. Where there is a choice between the counterflow and parallel flow configurations, the counterflow design is usually preferred for the following reasons:

1. The exchange of heat may raise the temperature of the cold fluid to more nearly the initial temperature of the hot fluid.
2. To transfer the same amount of heat, a smaller surface area is required.

Example Problem 8-8

(Same conditions as for the parallel flow sample problem.) Calculate the mean temperature difference in a double-pine counterflow heat exchanger where 500 kg/h of water enters at 80°C, flows through the outer annulus, and exits at 50°C. Inside the inner pipe 1000 kg/h of water enters at 20°C and exits at 35°C.

SOLUTION

SKETCH

Figure 8-15

EQUATION

$$\Delta T_{\text{lm}} = \frac{\Delta T_a - \Delta T_b}{\ln(\Delta T_a / \Delta T_b)} \quad [\Delta°C] \quad [\text{Eq. (8-12)}]$$

PARAMETERS

$$\Delta T_a = T_{h_{\text{in}}} - T_{c_{\text{out}}} \quad [\Delta°C]$$
$$T_{h_{\text{in}}} = 80°C \quad (\text{given})$$
$$T_{c_{\text{out}}} = 35°C \quad (\text{given})$$
$$\Delta T_a = 80 - 35 \quad [°C] - [°C] = [\Delta°C]$$
$$\Delta T_a = 45 \ \Delta°C$$
$$\Delta T_b = T_{h_{\text{out}}} - T_{c_{\text{in}}} \quad [\Delta°C]$$
$$T_{h_{\text{out}}} = 50°C \quad (\text{given})$$
$$T_{c_{\text{in}}} = 20°C \quad (\text{given})$$
$$\Delta T_b = 50 - 20 \quad [°C] - [°C] = [\Delta°C]$$
$$\Delta T_b = 30 \ \Delta°C$$

SUBSTITUTION

$$\Delta T_{\text{lm}} = \frac{45 - 30}{\ln(45/30)} \quad \frac{[\Delta°C] - [\Delta°C]}{[-]} = [\Delta°C]$$

$$= \frac{15}{\ln 1.500} = \frac{15}{0.4055}$$

ANSWER

$$\Delta T_{\text{lm}} = 37.0 \ \Delta°C$$

• *Cross Flow Configuration.* A cross flow heat exchanger is one in which the two fluids flow in paths essentially at right angles to each other as

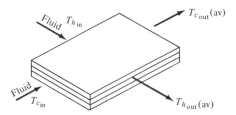

Figure 8-16   Cross flow heat exchanger configuration.

illustrated in Figure 8-16. In a cross flow heat exchanger, the fluids leaving the heat exchanger are not of a uniform temperature. The cold fluid filament adjacent to the hot fluid entrance will be heated to a higher temperature than the cold fluid filament adjacent to the hot fluid exit. A log mean temperature difference based analogously to counter flow requires a correction factor to obtain the effective mean temperature difference, as indicated by Eq. (8-15).

$$\Delta T_{lm} = F \frac{\left(T_{h_{in}} - T_{c_{out}}\right) - \left(T_{h_{out}} - T_{c_{in}}\right)}{\ln\left[\left(T_{h_{in}} - T_{c_{out}}\right) / \left(T_{h_{out}} - T_{c_{in}}\right)\right]} \quad [\Delta°C] \qquad (8\text{-}15)$$

where the temperatures are indicated in Figure 8-16. Thus, the heat transfer equation becomes

$$\dot{Q} = UAF\,\Delta T_{lm} \quad [W] \qquad (8\text{-}16)$$

where

$U =$ overall heat transfer coefficient $\left[W/m^2 \cdot \Delta_1°C\right]$

$A =$ heat transfer area $\left[m^2\right]$

$F =$ correction factor $[-]$

$\Delta T_{lm} =$ log mean temperature difference $\left[\Delta°C\right]$

Values of the correction factor (F) are usually obtained by using two parameters, (P) and (R). P is the ratio of the temperature drop in fluid No. 2 to the difference in the inlet temperatures of the two fluids. R is the ratio of the temperature drop in fluid No. 1 to the temperature drop in fluid No. 2. The correction factor plot for a single-pass cross flow heat exchanger with both fluids unmixed is given in Figure 8-17. The correction factor plot for a variation of a single-pass cross flow heat exchanger wherein one fluid is mixed throughout its path through the heat exchanger is shown in Figure 8-18.

· *Multiple Shell and Tube Passes.* The direction of flow for either or both fluids may change during its travel through the heat exchanger. Shell-and-

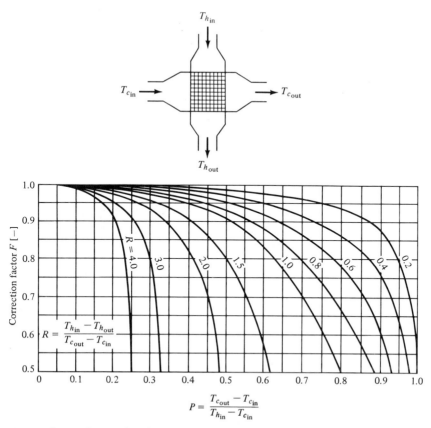

Figure 8-17 Correction-factor plot for single-pass cross flow heat exchanger with both fluids unmixed. (SOURCE: R. A. Bowman, A. E. Mueller, and W. M. Nagle, "Mean Temperature Difference in Design," *Trans. ASME* 62 (1940).)

tube heat exchangers are often provided with baffles in the heads to cause the fluid within the tubes to travel back and forth from one end to the other. In some cases, longitudinal baffles within the shell cause the fluid surrounding the tubes to travel the length of the shell a number of times. This configuration is often preferred to the strict counterflow design for the following reasons:

- lower cost of manufacture
- ease of disassembly for cleaning or repair
- reduced thermal stress due to expansion

Like the cross flow heat exchanger, the multipass designs must use a correction factor, F, which is a function of the specific configuration.

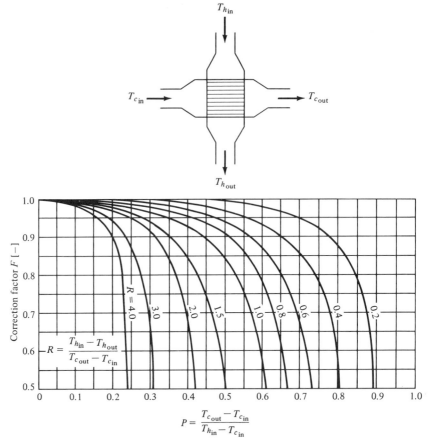

Figure 8-18   Correction-factor plot for single-pass cross flow exchanger, one fluid mixed, the other unmixed. (SOURCE: R. A. Bowman, A. E. Mueller, and W. M. Nagle, "Mean Temperature Difference in Design," *Trans. ASME* 62 (1940).)

Equation (8-16) must be used in conjunction with the charts shown in Figures 8-19 or 8-20 to determine the amount of heat transfer.

$$\dot{Q} = UAF\,\Delta T_m \quad [\text{W}] \quad [\text{Eq. (8-16)}]$$

8-2.4 Performance Analysis

In general, there are two approaches to the design and analysis of heat exchangers. The first is to establish a design for a particular set of requirements. The other is to determine what an existing heat exchanger with a certain performance for one set of conditions will do under another set of conditions. The mean effective temperature method, based upon Eq.

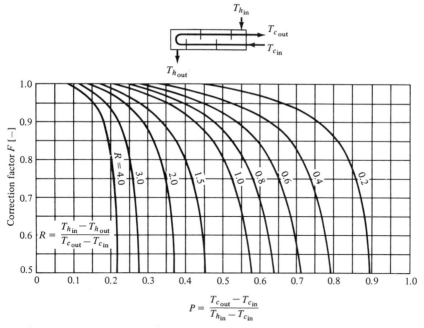

Figure 8-19 Correction-factor plot for exchanger with one shell pass and two, four, or any multiple of tube passes. (SOURCE: R. A. Bowman, A. E. Mueller, and W. M. Nagle, "Mean Temperature Difference in Design," *Trans. ASME* 62 (1940).)

(8-1) initially was the general method used for the design and analysis of heat exchangers. However, it was found to be a relatively cumbersome method to use for the analysis of a heat exchanger design, such as determining what would be the change in the amount of heat transferred by a heat exchanger if, for example, the mass flow rate of the cooling fluid were doubled, or if an inlet temperature were changed. To make this type of analysis of a heat exchanger easier, Kays and London[1] introduced what is known as the Effectiveness–NTU method. Both of these methods are discussed in the following sections, beginning with the mean effective temperature method.

THE MEAN EFFECTIVE TEMPERATURE METHOD

The mean effective temperature method is based upon the use of Eq. (8-1):

$$\dot{Q}_e = UA\,\Delta T_m \quad [\text{W}] \qquad [\text{Eq. (8-1)}]$$

[1]W. M. Kays, and A. L. London, *Compact Heat Exchangers.* 2nd ed., McGraw-Hill, New York, 1964.

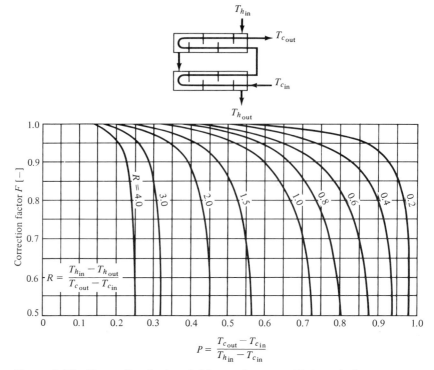

Figure 8-20   Correction-factor plot for exchanger with two shell passes and four, eight, or any multiple of tube passes. (SOURCE: R. A. Bowman, A. E. Mueller, and W. M. Nagle, "Mean Temperature Difference in Design," *Trans. ASME* 62 (1940).)

where

$\dot{Q}_e$ = the heat flow transferred through the heat exchanger structure   [W]

U = the overall heat transfer coefficient for the heat transfer path from one fluid, through the heat exchanger structure, and into the other fluid   [W/m²·Δ_1°C]

A = the direct heat transfer surface of the heat exchanger   [m²]

ΔT_m = the mean effective temperature difference between the two fluids in the heat exchanger   [Δ°C]

for parallel and counter flow—the log mean temperature difference ΔT_{lm} [Eq. (8-12)]

for cross flow and multipass—the log mean temperature difference for counter flow modified by a factor (F)

For determining design and performance parameters of a heat exchanger Eq. (8-1) can be used, and with the supplemental use of Eqs.

(8-4), (8-5), (8-6), and (8-12) all of the heat exchanger parameters can be established.

Example Problem 8-9

If a mass flow of hot oil of 1 kg/s were cooled from 150°C to 80°C by water entering at 25°C and leaving at 90°C in a counterflow heat exchanger, what would be the direct surface area (A) required if the value of U were 500 [W/m²·Δ_1°C]? The specific heat of the oil is 2 kJ/kg·Δ_1°C.

SOLUTION

SKETCH

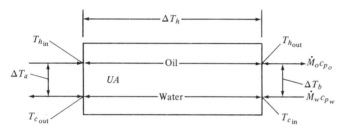

Figure 8-21

EQUATION

$$A = \frac{\dot{Q}_e}{U\Delta T_{lm}} \quad [\text{m}^2] \qquad [\text{Eq. (8-1)}]$$

PARAMETERS

$$\dot{Q}_e = \dot{Q}_h = \dot{M}_h c_{p_h} \Delta T_h \quad [\text{W}] \qquad [\text{Eqs. (8-4) and (8-6)}]$$

$$\dot{M}_h = 1 \text{ kg/s} \quad \text{(given)}$$

$$c_{p_h} = 2 \text{ kJ/kg·}\Delta_1\text{°C} \quad \text{(given)}$$

$$\Delta T_h = \left(T_{h_{in}} - T_{h_{out}}\right) \quad [\Delta\text{°C}]$$

$$T_{h_{in}} = 150\text{°C} \quad \text{(given)}$$

$$T_{h_{out}} = 80\text{°C} \quad \text{(given)}$$

$$\Delta T_h = 150 - 80 \quad [\text{°C}] - [\text{°C}] = [\Delta\text{°C}]$$

$$\Delta T_h = 70 \quad [\Delta\text{°C}]$$

$$\dot{Q}_e = (1)(2)(70) \quad [\text{kg/s}][\text{kJ/kg·}\Delta\text{°C}][\Delta\text{°C}] = [\text{kJ/s}] = [\text{kW}]$$

$$\dot{Q}_e = 140 \text{ kW} = 140{,}000 \text{ W}$$

$$U = 500 \text{ W/m}^2 \cdot \Delta_1 {}^\circ\text{C} \quad \text{(given)}$$

$$\Delta T_{\text{lm}} = \frac{\Delta T_a - \Delta T_b}{\ln(\Delta T_a / \Delta T_b)} \quad [\Delta^\circ\text{C}] \quad [\text{Eq. (8-12)}]$$

$$\Delta T_a = (T_{h_{\text{in}}} - T_{c_{\text{out}}}) \quad [\Delta^\circ\text{C}]$$

$$T_{h_{\text{in}}} = 150^\circ\text{C} \quad \text{(given)}$$

$$T_{c_{\text{out}}} = 90^\circ\text{C} \quad \text{(given)}$$

$$\Delta T_a = 150 - 90 \quad [^\circ\text{C}] - [^\circ\text{C}] = [\Delta^\circ\text{C}]$$

$$\Delta T_a = 60 \quad [\Delta^\circ\text{C}]$$

$$\Delta T_b = (T_{h_{\text{out}}} - T_{c_{\text{in}}}) \quad [\Delta^\circ\text{C}]$$

$$T_{h_{\text{out}}} = 80 \quad [^\circ\text{C}] \quad \text{(given)}$$

$$T_{c_{\text{in}}} = 25 \quad [^\circ\text{C}] \quad \text{(given)}$$

$$\Delta T_b = 80 - 25 \quad [^\circ\text{C}] - [^\circ\text{C}] = [\Delta^\circ\text{C}]$$

$$\Delta T_b = 55 \quad [\Delta^\circ\text{C}]$$

$$\Delta T_{\text{lm}} = \frac{60 - 55}{\ln(60/55)} = \frac{[\Delta^\circ\text{C}] - [\Delta^\circ\text{C}]}{[-]} = [\Delta^\circ\text{C}]$$

$$= \frac{5}{\ln 1.0909} = \frac{5}{0.0870}$$

$$\Delta T_{\text{lm}} = 57.47 \quad [\Delta^\circ\text{C}]$$

SUBSTITUTION

$$A = \frac{140{,}000}{(500)(57.47)} \quad \frac{[\text{W}]}{[\text{W/m}^2 \cdot \Delta_1 {}^\circ\text{C}][\Delta^\circ\text{C}]} = [\text{m}^2]$$

ANSWER

$$A = 4.87 \quad [\text{m}^2]$$

Example Problem 8-10

Using the heat exchanger in Example Problem 8-6 and keeping the inlet oil flow, inlet oil temperature, and the U value the same, to what temperature would the oil be cooled by doubling the cooling water mass flow rate with an inlet temperature of 25°C?

SOLUTION

The temperatures $T_{h_{out}}$ and $T_{c_{out}}$ can be related by the use of Eqs. (8-4), (8-5), (8-6), and (8-12).

$$\dot{Q}_h = \dot{Q}_c = \dot{Q}_e \quad [W] \qquad [\text{Eq. (8-6)}]$$

$$\dot{Q}_h = \dot{M}_h c_{p_h} (T_{h_{in}} - T_{h_{out}}) \quad [W]$$

$$\dot{Q}_c = \dot{M}_c c_{p_c} (T_{c_{out}} - T_{c_{in}}) \quad [W]$$

Equating $\dot{Q}_h = \dot{Q}_c$,

$$\dot{M}_h c_{p_h} (T_{h_{in}} - T_{h_{out}}) = \dot{M}_c c_{p_c} (T_{c_{out}} - T_{c_{in}})$$

Two equations can be obtained:

1. $\dot{Q}_h = \dot{Q}_c$
2. $\dot{Q}_h = \dot{Q}_e$

Using the first equation ($\dot{Q}_h = \dot{Q}_c$) the temperatures $T_{h_{out}}$ and $T_{c_{out}}$ are related by a "basic equation"

$$\dot{Q}_h = \dot{M}_h c_{p_h} (T_{h_{in}} - T_{h_{out}}) \quad [W] \qquad [\text{Eq. (8-4)}]$$

$$\dot{Q}_c = \dot{M}_c c_{p_c} (T_{c_{out}} - T_{c_{in}}) \quad [W] \qquad [\text{Eq. (8-5)}]$$

BASIC EQUATION

$$\dot{M}_h c_{p_h} (T_{h_{in}} - T_{h_{out}}) = \dot{M}_c c_{p_c} (T_{c_{out}} - T_{c_{in}}) \quad [W]$$

PARAMETER EVALUATION

$$\dot{M}_h = 1 \text{ kg/s} \quad \text{(given)}$$

$$c_{p_h} = 2000 \text{ J/kg} \cdot \Delta_1{}^\circ\text{C} \quad \text{(given)}$$

$$T_{h_{in}} = 150^\circ\text{C} \quad \text{(given)}$$

$$\dot{M}_c = 2 \text{ times } \dot{M}_c \text{ for Example Problem 8-9}$$

Using Eq. (8-6)

$$\dot{Q}_e = \dot{Q}_c \quad [W]$$

$$\dot{Q}_e = \dot{M}_c c_{p_c} (T_{c_{out}} - T_{c_{in}}) \quad [W]$$

Therefore,

$$\dot{M}_c = c_{p_c} \frac{\dot{Q}_e}{(T_{c_{out}} - T_{c_{in}})} \quad [\text{kg/s}]$$

$$\dot{M}_c = 2 \left[\frac{\dot{Q}_e}{c_{p_c} (T_{c_{out}} - T_{c_{in}})} \right] \quad [\text{kg/s}]$$

$$\dot{Q}_e = 140,000 \quad [\text{W}] \qquad \text{(Example Problem 8-9)}$$

$$c_{p_c} = 4200 \text{ J/kg} \cdot \Delta_1°\text{C}$$

$$T_{c_{\text{out}}} = 90°\text{C} \qquad \text{(given)}$$

$$T_{c_{\text{in}}} = 25°\text{C} \qquad \text{(given)}$$

$$\dot{M}_c = 2\left[\frac{140,000}{4200(90-25)}\right] \quad \frac{[\text{W}]}{[\text{J/kg} \cdot \Delta_1°\text{C}][°\text{C}]} = [\text{kg/s}]$$

$$= 1.026 \text{ kg/s}$$

$$c_{p_c} = 4200 \text{ J/kg} \cdot \Delta_1°\text{C} \qquad \text{(Appendix B, Table B-4)}$$

$$T_{c_{\text{in}}} = 25°\text{C} \qquad \text{(given)}$$

Substituting in the "basic equation,"

$$(1)(2000)\left(150 - T_{h_{\text{out}}}\right) = (1.026)(4200)\left(T_{c_{\text{out}}} - 25\right)$$

$$T_{h_{\text{out}}} + 2.155 \, T_{c_{\text{out}}} = 203.9 \tag{1}$$

Using the second equation ($\dot{Q}_g = \dot{Q}_e$)

$$\dot{M}_h c_{p_h}\left(T_{h_{\text{in}}} - T_{h_{\text{out}}}\right) = UA \, \Delta T_{\text{lm}} \quad [\text{W}]$$

$$\dot{M}_h c_{p_h}\left(T_{h_{\text{in}}} - T_{h_{\text{out}}}\right) = UA\left[\frac{\left(T_{h_{\text{in}}} - T_{c_{\text{out}}}\right) - \left(T_{h_{\text{out}}} - T_{h_{\text{in}}}\right)}{\ln\left(\dfrac{150 - T_{c_{\text{out}}}}{T_{h_{\text{out}}} - 25}\right)}\right]$$

Substituting values listed above

$$(1)(2000)\left(150 - T_{h_{\text{out}}}\right) = (500)(4.87)\left[\frac{\left(150 - T_{c_{\text{out}}}\right) - \left(T_{h_{\text{out}}} - 25\right)}{\ln\left(\dfrac{150 - T_{c_{\text{out}}}}{T_{h_{\text{out}}} - 25}\right)}\right]$$

$$T_{h_{\text{out}}} + 1.218\left[\frac{\left(150 - T_{c_{\text{out}}}\right) - \left(T_{h_{\text{out}}} - 25\right)}{\ln\left(\dfrac{150 - T_{c_{\text{out}}}}{T_{h_{\text{out}}} - 25}\right)}\right] = 150 \tag{2}$$

Equations (1) and (2) are two equations with the two unknowns $T_{h_{\text{out}}}$ and $T_{c_{\text{out}}}$, which satisfy the mathematical requirements for solution. However, Eqs. (1) and (2) do not lend themselves readily to a direct mathematical solution. Consequently, the practical method for solution is to solve by iteration, as illustrated below:

FIRST ITERATION—ASSUME $T_{h_{out}} = 70°C$

Solve Eq. (1) for $T_{c_{out}}$ and substitute $T_{h_{out}}$ and $T_{c_{out}}$ in Eq. (2) to see if it is valid.

$$T_{c_{out}} = 94.62 - 0.464T_{h_{out}}$$

$$T_{c_{out}} = 94.62 - (0.464)(70) = 94.62 - 32.48$$

$$T_{c_{out}} = 62.14°C$$

Substituting $T_{h_{out}}$ and $T_{c_{out}}$ in Eq. (2).

$$70 + 1.218 \left[\frac{(150 - 62.14) - (70 - 25)}{\ln\left(\frac{150 - 62.14}{70 - 25}\right)} \right]$$

$$70 + 1.218 \left[\frac{87.86 - 45}{\ln 1.952} \right]$$

$$70 + 1.218 \left[\frac{42.86}{0.669} \right] = 70 + 78.03 = 148.03 \neq 150$$

SECOND ITERATION—ASSUME $T_{h_{out}} = 72°C$

$$T_{c_{out}} = 94.62 - (0.464)(72) = 94.62 - 33.41 \tag{1}$$

$$T_{c_{out}} = 61.21°C$$

$$72 + 1.218 \left[\frac{(150 - 61.21) - (72 - 25)}{\ln \frac{(150 - 61.21)}{(72 - 25)}} \right] \tag{2}$$

$$72 + 1.218 \frac{88.79 - 47}{\ln 1.889} = 72 + 1.218 \left(\frac{41.79}{0.636} \right)$$

$$72 + 80.03 = 152.03 \neq 150$$

THIRD ITERATION—ASSUME $T_{c_{out}}$ HALFWAY BETWEEN 70°C AND 72°C

ANSWER

$$T_{h_{out}} = 71°C$$

THE EFFECTIVENESS–NTU METHOD

The Effectiveness–NTU method (NTU method) offers many advantages for the evaluation of different types of heat exchangers. While the log mean temperature difference (LMTD) method of heat exchanger analysis is useful when all of the inlet and outlet fluid temperatures are known, the use of the LMTD method to predict the performance of a given heat exchanger design for other conditions involves an iterative procedure (due to the logarithmic function) as illustrated by Example Problem 8-10.

The Effectiveness–NTU method involves the following parameters:

- the heat exchanger effectiveness (ε)
- the fluid flow capacity rates (C_{min}, C_{max})
- the parameter (UA/C_{min}) called NTU, meaning Number of Transfer Units. This parameter is related to the size of the heat exchanger.

The effectiveness of a heat exchanger is defined as

$$\text{effectiveness } (\varepsilon) = \frac{\text{actual heat transfer}}{\text{maximum possible heat transfer}}$$

The *actual heat transfer* ($\dot{Q}_a$) for the parallel flow configuration shown in Figure 8-22 is

$$\dot{Q}_a = \dot{M}_h c_{p_h}\left(T_{h_{in}} - T_{h_{out}}\right) \qquad \left[\text{Eqs. (8-4) and (8-5)}\right]$$

$$= \dot{M}_c c_{p_c}\left(T_{c_{out}} - T_{c_{in}}\right)$$

The *maximum possible heat transfer* ($\dot{Q}_{mp}$) is the concept of the heat transferred if the fluid were heated or cooled (as the case may be) from its inlet temperature to the inlet temperature of the other fluid. Hence

$$\dot{Q}_{mp_h} = \dot{M}_h c_{p_h}\left(T_{h_{in}} - T_{c_{in}}\right) \tag{8-17}$$

$$\dot{Q}_{mp_c} = \dot{M}_c c_{p_c}\left(T_{c_{in}} - T_{h_{in}}\right) \tag{8-18}$$

The maximum possible heat transfer ($\dot{Q}_{mp}$) is different for the two fluid flows in the heat exchanger when the product of the mass flow rate (M)

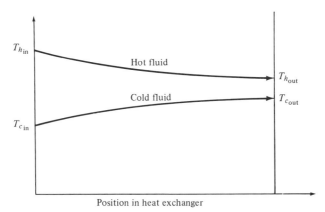

Figure 8-22 Fluid temperature profiles in a parallel flow heat exchanger.

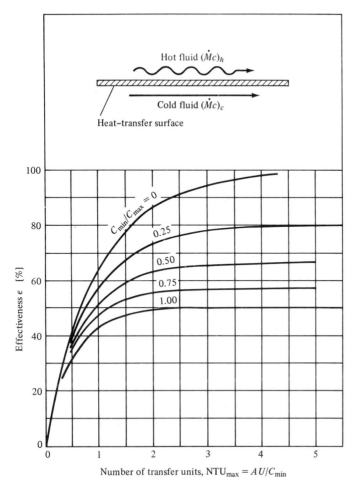

Figure 8-23 Effectiveness for parallel flow exchanger performance. (SOURCE: W. M. Kays and A. L. London, *Compact Heat Exchangers*, 2nd ed., McGraw-Hill, New York, 1964.)

and the specific heat (c_p) of the two flows are different. In this case the effectiveness (ε) of one fluid will be different than the effectiveness of the other fluid flow.

In the Effectiveness–NTU method, the *effectiveness of the fluid stream having the smaller heat capacity* $(\dot{M}c_p)$ *is used*. The effectiveness is calculated using the larger of the temperature changes of the two fluids passing through the heat exchanger (ΔT_{f_l}). The temperature changes for the fluids are:

for the hot fluid $\quad \Delta T_f = T_{h_{in}} - T_{h_{out}} \quad [\Delta°C] \quad\quad$ (8-19)

for the cold fluid $\quad \Delta T_f = T_{c_{out}} - T_{c_{in}} \quad [\Delta°C] \quad\quad$ (8-20)

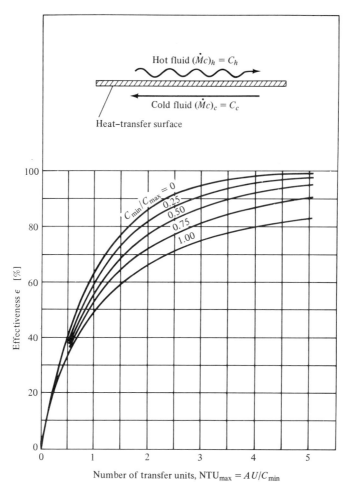

Figure 8-24 Effectiveness for counterflow exchanger performance.
(SOURCE: W. M. Kays and A. L. London, *Compact Heat Exchangers*, 2nd ed., McGraw-Hill, New York, 1964.)

The effectiveness equation is

$$\varepsilon_c = \frac{\Delta T_{f_l}}{\left(T_{h_{in}} - T_{c_{in}}\right)} \quad [-] \tag{8-21}$$

where the subscript h refers to the hot fluid flow, subscript c refers to the cold fluid flow and ΔT_{f_l} is the larger value from Eqs. (8-19) and (8-20).

The effectiveness is then simply the ratio of the temperature change of the fluid with the smaller heat capacity to the maximum temperature difference in the heat exchanger.

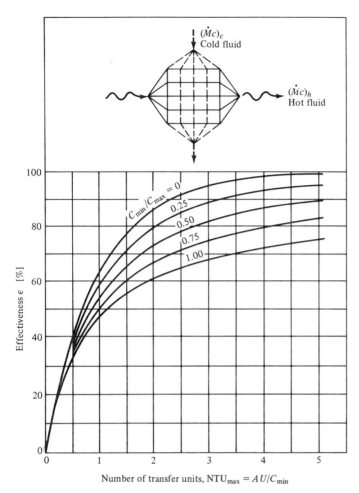

Figure 8-25 Effectiveness for cross flow exchanger with fluids unmixed. (SOURCE: W. M. Kays and A. L. London, *Compact Heat Exchangers*, 2nd ed., McGraw-Hill, New York, 1964.)

The heat capacity rate (C) of the fluid flow is defined as

$$C = \dot{M} c_p \quad [\mathrm{W}/\Delta_1{}^\circ\mathrm{C}] \tag{8-22}$$

where

$\dot{M}$ = mass flow rate $\quad [\mathrm{kg/s}]$

c_p = specific heat $\quad [\mathrm{J/kg} \cdot \Delta_1{}^\circ\mathrm{C}]$

Each of the two fluids has its own heat capacity rate:

$$C_h = \dot{M}_h c_{p_h}, \qquad C_c = \dot{M}_c c_{p_c}$$

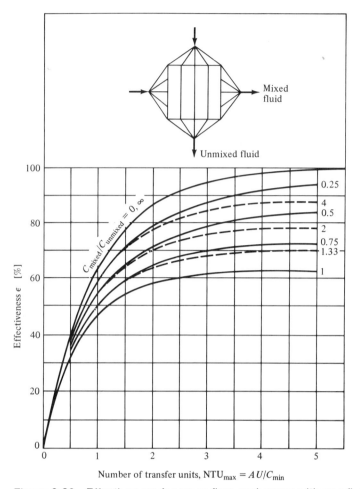

Figure 8-26 Effectiveness for cross flow exchanger with one fluid mixed. (SOURCE: W. M. Kays and A. L. London, *Compact Heat Exchangers*, 2nd ed., McGraw-Hill, New York, 1964.)

where the subscripts c and h refer to the cold and hot fluid streams, respectively. The higher value of C, regardless of whether it is for the hot or cold fluid stream becomes C_{max}, and the lower value becomes C_{min}. The Effectiveness–NTU method makes use of both

- the ratio C_{min}/C_{max}
- C_{min}

Kays and London[1] have published graphs of effectiveness versus the NTU parameter (AU/C_{min}) for various heat exchanger configurations. Graphs for some of the more common heat exchanger configurations are shown in Figures 8-23 through 8-28.

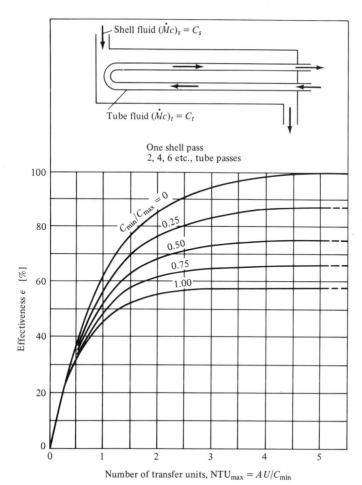

Figure 8-27   Effectiveness for 1–2 parallel counterflow exchanger performance. (SOURCE: W. M. Kays and A. L. London, *Compact Heat Exchangers*, 2nd ed., McGraw-Hill, New York, 1964.)

These curves give a relationship between:

- effectiveness
- C_{min}/C_{max}
- AU/C_{min}

With any two of these three parameters known, the third can be determined from these graphs. The three parameters serve to completely define the heat exchanger and its heat transfer performance.

8-3 HEAT PIPES

A heat pipe is a device for providing a high thermal conductance path for heat transfer. A heat pipe is in the form of a sealed cavity wherein heat is

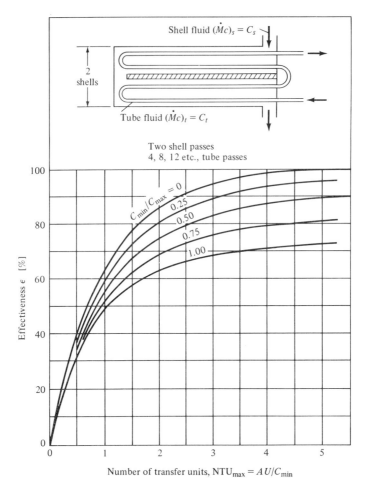

Figure 8-28 Effectiveness for 2–4 multipass counterflow exchanger performance. (SOURCE: W. M. Kays and A. L. London, *Compact Heat Exchangers*, 2nd ed., McGraw-Hill, New York, 1964.)

transported by the flow of vapor from an evaporating region to a condensing region. A significant distinction between a heat pipe and a natural circulation loop or reflux condenser is the utilization of capillary wicking in the heat pipe to return the condensate to the evaporation region in contrast to the return by gravity in natural circulation loops or reflux condensers. Consequently, heat pipes do not require a gravitational field for operation. The elements of a heat pipe are illustrated in the diagram shown in Figure 8-29.

This diagram shows a sealed cavity with an evaporating region and a condensing region. The vapor flows from the evaporating region to the condensing region. A capillary wick, with an entrance capillary radius r_c at

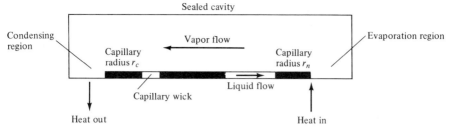

Figure 8-29 Diagram of a heat pipe.

the condensing region and an exit capillary radius r_h at the evaporating region conveys the condensed liquid back to the evaporating region. Because the heat pipe has a minimum of structure and the heat is transported by a flow of vapor, the heat pipe has a low weight associated with its high heat transport capability (high equivalent thermal conductance).

Interest in heat pipes has been increasing largely due to their potential attractiveness in space power applications, such as thermionic converters. In these applications the temperature level is relatively high (such as 1000 K) resulting in the use of liquid metals as the heat transfer fluid in the heat pipe.

The most significant properties of the heat transfer fluid with respect to heat pipe performance include:

- vapor pressure level
- slope of the vapor pressure curve with respect to temperature
- latent heat of vaporization
- surface tension
- cavity surface wetting ability
- density of vapor and liquid
- viscosity of vapor and liquid
- thermal conductivity of vapor and liquid

The maximum heat transport capability of a heat pipe is limited by

- the available capillary pumping
- the maximum heat fluxes that can be tolerated in the evaporating and condensing surfaces

The available capillary pumping must equal or exceed the total pressure drop in the fluid circulating system. The onset of boiling within the liquid in the capillary wick can disrupt the wicking action and the circulation of the liquid.

Other performance limits of heat pipes are related to life, reliability, and start-up transients. Liquid metals can dissolve the wick or cavity wall materials, resulting in depositions on the evaporating surface. Such deposits can eventually clog the wick pores and impair the heat transport capability of the heat pipe.

Heat pipe design is a relative complex combination of elements of heat transfer, geometry and mechanical design, and material selections. A discussion of such design factors is beyond the scope of this text.

PROBLEMS

8-1. For calculating the heat transfer into (cooling) or out of (heating) a house, the various heat transfer paths are characterized by the following surface (heat transfer path) areas and associated overall heat transfer coefficients (U):

SURFACE	AREA (A) [m^2]	OVERALL HEAT TRANSFER COEFFICIENT (U) W/m$^2 \cdot$°C
outside wall	150	1.0
windows	30	6.0
outside doors	5	3.0
floor	200	2.8
ceiling	200	3.5

Calculate the heat flow from the air inside the house to the outside air when the temperature difference is 30Δ°C.

8-2. In Problem 8-1, what area is the major loss of heat?

8-3. Characterize a house or building of your interest and calculate the heating load for a design temperature difference you select.

8-4. Using the overall heat transfer coefficient (U) from Appendix B, Table B-10 and the outside surface conductance (h) from Appendix B, Table B-11, calculate the overall heat transfer coefficient (U) for:
(a) a double paned window when the wind velocity is (1) 48 km/h; (2) 0 km/h.
(b) a frame wall, clapboard, 2.5-cm (1-in.) wood sheathing, lath and plaster, and 9.2-cm ($3\frac{5}{8}$-in.) rock wool fill between studying when the wind velocity is: (1) 48 km/h; (2) 0 km/h.

8-5. What would be the reduction in the heat load in Problem 8-1 if 15 cm of glass wool insulation (1.5 lb/ft^3) were placed over the ceiling area on top of the existing ceiling?

8-6. What would be the reduction in the heat load of Problem 8-1 if the windows were double paned?

8-7. A house has a double glass door (total size 3×2.5 m) facing south. The glass has a total reflection loss from both surfaces of 5% and an absorption of 10% for the incident solar radiations. At local noon when the zenith angle of the sun is 45°, how much solar energy [W] enters the house through the glass? Consider only the direct beam.

8-8. On a still clear night when the room temperature is 25°C and the outside air temperature is 0°C, what would be the heat loss through the glass doors (single pane) in Problem 8-7? Consider both convection and radiation.

8-9. A south facing wall of a house 3 m high and 8 m long is adjacent to a cement area extending 5 m out from the house for the full length of the wall. The concrete has a reflectivity of 15%, and at noon reaches a surface temperature of 45°C. The construction of the house wall is frame with 2.5 cm (1 in.) wood sheathing, lath and plaster painted with brown paint. If the outside air temperature is 30°C with a wind velocity of 25 km/h, and the room temperature is maintained at 25°C, how much heat [W] would enter the house at local noon when the sun's zenith angle is 30°?

8-10. How much would the heat load from the wall in Problem 8-9 be reduced if the wall were insulated with rock wool?

8-11. How much would the heat load from the wall in Problem 8-9 be reduced if the wall were painted white?

8-12. How much would the heat load from the wall in Problem 8-10 be reduced if the insulated wall were painted white?

8-13. Heat is transferred from hot water to an oil in a counterflow single-pass concentric pipe heat exchanger. The water enters the annulus at 90°C and exits at 40°C, while the oil enters the inner pipe at 25°C and exits at 45°C. Calculate the log mean temperature difference.

8-14. Determine the amount of heat being transferred by the heat exchanger in Problem 8-13 if the water flow is 600 kg/h.

8-15. A single-pass concentric pipe heat exchanger is 3 m long with a 5-cm diameter inner pipe. The overall heat transfer coefficient (U) is 300 [W/m$^2 \cdot \Delta_1$°C]. What flow rate of hot oil at 100°C could be cooled to 40°C (oil $c_p = 2093$ [J/kg·Δ_1°C]) with a 1000 kg/h water supply at a temperature of 20°C with parallel flow?

8-16. What flow rate of the hot oil could be cooled if the heat exchanger in Problem 8-15 were being used with counter flow?

8-17. If the outlet temperature of the oil in Problem 8-15 could be increased to 50°C, what increased oil flow rate could be cooled?

8-18. If the outlet temperature of the oil in Problem 8-16 could be increased to 50°C, what increased oil flow rate could be cooled?

8-19. What is the effectiveness of the heat exchanger in Problem 8-13?

8-20. What is the effectiveness of the heat exchanger in Problem 8-15?

8-21. What is the NTU value for the heat exchanger in Problem 8-15?

8-22. Using the Effectiveness–NTU method, (a) what would be the outlet temperature of the oil, and (b) what oil flow rate could be cooled to the 40°C with the same coolant flow if the heat exchanger in Problem 8-15 were being used in counter flow?

8-23. If the cooling water flow rate through the heat exchanger in Problem 8-15 were doubled with counter flow, what flow rate of hot oil could be cooled?

8-24. If the oil outlet temperature were increased to $50°C$ in Problem 8-15 (with parallel flow) what flow rate of hot oil could be cooled?

8-25. A cross flow heat exchanger (with unmixed fluid streams) is cooling oil. If the oil enters at $95°C$ and leaves at $45°C$, and the cooling air enters at $25°C$ and leaves at $45°C$, what is the log mean temperature difference?

8-26. What would be the NTU value for this heat exchanger?

8-27. If the oil flow rate were 10 kg/min, what AU product value would be required for the cross flow heat exchanger?

8-28. A tube in shell heat exchanger (1–2 parallel counter flow) is cooling oil. If the oil flow enters at $95°C$ and leaves at $45°C$ and the cooling water enters at $25°C$ and leaves at $45°C$. The oil is flowing through the tubes. What would be the NTU value for this heat exchanger?

8-29. If the oil flow rate were 10 kg/min what AU product value would be required for the tube-in-shell heat exchanger of Problem 8-28?

8-30. If the cooling water flow rate were doubled in Problem 8-25, with the same cooling water inlet temperature and same oil temperatures, what increase in the hot oil flow rate would be obtained?

Chapter 9
Application of Heat Transfer Principles to Heat Transfer Equipment

The purpose of this chapter is to illustrate the application of the heat transfer principles presented in this text to the design of heat transfer equipment. The modern boiler was selected as the vehicle for this discussion because, in addition to being one of the major types of heat transfer equipment, the design of a boiler includes a wide range of heat transfer principles.

9-1 AN OVERVIEW OF BOILERS

Basically a boiler transfers the heat generated by combustion of a fuel-air mixture into water to produce hot water or steam. The combustion is accomplished in a furnace, whose volume is designed to obtain complete combustion. Heat is transferred in the furnace section primarily by radiation to water cooled walls, with some additional convective heat transfer as the combustion gases pass over the surface of the furnace walls. The hot gases after leaving the furnace are then subjected to convective heat transfer to water cooled surfaces of the boiler convective section. When the temperature of the gases leaving the boiler convective section are suffi-

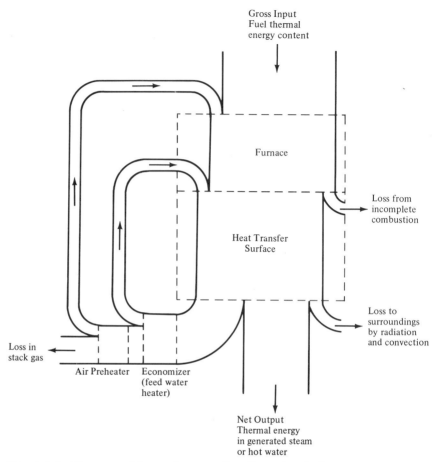

Figure 9-1   Diagram of thermal energy flow through a boiler.

ciently high, the heat loss in these stack gases can be reduced by the use of heat exchangers to heat the boiler feed water (economizers) and/or the air for combustion (air heaters).

The thermal energy flow through a boiler is illustrated by the diagram shown in Figure 9-1. This shows the input fuel thermal energy is distributed between three types of losses and the output thermal energy in the steam of hot water.

The overall boiler efficiency is usually called the "fuel-to-steam" efficiency to distinguish it from other boiler efficiency terms. The fuel-to-steam efficiency is the ratio of the thermal energy in the generated steam (or hot water) to the thermal energy in the input fuel. This overall efficiency is affected by three major categories of thermal losses, which are

- incomplete combustion
- heat loss in the stack gas
- heat loss to surroundings

The combustion efficiency is a measure of the performance of the fuel burner and the furnace volume design and relates to the ability of the two to burn the fuel completely.

The heat loss in the stack gas is related to the temperature and mass flow of the gases leaving the boiler and entering the stack; the higher the temperature, the greater the loss; the greater the mass flow rate, the greater the loss.

There is a minimum required temperature for the stack gas, which relates to the dew point of the stack gas. If condensation occurs on the surface of the exhaust stack, corrosion is encountered caused by the action of chemicals formed by the combination of some flue gas constituents with the condensed water. As boiler pressures increase, the saturation temperature also increases, resulting in higher flue gas temperatures leaving the boiler. The heat in such flue gases above the minimum flue gas temperature requirements can be recovered by transfer into the low temperature boiler feed water and/or into the air for fuel combustion. The feed water heater is usually called an economizer, and the air heater is called an air heater.

The mass flow rate of the flue gas for a particular fuel combustion rate is a function of the air fuel mixture. When the oxygen in the air exactly meets the requirements of the fuel combustion chemical reaction, the mixture ratio is termed "stoichiometric." When there is excess air the combustion mixture ratio is often termed "lean," and when there is insufficient air for complete combustion the mixture ratio is often termed "rich."

The heat loss from the boiler to its surroundings is by radiation and convection into the surrounding air from the exposed external surfaces of the boiler. Usually boilers are well insulated and this heat loss has a minor effect upon the overall boiler efficiency (1–2%).

A properly designed and operated modern boiler can given overall fuel-to-steam efficiencies in the range of 80–90%. The design of the boiler includes both combustion, mechanical/structural, and heat transfer considerations.

Through the years of boiler development, codes have been established to guide the design of boilers to assure the construction of safe and reliable boilers. Compliance with code requirements is required by government safety departments and insurance underwriters. The boiler codes give design ranges for mechanical design parameters, heat transfer flux rates, furnace volumetric firing (heat release) rates, and so forth, associated with various boiler pressures and temperatures. The primary boiler codes include the following.

• *ASME (American Society of Mechanical Engineers) Code.* This boiler code was developed for general boiler design, including power plant boilers, industrial boilers, and fired pressure vessels of all kinds. It has become a widely accepted code, and is often referred to or incorporated into other codes as applicable.

• *Code of Federal Regulations, Marine Engineering Regulations Title 46 CFR 50-64.* This boiler code applies to marine boilers. Because the reliability of boilers, including environments of severe rolling and pitching, is an important facet of ship safety, the Marine Engineering Regulations include considerations outside the scope of normal stationary boiler design. Sections of the ASME boiler code are cited and made a part of this marine boiler code.

• *NAV-SHIP Technical Manuals.* The naval requirements for boiler design are set forth in detail in a series of Naval Technical Manuals, which guide the design, construction, and inspection of naval boilers. Many of the details covered by the ASME code and the Federal Marine Engineering Regulations are adapted as appropriate into the naval requirements.

The size of a boiler is often described in units of a "boiler horsepower." This came from the early development of the steam engine and its associated boiler. For example, a 10 horsepower steam engine required a 10 horsepower boiler. As the steam engine became more efficient, a 10 horsepower boiler could supply the steam requirements of a steam engine with a larger power output. The boiler horsepower unit became defined as the heat transfer rate that would evaporate 34.5 lb/hr of water saturated at 212°F, converting it into saturated vapor at 212°F. This is equal to 33,472 Btu/hr or 9.81 kW. When the heating of subcooled feed water and/or adding pressure or superheat to the steam is involved, the steam evaporation rate is reduced commensurate with maintaining the heat transfer rate of 33,472 Btu/hr (per each boiler horsepower).

The application of boilers has developed into two general categories of current use. One is for industrial applications for producing steam or hot water. The other is for marine propulsion and central station electric power generation. The industrial boilers generally produce saturated steam or hot water with steam flow rates up to 12,250 kg/h (27,000 lb/hr) at pressures up to 2.0 MPa (300 psi). A packaged industrial boiler is shown in Figure 9-2.

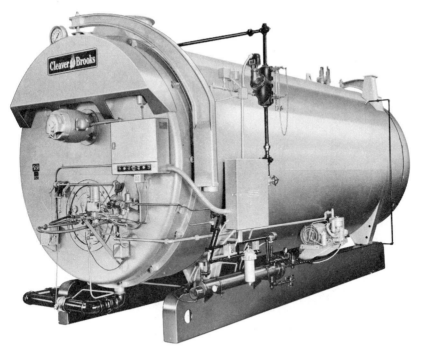

Figure 9-2 A modern packaged fire tube industrial boiler rated at 150 horsepower for oil or gas firing. (Courtesy of Cleaver Brooks, Division of Aqua-Chem, Inc.)

Marine propulsion boilers are much larger, typically operating around 6 MPa (900 psi) producing superheated steam around 550°C (1000°F). Because of their size and operating pressure modern marine boilers are water tube. These boilers typically include a steam superheater section as well as an economizer. A sectional view of a modern marine boiler is shown in Figure 9-3.

Central station electric power generation boilers operate at still higher pressures and temperatures with steam flow rates of an order of magnitude of 1,000,000 to 2,000,000 kg/h (2 to 4 million lb/hr). This category of a boiler typically includes steam superheating and expansion stage reheaters, economizers for feed water heating, and combustion air preheaters. Some supercritical boilers have pressures as high as 30 MPa (4500 psi) with steam temperatures of 650°C (1175°F). A sectional view of a central power station boiler is shown in Figure 9-4.

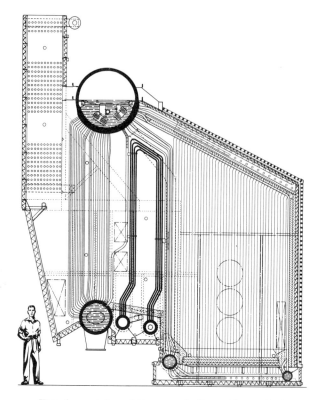

Two-drum Integral-Furnace boiler with vertical super-heater.

Figure 9-3 Sectional view of a modern marine boiler. (Courtesy of Babcock & Wilcox.)

9-2 A CHRONOLOGICAL SKETCH OF BOILER DEVELOPMENT

9-2.1 Stationary Boilers

The initial need for boilers was to produce steam for pumping water out of mines. At first atmospheric pressure steam was used, and the pumping action was obtained from the vacuum created by the condensation of the steam in an appropriate chamber. The early boilers were simple vessels positioned over a fire as illustrated in Figure 9-5.

As the Industrial Revolution progressed, the demand for steam increased, both as to the amount (steam flow rates) and for above atmospheric pressures, along with increased fuel-to-steam efficiency. The sta-

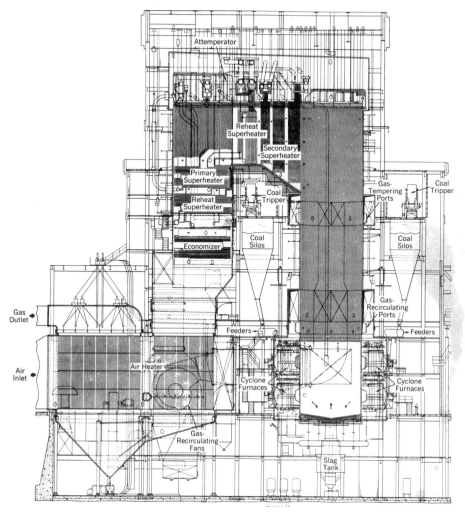

Figure 9-4   Sectional view of a central power station boiler. (Courtesy of Babcock & Wilcox.)

tionary boilers took the general form of some type of a pressure vessel set upon or surrounded by brickwork, with the fire underneath. One step upon the development path was the introduction of a large diameter furnace tube passing the length of a horizontal cylindrical shell. The combustion gases leaving the furnace tube were ducted along the outside of the boiler shell in one or more passes by the use of brick work. An example of this configuration, known as the Cornish boiler, is illustrated in

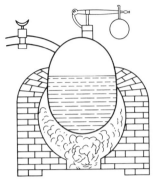

Figure 9-5 Papin's boiler, 1705. (SOURCE: *The Encyclopedia Britannica*, 11th ed., University Press, Cambridge, England, 1911.)

Figure 9-6(a). A modification known as the Lancaster boiler, which has two furnace tubes instead of one, is illustrated in Figure 9-6(b).

Further need for increased steam flow rates and higher pressures led to the development of water tube boiler configurations, as illustrated in Figure 9-7.

As the electrical networks expanded, the use of small electrical power plants diminished, and the need for large boilers for central electric generating plants increased. This has led to the construction of boilers, such as shown in Figure 9-8.

9-2.2 Propulsion System Boilers

With the application of the steam engine to locomotives and to steam ships, additional requirements were imposed upon the boiler design and

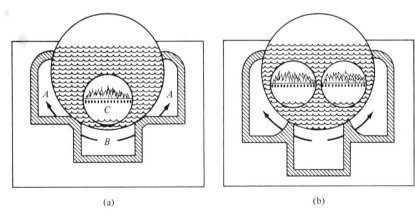

| (a) | (b) |

Figure 9-6 Early British boilers. (a) The Cornish boiler. (b) The Lancaster boiler. (SOURCE: *A Textbook on Mechanical Engineering*, International Textbook Co., Scranton, Pa., 1900.)

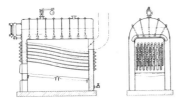

Inclined water tubes connecting front and rear water spaces complete circuit with steam space above. Stephen Wilcox, 1856.

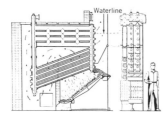

First Babcock & Wilcox boiler, patented in 1867.

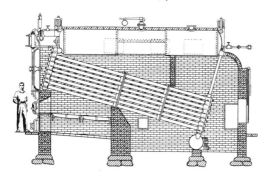

Babcock & Wilcox boiler developed in 1877.

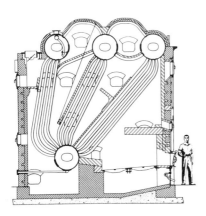

Early Stirling boiler arranged for hand firing.

Figure 9-7   Development of water tube boilers by Babcock & Wilcox. (Courtesy of Babcock & Wilcox.)

geometrical configurations. The boilers needed to be self-contained without external brickwork. Reduced weight became important; dimensional size limitations were imposed; and increased steam output for the boiler size was required.

RAILROAD LOCOMOTIVE BOILERS

The railroad locomotive boiler had numerous design restrictions, dictated by the distance between the rails, the size of the tunnels, bridges, and so forth, as well as requirements for a large steam flow rate from a relatively small boiler volume in order to obtain the needed power output of the locomotive. Initial attempts to build steam locomotives resulted in unacceptable low power outputs due to insufficient heat transfer rates into the boiler. The first practical locomotive (with sufficiently high heat

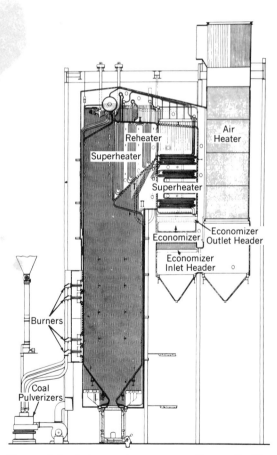

Figure 9-8 A large steam generating unit. (Courtesy of Babcock & Wilcox.)

transfer rate) was George Stevenson's "Rocket," which is illustrated in Figure 9-9. Its success was primarily due to the boiler design. In order to obtain a high heat transfer rate within the geometrical restrictions, the boiler was basically a cylindrical shell. A relatively large heat transfer direct surface area (A) was obtained by the use of a multiplicity of small tubes running the length of the boiler shell. A relatively high overall heat transfer coefficient (U) was obtained both by the relatively small diameter of the tubes and a forced flow (convection) of the combustion gases through the tubes pumped by the exhausting steam being discharged through jets into the smoke stack. A relatively high mean temperature difference (ΔT_m) was obtained by effecting the combustion in an external brick lined firebox so only the hot combustion gases were drawn into the boiler fire tubes.

To meet the requirements for increased locomotive power outputs, the size of the boilers had to be increased. Because little could be done to increase the overall heat transfer coefficient (U) between the flue gas and the water, or the mean temperature difference, the desired increase in heat transfer rate could primarily be obtained by increasing the heat transfer surface (see Section 8-3). The locomotive boiler construction became complex, taxing both the skill of the builder and the ingenuity of the designer to handle the stresses and thermal expansions encountered in the design. The firebox was enclosed by a water cooled shell, the boiler

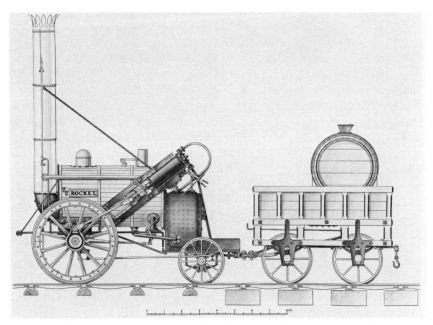

Figure 9-9 George Stevenson's "Rocket." British Crown Copyright Science Museum, London.

diameter was increased as much as possible, and the length was increased. To support the increased heat transfer rate, the firebox size was increased to accommodate a larger fire. Both the grate area was enlarged and the combustion path before the gases entered the fire tubes was lengthened. The typical design features of a large locomotive boiler of circa 1920 are illustrated in Figure 9-10.

MARINE BOILERS

The early marine boilers were in the form of a horizontal cylindrical shell with a large furnace tube(s) extending their length with a multiplicity of fire tubes to pass the combustion gases leaving the furnace tube back through the boiler for an increased amount of heat transfer. This family of boilers was commonly called "Scotch Boilers." They developed into relatively complex configurations as illustrated in Figures 9-11 and 9-12.

The Scotch boilers for large steam ships (circa 1900) were typically about 5 m in diameter with four furnace flues, with a length of 3 m for a single ended boiler or 6 m for a double ended boiler. Pressures were around 9 atm (900 kPa), and the required thickness of the shell and header plates made the boilers very heavy.

With the advent of the de Laval steam turbine, higher steam temperatures and pressures could significantly reduce the fuel consumption, and the cost of steamship operation.

This led to the development of the water tube boiler. The relatively large diameter cylindrical shell of the fire tube boiler was replaced by three relatively small diameter drums, or headers, between which tubes filled with water were run. In the marine applications the most effective use of the available space, with the use of short drums to reduce water level sloshing as the ship pitched and rolled, was favorable for the water tube boiler. A large combustion chamber could be provided with adequate wall cooling and subsequent flue gas cooling by a multiplicity of tubes filled with water. An early water tube marine boiler with three drums is illustrated in Figure 9-13. A sectional view of a modern marine boiler is shown in Figure 9-14.

The water tube boiler could produce higher pressure and temperature steam with less weight and volume (size) for boiler horsepower output. Heat transfer fluxes were increased by higher flue gas to tube surface film coefficients because the flue gas passed over the outside tube surface instead of flowing inside a long tube (see page 183). Due to the large size of the combustion chamber, significant radiation heat transfer was also obtained (see page 310 and Section 7-3). The weight of the boiler was reduced because the shell and drums were of smaller diameter and thinner walls could be used. Also the water tube walls could be thinner because tubes with internal pressure required thinner walls than tubes subjected to external pressure (as in the case of the fire tube boiler).

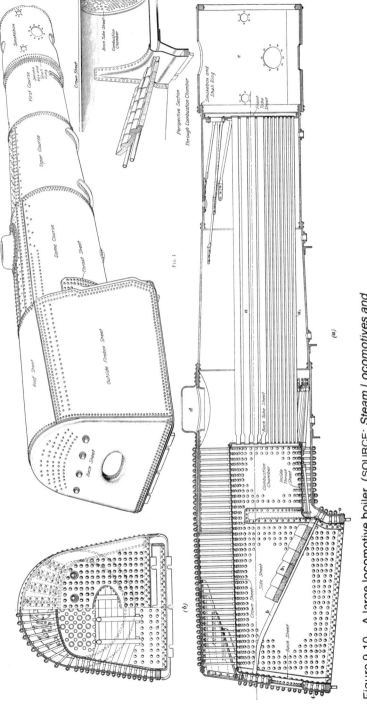

Figure 9-10   A large locomotive boiler. (SOURCE: *Steam Locomotives and Boilers*, International Textbook Co., Scranton, Pa., 1924.)

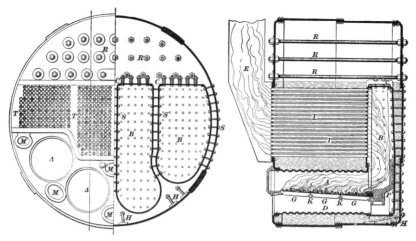

Figure 9-11 Single ended Scotch marine boiler. (SOURCE: *A Textbook on Mechanical Engineering*, International Textbook Co., Scranton, Pa., 1900.)

Improvements have been made in the heat transfer design. Heat transfer surface geometries have been developed to give increased heat transfer film coefficients, and the combustion pressure has been increased. This has resulted in a greater boiler output for a given volume, and reduced weight per boiler horsepower. A modern steam boiler producing steam with pressures around 6 MPa (900 psi) and 500°C (900°F) is shown in Figure 9-8.

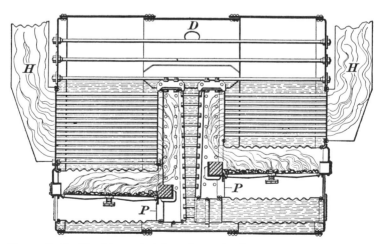

Figure 9-12 Double ended Scotch marine boiler. (SOURCE: *A Textbook on Mechanical Engineering*, International Textbook Co., Scranton, Pa., 1900.)

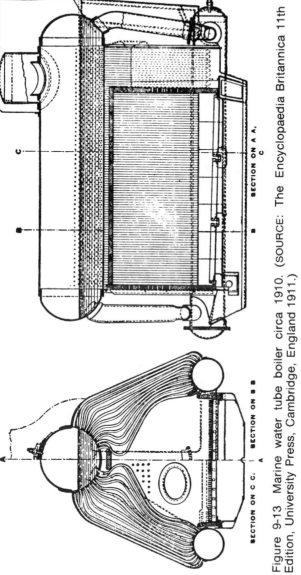

Figure 9-13 Marine water tube boiler circa 1910. (SOURCE: The Encyclopaedia Britannica 11th Edition, University Press, Cambridge, England 1911.)

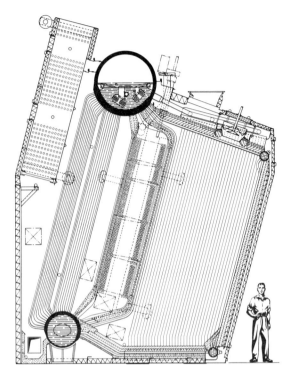

Top-fired two-drum Integral-Furnace boiler with horizontal superheater.

Figure 9-14   Sectional view of a modern marine boiler. (Courtesy of Babcock & Wilcox.)

The boiler shown in Figure 9-8 is described as a two-drum integral-furnace boiler with a vertical superheater. The furnace is water cooled by walls of closely spaced tubes. The combustion gases are passed through rows of tubes for steam generation and superheating. The combustion gases then pass through heat exchangers to reduce the discharged combustion gas temperature by transferring heat to the boiler feed water, combustion air preheating, or for other purposes.

Combustion furnaces can be gauged by several criteria. The criteria most frequently used are:

- combustion heat release (firing rate) per furnace unit volume;
- combustion heat release (firing rate) per unit area of radiant heat absorbing area;
- heat absorption rate per unit area of radiant heat absorbing surface.

The greater the combustion heat release per unit volume of the furnace, the smaller is the boiler size and weight for a given boiler heat output rate. It is important that the fuel be burned completely and cleanly within the furnace volume. Naval and maritime specifications place limits on the generally acceptable maximum volumetric furnace heat release rates, as well as on other furnace performance criteria.

The combustion heat release (firing rate or heat input) per unit area of radiant heat absorbing area is a measure of the "basic" heat transfer flux loading on the radiation absorbing heat transfer surfaces. The radiation absorbing heat transfer surfaces are the water tubes lining the walls of the furnace. Several rows of "screen tubes" are usually provided across the combustion gas exit from the furnace, acting as radiation absorbing heat transfer surfaces, and shielding the convection tubes from any significant amount of radiation heat transfer flux, which, if incident upon the convection tubes, would overheat them.

The heat absorption rate or output per unit area of radiant heat absorbing surface is a measure of the heat transfer into the water tubes lining the walls of the furnace. As pointed out earlier, the furnace wall is the area of the highest temperatures and the largest heat transfer fluxes. Combustion temperatures are typically 1500°C (2700°F). In coal fired boilers the combustion temperature must be kept below the melting point of the ash to avoid forming molten slag.

It is important that the circulation of the water in the tube be sufficient to maintain the boiling in the stable nucleate boiling regime III (see Section 5-1), maintaining the tube wall temperature within the operating temperature limits of the tube material. Generally the flow volume of the steam (vapor) exiting a water tube must be less than 70% of the total flow volume.

The temperature of the combustion gases leaving the furnace is controlled by the amount of heat that is transferred to the furnace walls. The more the amount of heat transferred in the furnace, the less is the amount of heat left in the gases leaving the furnace, and the lower the gas temperature. The gases exiting the furnace are typically 1080°C to 1170°C (1950°F to 2100°F). Because the heat transfer in the superheater section is from gas to steam (gas), the overall convection heat transfer coefficient will be lower than for the water filled tubes and the tube wall temperatures will be significantly higher due to the fact the heat transfer film coefficient inside the tube is much higher for boiling water than for steam (gas). Consequently the superheater tubes must be fabricated from appropriate high temperature materials.

The saturation temperatures of power boilers are sufficiently high so that heat exchangers (economizers and air preheaters) in the flue gas leaving the boiler can improve the steam generation efficiency by reducing the stack gas temperature and the associated stack heat losses. The economizer returns the heat to the boiler by heating the boiler feed water, and the air preheater stack heat losses. The economizer returns the heat transferred from the stack gases to the boiler by heating the boiler feed water, and the air preheater returns the heat transferred from the stack gases to the furnace combustion. The effect of these heat exchangers can vary significantly from one steam generator design or application to another.

9-3 THE PACKAGED INDUSTRIAL BOILER

The packaged industrial boiler is almost universally used for supplying process steam and hot water for industry and building heating. This category of boiler will be used for a discussion of the application of heat transfer principles, including both the fire tube and water tube boiler configurations. The heat transfer aspects of the combustion furnace and the convection inside or outside the tubes will be discussed, as well as other considerations.

9-3.1 The Fire Tube Packaged Industrial Boiler

The current fire tube packaged boiler design configuration is usually based on a horizontal cylindrical shell with a relatively large furnace tube extending the length of the boiler near the bottom of the shell with a series of fire tubes to provide return passes for the combustion gases exiting from the furnace tube. This configuration is illustrated in Figure 9-15. The intensity of the radiant energy transfer to the walls of the furnace tube is affected by the type of flame, flame temperature, and the geometry and size of the furnace tube. The gas temperature leaving the furnace at full load is typically 1100°C to 1200°C (1950°F to 2150°F). The emissivities of flames were discussed on page 310 and the radiant heat transfer in Section 7-3. Some convective heat transfer between the combustion gases and the surface of the furnace tube will occur as discussed on page 168.

The transfer of heat from the combustion gases to make the temperature of the flue gases leaving the boiler close to the boiler saturation temperature is a heat exchanger problem, with an objective of achieving an adequate effectiveness as discussed in Section 8-2. There are trade-offs between the tube diameter, length, and the flue gas mass flow rate through the tube. The UA parameter needs to be established in order to use the heat exchanger effectiveness equations or graphs in Section 8-2.

To obtain the overall heat transfer coefficient (U), both the heat transfer film coefficients on the inside and the outside of the tube must be evaluated. The film coefficient for gas flowing inside a tube is discussed on page 168 and boiling is discussed in Chapter 5. The flue gas mass flow rate is controlled by the combustion rate and air mixture, which establishes the volumetric flow rate. The selection of the number of fire tubes and their diameter (for each pass) then establishes the mass flow rate in the tubes.

Figure 9-15   Packaged fire tube industrial boiler. (Courtesy of Cleaver Brooks, Division of Aqua-Chem, Inc.)

The overall heat transfer coefficient for the heat transfer path from the flue gas through the tube wall and into the boiling water can be evaluated as discussed on page 413. The direct heat transfer surface (A) is determined by the total tube wall area.

A design can then be made to identify the number and sizes of the fire tubes and the number of passes through the boiler. For a given tube size and flue gas mass flow rate, the higher the effectiveness, the longer the tube must be. A cutaway view of a four-pass fire tube boiler is shown in Figure 9-16.

The flow of the combustion gases through the boiler is driven by a blower which supplies the "draft." The air flow must be controlled to provide the proper air-fuel mixture for combustion. As the load on the boiler varies, so must the fuel combustion rate and the air flow rate.

At boiler pressures of 1 MPa absolute (145 psig) the boiler saturation temperature is 180°C (355°F). Stack gases need to be kept at temperatures in this range to avoid condensation of sulfur combustion products upon the surfaces of the stack to minimize corrosion of the stack. At higher stack gas temperatures, heat can be recovered by the use of an economizer, thereby improving the fuel-to-steam efficiency. A typical economizer is a tubular heat exchanger with the stack gas passing over the outside, which

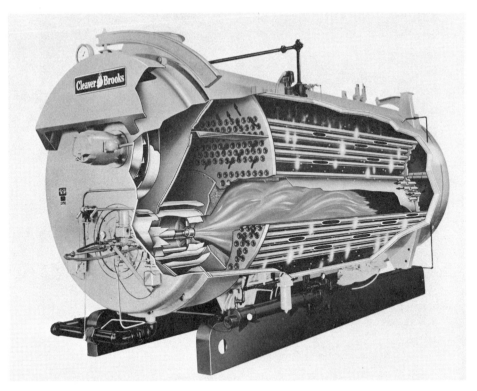

Figure 9-16   Cutaway view of a four-pass fire tube boiler. (Courtesy of Cleaver Brooks, Division of Aqua-Chem, Inc.)

is finned to improve the effective gas side heat transfer film coefficient, with the boiler feed water passing through the inside of the tubes. The use of extended surfaces (fins) on heat transfer surfaces to gas flow was discussed in Section 2-3.2. A finned tube economizer is shown in Figure 9-17.

The heat losses from a well-designed boiler to its surroundings by radiation and convection are minor, being in the range of 1.5 to 3% of the fuel heat input. All external surfaces need to be insulated with a low emissivity external surface. Emitted radiant energy from surfaces is discussed in Section 7-1, and free convection into air is discussed in Section 3-4.

In the operation of an industrial boiler it is important that the water be treated to prevent buildup of scale on the boiling surfaces. The scale has a high thermal resistance (low thermal conductivity) which results in a reduction in the overall heat transfer coefficient (U) as well as increasing the temperature of the tube wall, as discussed on page 57. The flue gas

Figure 9-17 A finned tube economizer. (Courtesy of Applied Engineering Co., Orangeburg, S.C.)

temperature leaving the boiler then goes up, the fuel-to-steam efficiency goes down, and when the tube wall temperature reaches its operating limit under its stressed condition, a failure of the boiler will occur.

FIRE TUBE INDUSTRIAL WATER HEATER
The general design of a fire tube water heater is similar to the fire tube steam generating boiler design, as illustrated by Figure 9-18. In the hot water heater design consideration must be given to the fact that:

- the overall heat transfer coefficient (U) is lower for the water heater because the film coefficient for heat transfer into water is much lower than the film coefficient for boiling;
- the average temperature of the water is lower, and controlled circulation of the inlet water may be necessary to prevent condensation and associated corrosion on surfaces exposed to the flue gases.

This requires a larger direct heat transfer surface for the same thermal

Figure 9-18   A fire tube industrial water heater. (Courtesy of Cleaver Brooks, Division of Aqua-Chem, Inc.)

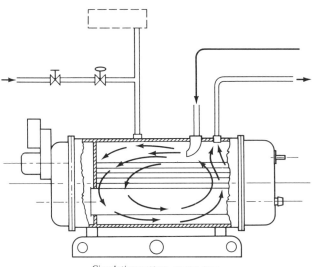

Circulation pattern creates even
temperature throughout the boiler

Figure 9-19   Return water forced circulation in a fire tube water heater. (Courtesy of Cleaver Brooks, Division of Aqua-Chem, Inc.)

output, and a water circulation and mixing means inside the heater, as illustrated in Figure 9-19.

9-3.2 The Water Tube Packaged Boiler

The performance of water tube and fire tube boilers for many applications is quite comparable. Their outline configuration for installation and maintenance requirements is different. The water tube boiler responds to load fluctuations faster. The fire tube boiler usually holds more water and requires more heat transfer surface for the same output. The water tube boiler is applicable to higher pressures than the fire tube, and to larger steam flow rates (boiler horsepower). A photograph of a water tube industrial boiler is shown in Figure 9-20.

The furnace volume is relatively large, as shown by the drum and tube arrangement of a boiler being fabricated in Figure 9-21. The furnace volume extends the length of the boiler (on the right-hand side of the picture), and is lined with one layer (row) of closely spaced tubes. The combustion gases then enter the bank of convective heat transfer tubes at

Figure 9-20   A modern packaged water tube industrial boiler. (Courtesy of Cleaver Brooks, Division of Aqua-Chem, Inc.)

Figure 9-21   Drum and tube arrangement for a water tube industrial boiler. (Courtesy of Cleaver Brooks, Division of Aqua-Chem, Inc.)

the far end, and by a series of vertical baffles are forced to flow up and down through the tube bank until they exit on the side at the front.

The radiant energy from the combustion is more intense (a higher radiant energy flux density at the heat transfer surface) than in the fire tube boiler because the radiation path length through the combustion gas is longer. To transfer this radiant energy heat flux into the water, an adequate water flow velocity must be maintained in each and every tube so as to maintain the tube wall temperature within operating limits. In this type of boiler the required circulation is usually provided by the free convection water flow resulting from the density change in the water as steam is generated. In other cases pumping is required to assure the required circulation. The radiant heat fluxes and heat transfer to the tube

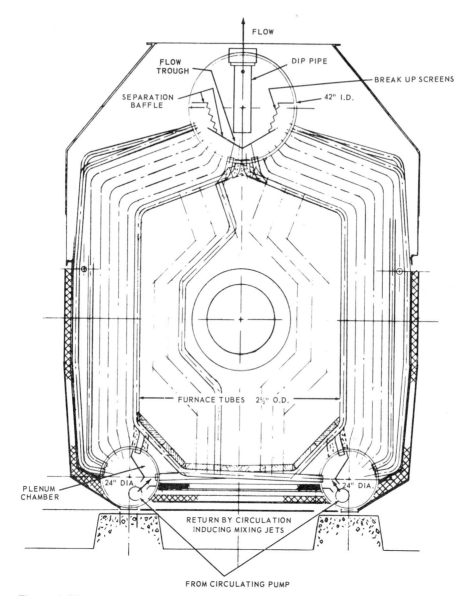

Figure 9-22   A sectional view of a large water tube industrial high-temperature water heater. (Courtesy of Cleavery Brooks, Division of Aqua-Chem, Inc.)

wall are discussed on page 310 and in Section 7-3.1, and boiling is discussed in Chapter 5.

The heat transfer in the convective tube banks generally follows the trends in tube and shell heat exchangers, discussed in Section 8-2 and gas flow through tube banks discussed on page 183. Each boiler manufacturing company has developed design information for the specific geometric configuration it uses, and usually regards this information as proprietary.

A water tube water heater is generally similar in appearance to a boiler of the same thermal output. One major design detail is the forced circulation of the water in the tubes. This is necessary to provide a sufficiently high heat transfer film coefficient into the water along each and every tube in order to maintain sufficiently low temperature of the water tubes lining the furnace wall for safe operation. A sectional view of a large high temperature water heater is shown in Figure 9-22.

APPENDIXES

APPENDIX C   CONVERSION FACTORS AND EQUATIONS OF EQUALITY

APPENDIX D   AN APPROACH TO PROBLEM SOLVING

APPENDIX E   MATHEMATICAL TABLES AND DIMENSIONAL DATA

Appendix A
Physical and Thermodynamic Data

Table A-1   ARDC STANDARD ATMOSPHERE (1959)

ALTITUDE, Z [m]	TEMPERATURE, T [K]	PRESSURE, p [kPa]	DENSITY, ρ [kg/m^3]
$-5,000$	320.69	177.610	1.9296
$-4,000$	314.18	159.600	1.7698
$-3,000$	307.67	142.970	1.6189
$-2,000$	301.16	127.780	1.4782
$-1,000$	294.66	113.930	1.3470
0	288.16	101.325	1.2250
500	284.91	95.461	1.1673
1,000	281.66	89.876	1.1117
1,500	278.41	84.560	1.0581
2,000	275.16	79.501	1.0066
2,500	271.92	74.692	0.9570
3,000	268.67	70.121	0.9093
3,500	265.42	65.780	0.8634
4,000	262.18	61.660	0.8194
4,500	258.93	57.752	0.7770
5,000	255.69	54.048	0.7364
6,000	249.20	47.217	0.6601
7,000	242.71	41.105	0.5900

Table A-1 (Continued)

ALTITUDE, Z [m]	TEMPERATURE, T [K]	PRESSURE, p [kPa]	DENSITY, ρ [kg/m^3]
8,000	236.23	35.651	0.5258
9,000	229.74	30.800	0.4671
10,000	223.26	26.500	0.4135
11,000	216.78	22.700	0.3648
12,000	216.66	19.399	0.3119
13,000	216.66	16.579	0.2666
14,000	216.66	14.170	0.2279
15,000	216.66	12.112	0.1948
16,000	216.66	10.353	0.1665
17,000	216.66	8.850	0.1423
18,000	216.66	7.565	0.1217
19,000	216.66	6.467	0.1040
20,000	216.66	5.529	0.0889
22,000	216.66	4.042	0.0650
24,000	216.66	2.955	0.0475
26,000	219.34	2.163	0.0344
28,000	225.29	1.595	0.0247
30,000	231.24	1.186	0.0179
35,000	246.09	0.584	0.0083
40,000	260.91	0.300	0.0040
45,000	275.71	0.160	0.0020
50,000	282.66	0.088	0.0011

Table A-2 GRAVITATIONAL ACCELERATION NEAR THE EARTH

ALTITUDE ABOVE SEA LEVEL [m]	LOCATION ON EARTH'S SURFACE, DEGREES LATITUDE					
	0°	20°	40° [m/s^2]	60°	80°	90°
0	9.780	9.787	9.802	9.819	9.831	9.832
250	9.780	9.786	9.801	9.818	9.830	9.831
500	9.779	9.785	9.800	9.817	9.829	9.830
1000	9.777	9.784	9.799	9.816	9.828	9.829
2000	9.774	9.780	9.796	9.814	9.826	9.827
3000	9.771	9.777	9.793	9.810	9.822	9.823
30,000	9.686	9.695	9.710	9.727	9.739	9.740

SOURCE: *U.S. Coast and Geodetic Survey*, 1912.

Table A-3   THERMODYNAMIC PROPERTIES OF SATURATED STEAM

TEMP., T [°C]	ABS. PRESS., P [kPa]	SPECIFIC VOLUME [m³/kg] SAT. LIQUID, v_f	SAT. VAPOR, v_g	INTERNAL ENERGY [kJ/kg] SAT. LIQUID, u_f	EVAP., u_{fg}	SAT. VAPOR, u_g	ENTHALPY [kJ/kg] SAT. LIQUID, h_f	EVAP., h_{fg}	SAT. VAPOR, h_g	ENTROPY [kJ/kg·K] SAT. LIQUID, s_f	EVAP., s_{fg}	SAT. VAPOR, s_g
0.01	0.6113	0.001 000	206.14	.00	2375.3	2375.3	.01	2501.3	2501.4	.0000	9.1562	9.1562
5	0.8721	0.001 000	147.12	20.97	2361.3	2382.3	20.98	2489.6	2510.6	.0761	8.9496	9.0257
10	1.2276	0.001 000	106.38	42.00	2347.2	2389.2	42.01	2477.7	2519.8	.1510	8.7498	8.9008
15	1.7051	0.001 001	77.93	62.99	2333.1	2396.1	62.99	2465.9	2528.9	.2245	8.5569	8.7814
20	2.339	0.001 002	57.79	83.95	2319.0	2402.9	83.96	2454.1	2538.1	.2966	8.3706	8.6672
25	3.169	0.001 003	43.36	104.88	2304.9	2409.8	104.89	2442.3	2547.2	.3674	8.1905	8.5580
30	4.246	0.001 004	32.89	125.78	2290.8	2416.6	125.79	2430.5	2556.3	.4369	8.0164	8.4533
35	5.628	0.001 006	25.22	146.67	2276.7	2423.4	146.68	2418.6	2565.3	.5053	7.8478	8.3531
40	7.384	0.001 008	19.52	167.56	2262.6	2430.1	167.57	2406.7	2574.3	.5725	7.6845	8.2570
45	9.593	0.001 010	15.26	188.44	2248.4	2436.8	188.45	2394.8	2583.2	.6387	7.5261	8.1648
50	12.349	0.001 012	12.03	209.32	2234.2	2443.5	209.33	2382.7	2592.1	.7038	7.3725	8.0763
55	15.758	0.001 015	9.568	230.21	2219.9	2450.1	230.23	2370.7	2600.9	.7679	7.2234	7.9913
60	19.940	0.001 017	7.671	251.11	2205.5	2456.6	251.13	2358.5	2609.6	.8312	7.0784	7.9096
65	25.03	0.001 020	6.197	272.02	2191.1	2463.1	272.06	2346.2	2618.3	.8935	6.9375	7.8310
70	31.19	0.001 023	5.042	292.95	2176.6	2469.6	292.98	2333.8	2626.8	.9549	6.8004	7.7553
75	38.58	0.001 026	4.131	313.90	2162.0	2475.9	313.93	2321.4	2635.3	1.0155	6.6669	7.6824
80	47.39	0.001 029	3.407	334.86	2147.4	2482.2	334.91	2308.8	2643.7	1.0753	6.5369	7.6122
85	57.83	0.001 033	2.828	355.84	2132.6	2488.4	355.90	2296.0	2651.9	1.1343	6.4102	7.5445
90	70.14	0.001 036	2.361	376.85	2117.7	2494.5	376.92	2283.2	2660.1	1.1925	6.2866	7.4791
95	84.55	0.001 040	1.982	397.88	2102.7	2500.6	397.96	2270.2	2668.1	1.2500	6.1659	7.4159
	[MPa]											
100	0.101 35	0.001 044	1.6729	418.94	2087.6	2506.5	419.04	2257.0	2676.1	1.3069	6.0480	7.3549
105	0.120 82	0.001 048	1.4194	440.02	2072.3	2512.4	440.15	2243.7	2683.8	1.3630	5.9328	7.2958
110	0.143 27	0.001 052	1.2102	461.14	2057.0	2518.1	461.30	2230.2	2691.5	1.4185	5.8202	7.2387
115	0.169 06	0.001 056	1.0366	482.30	2041.4	2523.7	482.48	2216.5	2699.0	1.4734	5.7100	7.1833

Table A-3 (Continued)

TEMP., T [°C]	ABS. PRESS., P [MPa]	SPECIFIC VOLUME [m³/kg] SAT. LIQUID, v_f	SAT. VAPOR, v_g	INTERNAL ENERGY [kJ/kg] SAT. LIQUID, u_f	EVAP., u_{fg}	SAT. VAPOR, u_g	ENTHALPY [kJ/kg] SAT. LIQUID, h_f	EVAP., h_{fg}	SAT. VAPOR, h_g	ENTROPY [kJ/kg·K] SAT. LIQUID, s_f	EVAP., s_{fg}	SAT. VAPOR, s_g
120	0.198 53	0.001 060	0.8919	503.50	2025.8	2529.3	503.71	2202.6	2706.3	1.5276	5.6020	7.1296
125	0.2321	0.001 065	0.7706	524.74	2009.9	2534.6	524.99	2188.5	2713.5	1.5813	5.4962	7.0775
130	0.2701	0.001 070	0.6685	546.02	1993.9	2539.9	546.31	2174.2	2720.5	1.6344	5.3925	7.0269
135	0.3130	0.001 075	0.5822	567.35	1977.7	2545.0	567.69	2159.6	2727.3	1.6870	5.2907	6.9777
140	0.3613	0.001 080	0.5089	588.74	1961.3	2550.0	589.13	2144.7	2733.9	1.7391	5.1908	6.9299
145	0.4154	0.001 085	0.4463	610.18	1944.7	2554.9	610.63	2129.6	2740.3	1.7907	5.0926	6.8833
150	0.4758	0.001 091	0.3928	631.68	1927.9	2559.5	632.20	2114.3	2746.5	1.8418	4.9960	6.8379
155	0.5431	0.001 096	0.3468	653.24	1910.8	2564.1	653.84	2098.6	2752.4	1.8925	4.9010	6.7935
160	0.6178	0.001 102	0.3071	674.87	1893.5	2568.4	675.55	2082.6	2758.1	1.9427	4.8075	6.7502
165	0.7005	0.001 108	0.2727	696.56	1876.0	2572.5	697.34	2066.2	2763.5	1.9925	4.7153	6.7078
170	0.7917	0.001 114	0.2428	718.33	1858.1	2576.5	719.21	2049.5	2768.7	2.0419	4.6244	6.6663
175	0.8920	0.001 121	0.2168	740.17	1840.0	2580.2	741.17	2032.4	2773.6	2.0909	4.5347	6.6256
180	1.0021	0.001 127	0.194 05	762.09	1821.6	2583.7	763.22	2015.0	2778.2	2.1396	4.4461	6.5857
185	1.1227	0.001 134	0.174 09	784.10	1802.9	2587.0	785.37	1997.1	2782.4	2.1879	4.3586	6.5465
190	1.2544	0.001 141	0.156 54	806.19	1783.8	2590.0	807.62	1978.8	2786.4	2.2359	4.2720	6.5079
195	1.3978	0.001 149	0.141 05	828.37	1764.4	2592.8	829.98	1960.0	2790.0	2.2835	4.1863	6.4698
200	1.5538	0.001 157	0.127 36	850.65	1744.7	2595.3	852.45	1940.7	2793.2	2.3309	4.1014	6.4323
205	1.7230	0.001 164	0.115 21	873.04	1724.5	2597.5	875.04	1921.0	2796.0	2.3780	4.0172	6.3952
210	1.9062	0.001 173	0.104 41	895.53	1703.9	2599.5	897.76	1900.7	2798.5	2.4248	3.9337	6.3585
215	2.104	0.001 181	0.094 79	918.14	1682.9	2601.1	920.62	1879.9	2800.5	2.4714	3.8507	6.3221
220	2.318	0.001 190	0.086 19	940.87	1661.5	2602.4	943.62	1858.5	2802.1	2.5178	3.7683	6.2861
225	2.548	0.001 199	0.078 49	963.73	1639.6	2603.3	966.78	1836.5	2803.3	2.5639	3.6863	6.2503
230	2.795	0.001 209	0.071 58	986.74	1617.2	2603.9	990.12	1813.8	2804.0	2.6099	3.6047	6.2146
235	3.060	0.001 219	0.065 37	1009.89	1594.2	2604.1	1013.62	1790.5	2804.2	2.6558	3.5233	6.1791

240	3.344	0.001 229	0.059 76	1033.21	1570.8	2604.0	1037.32	1766.5	2803.8	2.7015	3.4422	6.1437
245	3.648	0.001 240	0.054 71	1056.71	1546.7	2603.4	1061.23	1741.7	2803.0	2.7472	3.3612	6.1083
250	3.973	0.001 251	0.050 13	1080.39	1522.0	2602.4	1085.36	1716.2	2801.5	2.7927	3.2802	6.0730
255	4.319	0.001 263	0.045 98	1104.28	1496.7	2600.9	1109.73	1689.8	2799.5	2.8383	3.1992	6.0375
260	4.688	0.001 276	0.042 21	1128.39	1470.6	2599.0	1134.37	1662.5	2796.9	2.8838	3.1181	6.0019
265	5.081	0.001 289	0.038 77	1152.74	1443.9	2596.6	1159.28	1634.4	2793.6	2.9294	3.0368	5.9662
270	5.499	0.001 302	0.035 64	1177.36	1416.3	2593.7	1184.51	1605.2	2789.7	2.9751	2.9551	5.9301
275	5.942	0.001 317	0.032 79	1202.25	1387.9	2590.2	1210.07	1574.9	2785.0	3.0208	2.8730	5.8938
280	6.412	0.001 332	0.030 17	1227.46	1358.7	2586.1	1235.99	1543.6	2779.6	3.0668	2.7903	5.8571
285	6.909	0.001 348	0.027 77	1253.00	1328.4	2581.4	1262.31	1511.0	2773.3	3.1130	2.7070	5.8199
290	7.436	0.001 366	0.025 57	1278.92	1297.1	2576.0	1289.07	1477.1	2766.2	3.1594	2.6227	5.7821
295	7.993	0.001 384	0.023 54	1305.2	1264.7	2569.9	1316.3	1441.8	2758.1	3.2062	2.5375	5.7437
300	8.581	0.001 404	0.021 67	1332.0	1231.0	2563.0	1344.0	1404.9	2749.0	3.2534	2.4511	5.7045
305	9.202	0.001 425	0.019 948	1359.3	1195.9	2555.2	1372.4	1366.4	2738.7	3.3010	2.3633	5.6643
310	9.856	0.001 447	0.018 350	1387.1	1159.4	2546.4	1401.3	1326.0	2727.3	3.3493	2.2737	5.6230
315	10.547	0.001 472	0.016 867	1415.5	1121.1	2536.6	1431.0	1283.5	2714.5	3.3982	2.1821	5.5804
320	11.274	0.001 499	0.015 488	1444.6	1080.9	2525.5	1461.5	1238.6	2700.1	3.4480	2.0882	5.5362
330	12.845	0.001 561	0.012 996	1505.3	993.7	2498.9	1525.3	1140.6	2665.9	3.5507	1.8909	5.4417
340	14.586	0.001 638	0.010 797	1570.3	894.3	2464.6	1594.2	1027.9	2622.0	3.6594	1.6763	5.3357
350	16.513	0.001 740	0.008 813	1641.9	776.6	2418.4	1670.6	893.4	2563.9	3.7777	1.4335	5.2112
360	18.651	0.001 893	0.006 945	1725.2	626.3	2351.5	1760.5	720.5	2481.0	3.9147	1.1379	5.0526
370	21.03	0.002 213	0.004 925	1844.0	384.5	2228.5	1890.5	441.6	2332.1	4.1106	.6865	4.7971
374.14	22.09	0.003 155	0.003 155	2029.6	0	2029.6	2099.3	0	2099.3	4.4298	0	4.4298

SOURCE: Adapted from Joseph H. Keenan, Frederick G. Keyes, Philip G. Hill, and Joan G. Moore, Steam Tables, Wiley, New York, 1969.

Table A-4 THERMODYNAMIC PROPERTIES OF SATURATED FREON-12

TEMP. [°C]	ABS. PRESS., P [MPa]	SPECIFIC VOLUME [m³/kg]			ENTHALPY [kJ/kg]			ENTROPY [kJ/kg·K]		
		SAT. LIQUID, v_f	EVAP., v_{fg}	SAT. VAPOR, v_g	SAT. LIQUID, h_f	EVAP., h_{fg}	SAT. VAPOR, h_g	SAT. LIQUID, s_f	EVAP., s_{fg}	SAT. VAPOR, s_g
−90	0.0028	0.000 608	4.414 937	4.415 545	−43.243	189.618	146.375	−0.2084	1.0352	0.8268
−85	0.0042	0.000 612	3.036 704	3.037 316	−38.968	187.608	148.640	−0.1854	0.9970	0.8116
−80	0.0062	0.000 617	2.137 728	2.138 345	−34.688	185.612	150.924	−0.1630	0.9609	0.7979
−75	0.0088	0.000 622	1.537 030	1.537 651	−30.401	183.625	153.224	−0.1411	0.9266	0.7855
−70	0.0123	0.000 627	1.126 654	1.127 280	−26.103	181.640	155.536	−0.1197	0.8940	0.7744
−65	0.0168	0.000 632	0.840 534	0.841 166	−21.793	179.651	157.857	−0.0987	0.8630	0.7643
−60	0.0226	0.000 637	0.637 274	0.637 910	−17.469	177.653	160.184	−0.0782	0.8334	0.7552
−55	0.0300	0.000 642	0.490 358	0.491 000	−13.129	175.641	162.512	−0.0581	0.8051	0.7470
−50	0.0391	0.000 648	0.382 457	0.383 105	−8.772	173.611	164.840	−0.0384	0.7779	0.7396
−45	0.0504	0.000 654	0.302 029	0.302 682	−4.396	171.558	167.163	−0.0190	0.7519	0.7329
−40	0.0642	0.000 659	0.241 251	0.241 910	−0.000	169.479	169.479	−0.0000	0.7269	0.7269
−35	0.0807	0.000 666	0.194 732	0.195 398	4.416	167.368	171.784	0.0187	0.7027	0.7214
−30	0.1004	0.000 672	0.158 703	0.159 375	8.854	165.222	174.076	0.0371	0.6795	0.7165
−25	0.1237	0.000 679	0.130 487	0.131 166	13.315	163.037	176.352	0.0552	0.6570	0.7121
−20	0.1509	0.000 685	0.108 162	0.108 847	17.800	160.810	178.610	0.0730	0.6352	0.7082
−15	0.1826	0.000 693	0.090 326	0.091 018	22.312	158.534	180.846	0.0906	0.6141	0.7046
−10	0.2191	0.000 700	0.075 946	0.076 646	26.851	156.207	183.058	0.1079	0.5936	0.7014
−5	0.2610	0.000 708	0.064 255	0.064 963	31.420	153.823	185.243	0.1250	0.5736	0.6986
0	0.3086	0.000 716	0.054 673	0.055 389	36.022	151.376	187.397	0.1418	0.5542	0.6960
5	0.3626	0.000 724	0.046 761	0.047 485	40.659	148.859	189.518	0.1585	0.5351	0.6937
10	0.4233	0.000 733	0.040 180	0.040 914	45.337	146.265	191.602	0.1750	0.5165	0.6916
15	0.4914	0.000 743	0.034 671	0.035 413	50.058	143.586	193.644	0.1914	0.4983	0.6897
20	0.5673	0.000 752	0.030 028	0.030 780	54.828	140.812	195.641	0.2076	0.4803	0.6879

25	0.6516	0.000 763	0.026 091	0.026 854	59.653	137.933	197.586	0.2237	0.4626	0.6863
30	0.7449	0.000 774	0.022 734	0.023 508	64.539	134.936	199.475	0.2397	0.4451	0.6848
35	0.8477	0.000 786	0.019 855	0.020 641	69.494	131.805	201.299	0.2557	0.4277	0.6834
40	0.9607	0.000 798	0.017 373	0.018 171	74.527	128.525	203.051	0.2716	0.4104	0.6820
45	1.0843	0.000 811	0.015 220	0.016 032	79.647	125.074	204.722	0.2875	0.3931	0.6806
50	1.2193	0.000 826	0.013 344	0.014 170	84.868	121.430	206.298	0.3034	0.3758	0.6792
55	1.3663	0.000 841	0.011 701	0.012 542	90.201	117.565	207.766	0.3194	0.3582	0.6777
60	1.5259	0.000 858	0.010 253	0.011 111	95.665	113.443	209.109	0.3355	0.3405	0.6760
65	1.6988	0.000 877	0.008 971	0.009 847	101.279	109.024	210.303	0.3518	0.3224	0.6742
70	1.8858	0.000 897	0.007 828	0.008 725	107.067	104.255	211.321	0.3683	0.3038	0.6721
75	2.0874	0.000 920	0.006 802	0.007 723	113.058	99.068	212.126	0.3851	0.2845	0.6697
80	2.3046	0.000 946	0.005 875	0.006 821	119.291	93.373	212.665	0.4023	0.2644	0.6667
85	2.5380	0.000 976	0.005 029	0.006 005	125.818	87.047	212.865	0.4201	0.2430	0.6631
90	2.7885	0.001 012	0.004 246	0.005 258	132.708	79.907	212.614	0.4385	0.2200	0.6585
95	3.0569	0.001 056	0.003 508	0.004 563	140.068	71.658	211.726	0.4579	0.1946	0.6526
100	3.3440	0.001 113	0.002 790	0.003 903	148.076	61.768	209.843	0.4788	0.1655	0.6444
105	3.6509	0.001 197	0.002 045	0.003 242	157.085	49.014	206.099	0.5023	0.1296	0.6319
110	3.9784	0.001 364	0.001 098	0.002 462	168.059	28.425	196.484	0.5322	0.0742	0.6064
112	4.1155	0.001 792	0.000 005	0.001 797	174.920	0.151	175.071	0.5651	0.0004	0.5655

SOURCE: Copyright 1955 and 1956, E. I. du Pont de Nemours & Company, Inc. Reprinted by permission. Adapted from English units.

Table A-5 THERMODYNAMIC PROPERTIES OF SATURATED AMMONIA

TEMP. [°C]	ABS. PRESS., P [kPa]	SPECIFIC VOLUME [m³/kg] SAT. LIQUID v_f	EVAP. v_{fg}	SAT. VAPOR v_g	ENTHALPY [kJ/kg] SAT. LIQUID h_f	EVAP. h_{fg}	SAT. VAPOR h_g	ENTROPY [kJ/kg·K] SAT. LIQUID s_f	EVAP. s_{fg}	SAT. VAPOR s_g
−50	40.88	0.001 424	2.6239	2.6254	−44.3	1416.7	1372.4	−0.1942	6.3502	6.1561
−48	45.96	0.001 429	2.3518	2.3533	−35.5	1411.3	1375.8	−0.1547	6.2696	6.1149
−46	51.55	0.001 434	2.1126	2.1140	−26.6	1405.8	1379.2	−0.1156	6.1902	6.0746
−44	57.69	0.001 439	1.9018	1.9032	−17.8	1400.3	1382.5	−0.0768	6.1120	6.0352
−42	64.42	0.001 444	1.7155	1.7170	−8.9	1394.7	1385.8	−0.0382	6.0349	5.9967
−40	71.77	0.001 449	1.5506	1.5521	0.0	1389.0	1389.0	0.0000	5.9589	5.9589
−38	79.80	0.001 454	1.4043	1.4058	8.9	1383.3	1392.2	0.0380	5.8840	5.9220
−36	88.54	0.001 460	1.2742	1.2757	17.8	1377.6	1395.4	0.0757	5.8101	5.8858
−34	98.05	0.001 465	1.1582	1.1597	26.8	1371.8	1398.5	0.1132	5.7372	5.8504
−32	108.37	0.001 470	1.0547	1.0562	35.7	1365.9	1401.6	0.1504	5.6652	5.8156
−30	119.55	0.001 476	0.9621	0.9635	44.7	1360.0	1404.6	0.1873	5.5942	5.7815
−28	131.64	0.001 481	0.8790	0.8805	53.6	1354.0	1407.6	0.2240	5.5241	5.7481
−26	144.70	0.001 487	0.8044	0.8059	62.6	1347.9	1410.5	0.2605	5.4548	5.7153
−24	158.78	0.001 492	0.7373	0.7388	71.6	1341.8	1413.4	0.2967	5.3864	5.6831
−22	173.93	0.001 498	0.6768	0.6783	80.7	1335.6	1416.2	0.3327	5.3188	5.6515
−20	190.22	0.001 504	0.6222	0.6237	89.7	1329.3	1419.0	0.3684	5.2520	5.6205
−18	207.71	0.001 510	0.5728	0.5743	98.8	1322.9	1421.7	0.4040	5.1860	5.5900
−16	226.45	0.001 515	0.5280	0.5296	107.8	1316.5	1424.4	0.4393	5.1207	5.5600
−14	246.51	0.001 521	0.4874	0.4889	116.9	1310.0	1427.0	0.4744	5.0561	5.5305
−12	267.95	0.001 528	0.4505	0.4520	126.0	1303.5	1429.5	0.5093	4.9922	5.5015
−10	290.85	0.001 534	0.4169	0.4185	135.2	1296.8	1432.0	0.5440	4.9290	5.4730
−8	315.25	0.001 540	0.3863	0.3878	144.3	1290.1	1434.4	0.5785	4.8664	5.4449
−6	341.25	0.001 546	0.3583	0.3599	153.5	1283.3	1436.8	0.6128	4.8045	5.4173

−4	368.90	0.001 553	0.3328	0.3343	162.7	1276.4	1439.1	0.6469	4.7432	5.3901
−2	398.27	0.001 559	0.3094	0.3109	171.9	1269.4	1441.3	0.6808	4.6825	5.3633
0	429.44	0.001 566	0.2879	0.2895	181.1	1262.4	1443.5	0.7145	4.6223	5.3369
2	462.49	0.001 573	0.2683	0.2698	190.4	1255.2	1445.6	0.7481	4.5627	5.3108
4	497.49	0.001 580	0.2502	0.2517	199.6	1248.0	1447.6	0.7815	4.5037	5.2852
6	534.51	0.001 587	0.2335	0.2351	208.9	1240.6	1449.6	0.8148	4.4451	5.2599
8	573.64	0.001 594	0.2182	0.2198	218.3	1233.2	1451.5	0.8479	4.3871	5.2350
10	614.95	0.001 601	0.2040	0.2056	227.6	1225.7	1453.3	0.8808	4.3295	5.2104
12	658.52	0.001 608	0.1910	0.1926	237.0	1218.1	1455.1	0.9136	4.2725	5.1861
14	704.44	0.001 616	0.1789	0.1805	246.4	1210.4	1456.8	0.9463	4.2159	5.1621
16	752.79	0.001 623	0.1677	0.1693	255.9	1202.6	1458.5	0.9788	4.1597	5.1385
18	803.66	0.001 631	0.1574	0.1590	265.4	1194.7	1460.0	1.0112	4.1039	5.1151
20	857.12	0.001 639	0.1477	0.1494	274.9	1186.7	1461.5	1.0434	4.0486	5.0920
22	913.27	0.001 647	0.1388	0.1405	284.4	1178.5	1462.9	1.0755	3.9937	5.0692
24	972.19	0.001 655	0.1305	0.1322	294.0	1170.3	1464.3	1.1075	3.9392	5.0467
26	1033.97	0.001 663	0.1228	0.1245	303.6	1162.0	1465.6	1.1394	3.8850	5.0244
28	1098.71	0.001 671	0.1156	0.1173	313.2	1153.6	1466.8	1.1711	3.8312	5.0023
30	1166.49	0.001 680	0.1089	0.1106	322.9	1145.0	1467.9	1.2028	3.7777	4.9805
32	1237.41	0.001 689	0.1027	0.1044	332.6	1136.4	1469.0	1.2343	3.7246	4.9589
34	1311.55	0.001 698	0.0969	0.0986	342.3	1127.6	1469.9	1.2656	3.6718	4.9374
36	1389.03	0.001 707	0.0914	0.0931	352.1	1118.7	1470.8	1.2969	3.6192	4.9161
38	1469.92	0.001 716	0.0863	0.0880	361.9	1109.7	1471.5	1.3281	3.5669	4.8950
40	1554.33	0.001 726	0.0815	0.0833	371.7	1100.5	1472.2	1.3591	3.5148	4.8740
42	1642.35	0.001 735	0.0771	0.0788	381.6	1091.2	1472.8	1.3901	3.4630	4.8530
44	1734.09	0.001 745	0.0728	0.0746	391.5	1081.7	1473.2	1.4209	3.4112	4.8322
46	1829.65	0.001 756	0.0689	0.0707	401.5	1072.0	1473.5	1.4518	3.3595	4.8113
48	1929.13	0.001 766	0.0652	0.0669	411.5	1062.2	1473.7	1.4826	3.3079	4.7905
50	2032.62	0.001 777	0.0617	0.0635	421.7	1052.0	1473.7	1.5135	3.2561	4.7696

SOURCE: Adapted from National Bureau of Standards Circular No. 142, *Tables of Thermodynamic Properties of Ammonia.*

Appendix B
Heat Transfer Data

Table B-1   PROPERTIES OF AIR AT ATMOSPHERIC PRESSURE

(The values of μ, k, c_p and Pr are not strongly pressure-dependent and may be used over a fairly wide range of pressures.)

T [K]	ρ [kg/m³]	c_p [kJ/kg·Δ_1°C]	μ [Pa·s×10⁵]	ν [m²/s×10⁶]	k [W/m·Δ_1°C]	α_d [m²/s×10⁴]	Pr	α^* (1/m³·Δ_1°C)×10⁻⁶
100	3.6010	1.0266	0.6924	1.923	0.009246	0.02501	0.770	20396.
150	2.3675	1.0099	1.0283	4.343	0.013735	0.05745	0.753	2622.
200	1.7684	1.0061	1.3289	7.490	0.01809	0.10165	0.739	641.8
250	1.4128	1.0053	1.488	9.49	0.02227	0.13161	0.722	282.9
300	1.1774	1.0057	1.983	15.68	0.02624	0.22160	0.708	87.61
350	0.9980	1.0090	2.075	20.76	0.03003	0.2983	0.697	45.21
400	0.8826	1.0140	2.286	25.90	0.03365	0.3760	0.689	25.13
450	0.7833	1.0207	2.484	28.86	0.03707	0.4222	0.683	16.31
500	0.7048	1.0295	2.671	37.90	0.04038	0.5564	0.680	9.285
550	0.6423	1.0392	2.848	44.34	0.04360	0.6532	0.680	6.182
600	0.5879	1.0551	3.018	51.34	0.04659	0.7512	0.680	4.234
650	0.5430	1.0635	3.177	58.51	0.04953	0.8578	0.682	3.009
700	0.5030	1.0752	3.332	66.25	0.05230	0.9672	0.684	2.188
750	0.4709	1.0856	3.481	73.91	0.05509	1.0774	0.686	1.638
800	0.4405	1.0978	3.625	82.29	0.05779	1.1951	0.689	1.246
850	0.4149	1.1095	3.765	90.75	0.06028	1.3097	0.692	0.9734
900	0.3925	1.1212	3.899	99.3	0.06279	1.4271	0.696	0.7674
950	0.3716	1.1321	4.023	108.2	0.06525	1.5510	0.699	0.6132
1000	0.3524	1.1417	4.152	117.8	0.06752	1.6779	0.702	0.4961

Table B-1 (Continued)

T [K]	ρ [kg/m³]	c_p [kJ/kg·Δ₁°C]	μ [Pa·s×10⁵]	ν [m²/s×10⁶]	k [W/m·Δ₁°C]	α_d [m²/s×10⁴]	Pr	α^* (1/m³·Δ₁°C)×10⁻⁶
1100	0.3204	1.160	4.44	138.6	0.0732	1.969	0.704	0.3270
1200	0.2947	1.179	4.69	159.1	0.0782	2.251	0.707	0.2272
1300	0.2707	1.197	4.93	182.1	0.0837	2.583	0.705	0.1605
1400	0.2515	1.214	5.17	205.5	0.0891	2.920	0.705	0.1160
1500	0.2355	1.230	5.40	229.1	0.0946	3.262	0.705	0.0878
1600	0.2211	1.248	5.63	254.5	0.100	3.609	0.705	0.0672
1700	0.2082	1.267	5.85	280.5	0.105	3.977	0.705	0.0518
1800	0.1970	1.287	6.07	308.1	0.111	4.379	0.704	0.0407
1900	0.1858	1.309	6.29	338.5	0.117	4.811	0.704	0.0319
2000	0.1762	1.338	6.50	369.0	0.124	5.260	0.702	0.0253
2100	0.1682	1.372	6.72	399.6	0.131	5.715	0.700	0.0206
2200	0.1602	1.419	6.93	432.6	0.139	6.120	0.707	0.0167
2300	0.1538	1.482	7.14	464.0	0.149	6.540	0.710	0.0139
2400	0.1458	1.574	7.35	504.0	0.161	7.020	0.718	0.0116
2500	0.1394	1.688	7.57	543.5	0.175	7.441	0.730	0.0097

SOURCE: From *Natl. Bur. Stand.* (U.S.) *Circ.* 564, 1955.

$*\alpha = \left[\dfrac{g\beta\rho^2 c_p}{\mu k} \right] \left[\dfrac{1}{m^3 \cdot \Delta_1 °C} \right]$

Table B-2   PROPERTIES OF GASES AT ATMOSPHERIC PRESSURE

(Values of μ, k, c_p, and Pr are not Strongly Pressure-Dependent for He, H_2, O_2, and N_2 and may be used over a fairly wide range of pressures.)

T [K]	ρ [kg/m³]	c_p [kJ/kg·Δ_1°C]	μ [Pa·s]	ν [m²/s]	k [W/m·Δ_1°C]	α_d [m²/s]	Pr [—]
HELIUM							
144	0.3379	5.200	125.5×10^{-7}	37.11×10^{-6}	0.0928	0.5275×10^{-4}	0.70
200	0.2435	5.200	156.6	64.38	0.1177	0.9288	0.694
255	0.1906	5.200	181.7	95.50	0.1357	1.3675	0.70
366	0.13280	5.200	230.5	173.6	0.1691	2.449	0.71
477	0.10204	5.200	275.0	269.3	0.197	3.716	0.72
589	0.08282	5.200	311.3	375.8	0.225	5.215	0.72
700	0.07032	5.200	347.5	494.2	0.251	6.661	0.72
800	0.06023	5.200	381.7	634.1	0.275	8.774	0.72
HYDROGEN							
150	0.16371	12.602	5.595×10^{-6}	34.18×10^{-5}	0.0981	0.475×10^{-4}	0.718
200	0.12270	13.540	6.813	55.53	0.1282	0.772	0.719
250	0.09819	14.059	7.919	80.64	0.1561	1.130	0.713
300	0.08185	14.314	8.963	109.5	0.182	1.554	0.706
350	0.07016	14.436	9.954	141.9	0.206	2.031	0.697
400	0.06135	14.491	10.864	177.1	0.228	2.568	0.690
450	0.05462	14.499	11.779	215.6	0.251	3.164	0.682
500	0.04918	14.507	12.636	257.0	0.272	3.817	0.675
550	0.04469	14.532	13.475	301.6	0.292	4.516	0.668
600	0.04085	14.537	14.285	349.7	0.315	5.306	0.664
700	0.03492	14.574	15.89	455.1	0.351	6.903	0.659
800	0.03060	14.675	17.40	569	0.384	8.563	0.664
900	0.02723	14.821	18.78	690	0.412	10.217	0.676

Table B-2 (Continued)

T [K]	ρ [kg/m³]	c_p [kJ/kg·Δ_1°C]	μ [Pa·s]	ν [m²/s]	k [W/m·Δ_1°C]	α_d [m²/s]	Pr [—]
			OXYGEN				
150	2.6190	0.9178	11.490×10^{-6}	4.387×10^{-6}	0.01367	0.05688×10^{-4}	0.773
200	1.9559	0.9131	14.850	7.593	0.01824	0.10214	0.745
250	1.5618	0.9157	17.87	11.45	0.02259	0.15794	0.725
300	1.3007	0.9203	20.63	15.86	0.02676	0.22353	0.709
350	1.1133	0.9291	23.16	20.80	0.03070	0.2968	0.702
400	0.9755	0.9420	25.54	26.18	0.03461	0.3768	0.695
450	0.8682	0.9567	27.77	31.99	0.03828	0.4609	0.694
500	0.7801	0.9722	29.91	38.34	0.04173	0.5502	0.697
550	0.7096	0.9881	31.97	45.05	0.04517	0.6441	0.700
			NITROGEN				
200	1.7108	1.0429	12.947×10^{-6}	7.568×10^{-6}	0.01824	0.10224×10^{-4}	0.747
300	1.1421	1.0408	17.84	15.63	0.02620	0.22044	0.713
400	0.8538	1.0459	21.98	25.74	0.03335	0.3734	0.691
500	0.6824	1.0555	25.70	37.66	0.03984	0.5530	0.684
600	0.5687	1.0756	29.11	51.19	0.04580	0.7486	0.686
700	0.4934	1.0969	32.13	65.13	0.05123	0.9466	0.691
800	0.4277	1.1225	34.84	81.46	0.05609	1.1685	0.700
900	0.3796	1.1464	37.49	91.06	0.06070	1.3946	0.711
1000	0.3412	1.1677	40.00	117.2	0.06475	1.6250	0.724
1100	0.3108	1.1857	42.28	136.0	0.06850	1.8591	0.736
1200	0.2851	1.2037	44.50	156.1	0.07184	2.0932	0.748

CARBON DIOXIDE

220	2.4733	0.783	11.105×10^{-6}	4.490×10^{-6}	0.010805	0.05920×10^{-4}	0.818
250	2.1657	0.804	12.590	5.813	0.012884	0.07401	0.793
300	1.7973	0.871	14.958	8.321	0.016572	0.10588	0.770
350	1.5362	0.900	17.205	11.19	0.02047	0.14808	0.755
400	1.3424	0.942	19.32	14.39	0.02461	0.19463	0.738
450	1.1918	0.980	21.34	17.90	0.02897	0.24813	0.721
500	1.0732	1.013	23.26	21.67	0.03352	0.3084	0.702
550	0.9739	1.047	25.08	25.74	0.03821	0.3750	0.685
600	0.8938	1.076	26.83	30.02	0.04311	0.4483	0.668

AMMONIA, NH_3

273	0.7929	2.177	9.353×10^{-6}	1.18×10^{-5}	0.0220	0.1308×10^{-4}	0.90
323	0.6487	2.177	11.035	1.70	0.0270	0.1920	0.88
373	0.5590	2.236	12.886	2.30	0.0327	0.2619	0.87
423	0.4934	2.315	14.672	2.97	0.0391	0.3432	0.87
473	0.4405	2.395	16.49	3.74	0.0467	0.4421	0.84

WATER VAPOR

380	0.5863	2.060	12.71×10^{-6}	2.16×10^{-5}	0.0246	0.2036×10^{-4}	1.060
400	0.5542	2.014	13.44	2.42	0.0261	0.2338	1.040
450	0.4902	1.980	15.25	3.11	0.0299	0.307	1.010
500	0.4405	1.985	17.04	3.86	0.0339	0.387	0.996
550	0.4005	1.997	18.84	4.70	0.0379	0.475	0.991
600	0.3652	2.026	20.67	5.66	0.0422	0.573	0.986
650	0.3380	2.056	22.47	6.64	0.0464	0.666	0.995
700	0.3140	2.085	24.26	7.72	0.0505	0.772	1.000
750	0.2931	2.119	26.04	8.88	0.0549	0.883	1.005
800	0.2739	2.152	27.86	10.20	0.0592	1.001	1.010
850	0.2579	2.186	29.69	11.52	0.0637	1.130	1.019

SOURCE: Adapted to SI units from E. R. G. Eckert and R. M. Drake, *Heat and Mass Transfer*, 2d ed., McGraw-Hill, New York, 1959.

Table B-3 PROPERTIES OF STEAM

PRESSURE [MPa (psia)]	SATURATED VAPOR		SUPERHEATED VAPOR TEMPERATURE [°C (°F)]					
			93.3 (200)	204 (400)	316 (600)	426 (800)	538 (1000)	649 (1200)
0.007	ρ	0.0480	0.031	0.025	—	—	—	—
(1)	c_p	1.9	1.91	1.94	—	—	—	—
	μ	9.506	12.81	16.95	20.67	24.38	27.69	30.58
	k	0.016	0.023	0.0318	0.0412	0.0505	0.0600	—
1.724	ρ	8.69	—	—	6.487	5.446	4.661	4.085
(250)	c_p	2.9	—	—	2.23	2.18	2.22	2.27
	μ	—	—	—	—	—	—	—
	k	0.0365	—	—	0.0427	0.0512	0.0604	—
3.447	ρ	17.268	—	—	13.82	11.13	9.43	8.22
(500)	c_p	3.5	—	—	2.65	2.29	2.25	2.32
	μ	22.38	—	—	24.38	30.17	30.17	33.06
	k	0.0434	—	—	0.0450	0.0522	0.0609	—
6.895	ρ	35.91	—	—	31.19	23.29	19.32	16.64
(1000)	c_p	5.	—	—	3.56	2.57	2.38	2.40
	μ	28.93	—	—	28.52	30.58	33.06	35.54
	k	0.0547	—	—	0.0521	0.0543	0.0618	—
10.34	ρ	58.64	—	—	55.5	36.63	29.55	25.20
(1500)	c_p	7.	—	—	6.05	2.90	2.53	2.45
	μ	33.89	—	—	33.89	33.89	35.96	38.02
	k	0.0656	—	—	0.0660	0.0574	0.0630	—
13.79	ρ	85.07	—	—	—	52.14	40.64	34.23
(2000)	c_p	9.	—	—	—	3.51	2.66	2.53
	μ	38.85	—	—	—	35.54	38.85	40.09
	k	0.0770	—	—	—	0.0614	0.0644	—
17.24	ρ	122.6	—	—	—	69.86	52.20	43.39
(2500)	c_p	11.	—	—	—	4.4	2.85	2.61
	μ	44.64	—	—	—	41.74	41.74	42.98
	k	—	—	—	—	—	—	—
20.68	ρ	188.5	—	—	—	91.07	64.49	52.81
(3000)	c_p	14.	—	—	—	6.33	3.06	2.78
	μ	47.94	—	—	—	45.46	44.63	45.46
	k	—	—	—	—	—	—	—

UNITS: ρ [kg/m³]; c_p [kJ/kg·Δ_1°C]; μ [Pa·s×10^{-6}]; k[W/m·Δ_1°C]

SOURCE: Adapted from J. Keenan and F. Keyes, *Thermodynamic Properties of Steam*, Wiley, New York, 1955; *Thermodynamic and Transport Properties of Steam*, American Society of Mechanical Engineers, 1967; and J. Keenan, F. Keyes, P. Hill, and J. Moore, *Steam Tables*, Wiley, New York, 1969.

Table B-4   PROPERTIES OF WATER (SATURATED LIQUID)

TEMP.	c_p	ρ	μ	k	$\alpha^* = \dfrac{g\beta\rho^2 c_p}{\mu k}$	Pr	r	β
[°C]	[kJ/kg·Δ_1°C]	[kg/m³]	[kg/m·s]	[W/m·Δ_1°C]	[1/m³·Δ_1°C]	[—]	[kJ/kg]	[1/Δ_1°C]
0	4.194	1000.0	1.79×10^{-3}	0.566	0.0	13.26	2501	0.00000
10	4.202	1000.0	1.31	0.585	5.37×10^9	9.41	2478	0.00010
20	4.190	998.0	1.01	0.602	13.5	7.03	2454	0.00020
30	4.179	996.0	0.803	0.619	23.7	5.42	2431	0.00029
40	4.177	992.6	0.656	0.633	36.9	4.33	2407	0.00038
50	4.178	988.1	0.536	0.644	53.3	3.48	2383	0.00046
60	4.183	983.3	0.475	0.654	67.6	3.04	2359	0.00053
70	4.187	977.8	0.408	0.664	85.4	2.573	2334	0.00059
80	4.197	971.8	0.359	0.671	103	2.245	2309	0.00064
90	4.206	965.4	0.318	0.676	122	1.979	2283	0.00069
100	4.219	958.6	0.283	0.682	146	1.751	2257	0.00074
110	4.233	951.3	0.253	0.685	173	1.563	2230	0.00080
120	4.251	943.4	0.229	0.685	203	1.421	2203	0.00086
130	4.270	935.1	0.211	0.685	230	1.315	2174	0.00091
140	4.294	926.4	0.196	0.685	258	1.229	2145	0.00096
150	4.321	917.3	0.185	0.684	287	1.169	2114	0.00102
160	4.350	907.8	0.174	0.681	320	1.111	2083	0.00108
170	4.383	897.7	0.164	0.679	354	1.059	2050	0.00114
180	4.420	887.3	0.154	0.676	390	1.007	2015	0.00120

Table B-4 (Continued)

TEMP. [°C]	c_p [kJ/kg·Δ_1°C]	ρ [kg/m³]	μ [kg/m·s]	k [W/m·Δ_1°C]	α^* $\dfrac{g\beta\rho^2 c_p}{\mu k}$ [1/m³·Δ_1°C]	Pr [—]	r [kJ/kg]	β [1/Δ_1°C]
190	4.461	876.4	0.147	0.671	428	0.977	1979	0.00135
200	4.506	865.0	0.139	0.667	470	0.939	1941	0.00138
210	4.558	853.0	0.133	0.661	516	0.917	1901	0.00142
220	4.616	840.8	0.127	0.654	570	0.896	1859	0.00154
230	4.684	827.1	0.121	0.648	633	0.875	1814	0.00161
240	4.761	813.7	0.116	0.638	706	0.866	1767	0.00170
250	4.850	799.4	0.112	0.627	792	0.866	1716	0.00188
260	4.955	783.7	0.107	0.616	896	0.861	1663	0.00198
270	5.079	768.0	0.103	0.598	1.02×10^{12}	0.875	1605	0.00214
280	5.224	750.8	0.098	0.581	1.21	0.881	1544	0.00239
290	5.399	732.1	0.094	0.560	1.42	0.906	1477	0.00263
300	5.610	712.3	0.091	0.530	1.67	0.963	1405	0.00288

SOURCE: Adapted from Frank Kreith, *Principles of Heat Transfer*, 1st ed., International Textbook Co., Scranton, Pa., 1962; and Joan Moore, *Steam Tables*, Wiley, New York, 1969.

SOURCE: Values of the modulus, $\alpha = g\beta\rho^2 c / \mu k$, and k for ethyl alcohol, water, and transformer oil. (*Values for alcohol and oil by W. J. King Mech. Eng., May, 1932. Values for water from Table B-3.*)

*Evaluated at $g = 9.80$ m/s^2.

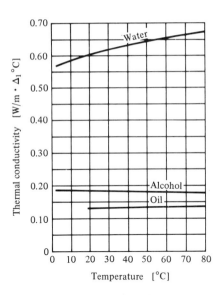
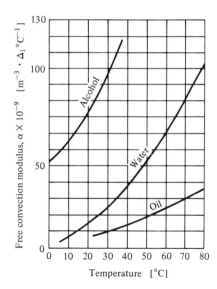

Properties of liquids. Values of the modulus, $\alpha = g\beta p^2 c_p / \mu k$ and k for ethyl alcohol, water, and transformer oil. (Values for alcohol and oil by W. J. King, *Mech. Eng.*, May, 1932. Values for water from Table B-3.)

Table B-5 PROPERTIES OF SATURATED LIQUIDS

T [°C]	ρ [kg/m³]	c_p [kg Δ_1 °C]	ν [m²/s]	k [W/m Δ_1 °C]	α_d [m²/s]	Pr [—]	r [kJ/kg]	β [1/Δ_1 °C]
AMMONIA, NH₃								
−50	703.69	4.463	0.435×10^{-6}	0.547	1.742×10^{-7}	2.60	1416.7	
−40	691.68	4.467	0.406	0.547	1.775	2.28	1389.0	
−30	679.34	4.476	0.387	0.549	1.801	2.15	1360.0	
−20	666.69	4.509	0.381	0.547	1.819	2.09	1329.5	
−10	653.55	4.564	0.378	0.543	1.825	2.07	1296.8	
0	640.10	4.635	0.373	0.540	1.819	2.05	1262.4	
10	626.16	4.714	0.368	0.531	1.801	2.04	1225.7	
20	611.75	4.798	0.359	0.521	1.775	2.02	1186.7	2.45
30	596.37	4.890	0.349	0.507	1.742	2.01	1145.0	$\times10^{-3}$
40	580.99	4.999	0.340	0.493	1.701	2.00	1100.5	
50	564.33	5.116	0.330	0.476	1.654	1.99	1052.0	
CARBON DIOXIDE, CO₂								
−50	1,156.34	1.84	0.119×10^{-6}	0.0855	0.4021×10^{-7}	2.96		
−40	1,117.77	1.88	0.118	0.1011	0.4810	2.46		
−30	1,076.76	1.97	0.117	0.1116	0.5272	2.22		
−20	1,032.39	2.05	0.115	0.1151	0.5445	2.12		
−10	983.38	2.18	0.113	0.1099	0.5133	2.20		
0	926.99	2.47	0.108	0.1045	0.4578	2.38		
10	860.03	3.14	0.101	0.0971	0.3608	2.80		
20	772.57	5.0	0.091	0.0872	0.2219	4.10		14.00
30	597.81	36.4	0.080	0.0703	0.0279	28.7		$\times10^{-3}$

SULFUR DIOXIDE, SO_2

−50	1,560.84	1.3595	0.484×10^{-6}	0.242	1.141×10^{-7}	4.24
−40	1,536.81	1.3607	0.424	0.235	1.130	3.74
−30	1,520.64	1.3616	0.371	0.230	1.117	3.31
−20	1,488.60	1.3624	0.324	0.225	1.107	2.93
−10	1,463.61	1.3628	0.288	0.218	1.097	2.62
0	1,438.46	1.3636	0.257	0.211	1.081	2.38
10	1,412.51	1.3645	0.232	0.204	1.066	2.18
20	1,386.40	1.3653	0.210	0.199	1.050	2.00 1.94×10^{-3}
30	1,359.33	1.3662	0.190	0.192	1.035	1.83
40	1,329.22	1.3674	0.173	0.185	1.019	1.70
50	1,299.10	1.3683	0.162	0.177	0.999	1.61

DICHLORODIFLUOROMETHANE (FREON-12), CCl_2F_2

−50	1,546.75	0.8750	0.310×10^{-6}	0.067	0.501×10^{-7}	6.2	173.61 2.63×10^{-3}
−40	1,518.71	0.8847	0.279	0.069	0.514	5.4	169.48
−30	1,489.56	0.8956	0.253	0.069	0.526	4.8	165.22
−20	1,460.57	0.9073	0.235	0.071	0.539	4.4	160.81
−10	1,429.49	0.9203	0.221	0.073	0.550	4.0	156.21
0	1,397.45	0.9345	0.214	0.073	0.557	3.8	151.38
10	1,364.30	0.9496	0.203	0.073	0.560	3.6	146.27
20	1,330.18	0.9659	0.198	0.073	0.560	3.5	140.81
30	1,295.10	0.9835	0.194	0.071	0.560	3.5	134.94
40	1,257.13	1.0019	0.191	0.069	0.555	3.5	128.53
50	1,215.96	1.0216	0.190	0.067	0.545	3.5	121.43

Table B-5 (Continued)

T [°C]	ρ [kg/m³]	c_p [kgΔ_1°C]	ν [m²/s]	k [W/mΔ_1°C]	α_d [m²/s]	Pr [—]	r [kJ/kg]	β [1/Δ_1°C]
GLYCERIN, $C_3H_5(OH)_3$								
0	1,276.03	2.261	0.00831	0.282	0.983×10^{-7}	84.7×10^3		
10	1,270.11	2.319	0.00300	0.284	0.965	31.0		
20	1,264.02	2.386	0.00118	0.286	0.947	12.5		0.50
30	1,258.09	2.445	0.00050	0.286	0.929	5.38		$\times 10^{-3}$
40	1,252.01	2.512	0.00022	0.286	0.914	2.45		
50	1,244.96	2.583	0.00015	0.287	0.893	1.63		
ETHYLENE GLYCOL, $C_2H_4(OH)_2$								
0	1,130.75	2.294	57.53×10^{-6}	0.242	0.934×10^{-7}	615		
20	1,116.65	2.382	19.18	0.249	0.939	204		0.65
40	1,101.43	2.474	8.69	0.256	0.939	93		$\times 10^{-3}$
60	1,087.66	2.562	4.75	0.260	0.932	51		
80	1,077.56	2.650	2.98	0.261	0.921	32.4		
100	1,058.50	2.742	2.03	0.263	0.908	22.4		

ENGINE OIL (UNUSED)

0	899.12	1.796	0.00428	0.147	0.911×10^{-7}	47,100	0.70×10^{-3}
20	888.23	1.880	0.00090	0.145	0.872	10,400	
40	876.05	1.964	0.00024	0.144	0.834	2,870	
60	864.04	2.047	0.839×10^{-4}	0.140	0.800	1,050	
80	852.02	2.131	0.375	0.138	0.769	490	
100	840.01	2.219	0.203	0.137	0.738	276	
120	828.96	2.307	0.124	0.135	0.710	175	
140	816.94	2.395	0.080	0.133	0.686	116	
160	805.89	2.483	0.056	0.132	0.663	84	

MERCURY, Hg

0	13,628.22	0.1403	0.124×10^{-6}	8.20	42.99×10^{-7}	0.0288	1.82×10^{-4}
20	13,579.04	0.1394	0.114	8.69	46.06	0.0249	
50	13,505.84	0.1386	0.104	9.40	50.22	0.0207	
100	13,384.58	0.1373	0.0928	10.51	57.16	0.0162	
150	13,264.28	0.1365	0.0853	11.49	63.54	0.0134	
200	13,144.94	0.1570	0.0802	12.34	69.08	0.0116	
250	13,025.60	0.1357	0.0765	13.07	74.06	0.0103	
315.5	12,847	0.134	0.0673	14.02	81.5	0.0083	

SOURCE: Adapted to SI Units from E. R. G. Eckert and R. M. Drake, *Heat and Mass Transfer*, 2d ed., McGraw-Hill, New York, 1959.

Table B-6 PROPERTY VALUES FOR METALS

METAL	PROPERTIES AT $20°C$			
	ρ [kg/m^3]	c_p [kJ/kg·$\Delta_1°C$]	k [W/m·$\Delta_1°C$]	α [m^2/s×10^5]
Aluminum:				
Pure	2,707	0.896	204	8.418
Al-Cu (Duralumin) 94–96% Al, 3–5% Cu, trace Mg	2,787	0.883	164	6.676
Al-Si (Silumin, copper-bearing) 86.5% Al, 1% Cu	2,659	0.867	137	5.933
Al-Si (Alusil) 78–80% Al, 20–22% Si	2,627	0.854	161	7.172
Al-Mg-Si 97% Al, 1% Mg, 1% Si, 1% Mn	2,707	0.892	177	7.311
Lead	11,373	0.130	35	2.343
Iron:				
Pure	7,897	0.452	73	2.034
Wrought iron 0.5% C	7,849	0.46	59	1.626
Steel (C max≈1.5%):				
Carbon steel				
C≈0.5%	7,833	0.465	54	1.474
1.0%	7,801	0.473	43	1.172
1.5%	7,753	0.486	36	0.970
Nickel steel				
Ni≈0%	7,897	0.452	73	2.026
20%	7,933	0.46	19	0.526
40%	8,169	0.46	10	0.279
80%	8,618	0.46	35	0.872
Invar 36% Ni	8,137	0.46	10.7	0.286
Chrome steel				
Cr 0%	7,897	0.452	73	2.026
1%	7,865	0.46	61	1.665
5%	7,833	0.46	40	1.110
20%	7,689	0.46	22	0.635
Cr-Ni (chrome-nickel): 15% Cr, 10% Ni	7,865	0.46	19	0.526
18% Cr, 8% Ni (V2A)	7,817	0.46	16.3	0.444
20% Cr, 15% Ni	7,833	0.46	15.1	0.415
25% Cr, 20% Ni	7,865	0.46	12.8	0.361
Tungsten steel				
W-0%	7,897	0.452	73	2.026
1%	7,913	0.448	66	1.858
5%	8,073	0.435	54	1.525
10%	8,314	0.419	48	1.391

THERMAL CONDUCTIVITY k [W/m·Δ_1°C]								
−100°C −148°F	0°C 32°F	100°C 212°F	200°C 392°F	300°C 572°F	400°C 752°F	600°C 1112°F	800°C 1472°F	1000°C 1832°F
215	202	206	215	228	249			
126	159	182	194					
119	137	144	152	161				
144	157	168	175	178				
	175	189	204					
36.9	35.1	33.4	31.5	29.8				
87	73	67	62	55	48	40	36	35
	59	57	52	48	45	36	33	33
	55	52	48	45	42	35	31	29
	43	43	42	40	36	33	29	28
	36	36	36	35	33	31	28	28
87	73	67	62	55	48	40	36	35
	62	55	52	47	42	36	33	33
	40	38	36	36	33	29	29	29
	22	22	22	22	24	24	26	29
	16.3	17	17	19	19	22	26	31

Table B-6 (Continued)

METAL	ρ [kg/m^3]	c_p [kJ/kg·Δ_1°C]	k [W/m·Δ_1°C]	α [m^2/s×10^5]
		PROPERTIES AT 20°C		
Copper:				
Pure	8,954	0.3831	386	11.234
Aluminum bronze				
95% Cu, 5% Al	8,666	0.410	83	2.330
Bronze 75% Cu,				
25% Sn	8,666	0.343	26	0.859
Red brass 85% Cu,				
9% Sn, 6% Zn	8,714	0.385	61	1.804
Brass 70% Cu,				
30% Zn	8,522	0.385	111	3.412
German silver 62%				
Cu, 15% Ni,				
22% Zn	8,618	0.394	24.9	0.733
Constantan 60%				
Cu, 40% Ni	8,922	0.410	22.7	0.612
Gold, pure	19,300	0.1289	315	12.662
Magnesium:				
Pure	1,746	1.013	171	9.708
Mg-Al (electroly-				
tic) 6–8% Al,				
1–2% Zn	1,810	1.00	66	3.605
Molybdenum	10,220	0.251	123	4.790
Nickel:				
Pure (99.9%)	8,906	0.4459	90	2.266
Ni-Cr 90% Ni,				
10% Cr	8,666	0.444	17	0.444
80% Ni, 20% Cr	8,314	0.444	12.6	0.343
Silver:				
Purest	10,524	0.2340	419	17.004
Pure (99.9%)	10,524	0.2340	407	16.563
Tin, pure	7,304	0.2265	64	3.884
Tungsten	19,350	0.1344	163	6.271
Zinc, pure	7,144	0.3843	112.2	4.106

SOURCE: Adapted to SI units from E. R. G. Eckert and R. M. Drake, *Heat and Mass Transfer*, 2d ed., McGraw-Hill, New York, 1959.

THERMAL CONDUCTIVITY k [W/m·°C]								
−100°C −148°F	0°C 32°F	100°C 212°F	200°C 392°F	300°C 572°F	400°C 752°F	600°C 1112°F	800°C 1472°F	1000°C 1832°F
407	386	379	374	369	363	353		
	59	71						
88		128	144	147	147			
19.2		31	40	45	48			
21		22.2	26					
331	318	312	310	305	299	286		
178	171	168	163	157				
	52	62	74	83				
138	125	118	114	111	109	106	102	99
104	93	83	73	64	59			
	17.1	18.9	20.9	22.8	24.6			
	12.3	13.8	15.6	17.1	18.0	22.5		
419	417	415	412					
419	410	415	374	362	360			
74	65.9	59	57					
	166	151	142	133	126	112	76	
114	112	109	106	100	93			

Table B-7 PHYSICAL PROPERTIES OF SOME COMMON LOW-MELTING-POINT METALS

METAL	MELTING POINT [°C]	NORMAL BOILING POINT [°C]	TEMPERATURE [°C]	DENSITY [kg/m³×10⁻³]	VISCOSITY [kg/m·s×10³]	HEAT CAPACITY [kJ/kg·Δ₁°C]	THERMAL CONDUCTIVITY [W/m·Δ₁°C]	PRANDTL NUMBER [—]
Bismuth	271	1477	316	10.01	1.62	0.144	16.4	0.014
			760	9.47	0.79	0.165	15.6	0.0084
Lead	327	1737	371	10.5	2.40	0.159	16.1	0.024
			704	10.1	1.37	0.155	14.9	0.016
Lithium	179	1317	204	0.51	0.60	4.19	38.1	0.065
			982	0.44	0.42	4.19		
Mercury	−39	357	10	13.6	1.59	0.138	8.1	0.027
			316	12.8	0.86	0.134	14.0	0.0084
Potassium	63.8	760	149	0.81	0.5	0.796	45.0	0.0066
			704	0.67	0.14	0.754	33.1	0.0031
Sodium	97.8	883	204	0.90	0.43	1.34	80.3	0.0072
			704	0.78	0.18	1.26	59.7	0.0038
Sodium potassium: 22% Na	19	826	93.3	0.848	0.49	0.946	24.4	0.019
			760	0.69	0.146	0.883		
56% Na	−11	784	93.3	0.89	0.58	1.13	25.6	0.026
			760	0.74	0.16	1.04	28.9	0.058
Lead bismuth, 44.5% Pb	125	1670	288	10.3	1.76	0.147	10.7	0.024
			649	9.84	1.15			

source: Adapted to SI Units from J. G. Knudsen and D. L. Katz, *Fluid Dynamics and Heat Transfer*, McGraw-Hill, New York, 1958.

Table B-8   PROPERTIES OF NONMETALS

SUBSTANCE	TEMPERATURE [°C]	k [W/m·Δ_1°C]	ρ [kg/m³]	c_p [kJ/kg·Δ_1°C]	α_d [m²/s×10⁷]
Asphalt	20–55	0.74–0.76			
Brick:		STRUCTURAL AND HEAT-RESISTANT MATERIALS			
Building brick, common	20	0.69	1600	0.84	5.2
Face		1.32	2000		
Carborundum brick	600	18.5			
	1400	11.1			
Chrome brick	200	2.32	3000	0.84	9.2
	550	2.47			9.8
	900	1.99			7.9
Diatomaceous earth, molded and fired	200	0.24			
	870	0.31			
Fireclay brick, burnt 2426°F	500	1.04	2000	0.96	5.4
	800	1.07			
	1100	1.09			
Burnt 2642°F	500	1.28	2300	0.96	5.8
	800	1.37			
	1100	1.40			
Missouri	200	1.00	2600	0.96	4.0
	600	1.47			
	1400	1.77			

Table B-8 (Continued)

SUBSTANCE	TEMPERATURE [°C]	k [W/m·Δ_1°C]	ρ [kg/m³]	c_p [kJ/kg·Δ_1°C]	α_d [m²/s×10⁷]
Magnesite	200	3.81		1.13	
	650	2.77			
	1200	1.90			
STRUCTURAL AND HEAT-RESISTANT MATERIALS					
Cement, portland		0.29			
Mortar	23	1.16			
Concrete, cinder	23	0.76	1500		
Stone 1-2-4 mix	20	1.37	1900–2300	0.88	8.2–6.8
Glass, window	20	0.78 (av)	2700	0.84	3.4
Corosilicate	30–75	1.09	2200		
Plaster, gypsum	20	0.48	1440	0.84	4.0
Metal lath	20	0.47			
Wood lath	20	0.28			
Stone:					
Granite	100–300	1.73–3.98	2640	0.82	8–18
Limestone		1.26–1.33	2500	0.90	5.6–5.9
Marble		2.07–2.94	2500–2700	0.80	10–13.6
Sandstone	40	1.83	2160–2300	0.71	11.2–11.9
Wood (across the grain):					
Balsa, 8.8 lb/ft³	30	0.055	140		
Cypress	30	0.097	460		
Fir	23	0.11	420		
Maple or oak	30	0.166	540	2.72	0.96
Yellow pine	23	0.147	640	2.4	1.28
White pine	30	0.112	430	2.8	0.82

INSULATING MATERIAL

Asbestos:					
Loosely packed	−45	0.149			
	0	0.154	470–570	0.816	3.3–4
	100	0.161			
Asbestos-cement boards	20	0.74			
Sheets	51	0.166			
Felt, 40 laminations/in.	38	0.057			
	150	0.069			
	260	0.083			
20 laminations/in.	38	0.078			
	150	0.095			
	260	0.112			
Corrugated, 4 plies/in.	38	0.087			
	93	0.100			
	150	0.119			
Asbestos cement	—	2.08			
Balsam wool, 2.2 lb/ft^3	32	0.04	35		
Cardboard, corrugated	—	0.064			
Celotex	32	0.048			
Corkboard, 10 lb/ft^3	30	0.043	160		
Cork, regranulated	32	0.045	45–120	1.88	2–5.3
Ground	32	0.043	150		
Diatomaceous earth (Sil-o-cel)	0	0.061	320		
Felt, hair	30	0.036	130–200		
Wool	30	0.052	330		

Table B-8 (Continued)

SUBSTANCE	TEMPERATURE [°C]	k [W/m·Δ_1°C]	ρ [kg/m³]	c_p [kJ/kg·Δ_1°C]	α_d [m²/s×10⁷]
INSULATING MATERIAL					
Fiber, insulating board	20	0.048	240		
Glass wool, 1.5 lb/ft³	23	0.038	24	0.7	22.6
Insulex, dry	32	0.064			
		0.144			
Kapok	30	0.035			
Magnesia, 85%	38	0.067	270		
	93	0.071			
	150	0.074			
	204	0.080			
Rock wool, 10 lb/ft³	32	0.040	160		
Loosely packed	150	0.067	64		
	260	0.087			
Sawdust	23	0.059			
Silica aerogel	32	0.024	140		
Wood shavings	23	0.059			

SOURCE: Adapted to SI Units from A. I. Brown and S. M. Marco, *Introduction to Heat Transfer*, 3d ed., McGraw-Hill, New York, 1958.

Table B-9   THE NORMAL TOTAL EMISSIVITY OF VARIOUS SURFACES

SURFACES	°C	ε
A. METALS AND THEIR OXIDES		
Aluminum:		
Highly polished plate, 98.3% pure	227, 577	0.039, 0.057
Polished plate	23	0.040
Rough plate	25	0.055
Oxidized at 600°C	200, 600	0.11, 0.19
Al-surfaced roofing	43	0.216
Al-treated surfaces, heated at 600°C		
Copper	200, 600	0.18, 0.19
Steel	200, 600	0.52, 0.57
Brass:		
Highly polished:		
73.2% Cu, 26.7% Zn, by weight	890, 1250	0.028, 0.031
62.4% Cu, 36.8% Zn, 0.4% Pb, 0.3% Al,		
by weight	920, 377	0.0388, 0.037
82.9% Cu, 17.0% Zn, by weight	277	0.030
Hard-rolled, polished, but direction of polishing		
visible	21	0.038
But somewhat attacked	23	0.043
But traces of stearin from polish left on	24	0.053
Polished	38, 315	0.096, 0.096
Rolled plate:		
Natural surface	22	0.06
Rubbed with coarse emery	22	0.20
Dull plate	50, 350	0.22
Oxidized by heating at 600°C	200, 600	0.61, 0.59
Chromium:		
See Nickel Alloys for Ni-Cr steels		
Copper:		
Carefully polished electrolytic Cu	80	0.018
Commercial, emeried, polished, but pits		
remaining	19	0.030
Scraped shiny, but not mirrorlike	22	0.072
Polished	116	0.023
Plate heated at 600°C	200, 600	0.57, 0.57
Cuprous oxide	800, 1100	0.66, 0.54
Plate, heated for a long time, covered with thick		
oxide layer	25	0.78
Molten copper	1077, 1277	0.16, 0.13
Gold:		
Pure, highly polished	227, 632	0.018, 0.035
Iron and steel:		
Metallic surfaces (or very thin oxide layer):		
Electrolytic iron, highly polished	177, 227	0.052, 0.074
Polished iron	427, 1027	0.144, 0.377
Iron freshly emeried	20	0.242
Cast iron, polished	200	0.21
Wrought iron, highly polished	38, 250	0.28
Cast iron, newly turned	22	0.435

Table B-9 (Continued)

SURFACES	°C	ε
A. METALS AND THEIR OXIDES		
Polished steel casting	770, 1037	0.52, 0.56
Ground sheet steel	937, 1100	0.55, 0.61
Smooth sheet iron	900, 1037	0.55, 0.60
Cast iron, turned on lathe	880, 987	0.60, 0.70
Oxidized surfaces:		
Iron plate, pickled, then rusted red	20	0.612
Then completely rusted	19	0.685
Rolled sheet steel	21	0.657
Oxidized iron	100	0.736
Cast iron, oxidized at 600°C	200, 600	0.64, 0.78
Steel oxidized at 600°C	200, 600	0.79, 0.79
Smooth, oxidized electrolytic iron	127, 527	0.78, 0.82
Iron oxide	500, 1200	0.85, 0.89
Rough ingot iron	927, 1115	0.87, 0.95
Sheet steel, strong rough oxide layer	24	0.80
Dense shiny oxide layer	24	0.82
Cast plate:		
Smooth	23	0.80
Rough	23	0.82
Cast iron, rough, strongly oxidized	38, 250	0.95
Wrought iron, dull-oxidized	21, 360	0.94
Steel plate, rough	38, 371	0.94, 0.97
High-temperature alloy steels: see Nickel alloys		
Molten metals:		
Molten cast iron	1300, 1400	0.29, 0.29
Molten mild steel	1600, 1800	0.28, 0.28
Lead:		
Pure (99.96%) unoxidized	127, 227	0.057, 0.075
Gray oxidized	24	0.281
Oxidized at 200°C	200	0.63
Mercury, pure clean	0, 100	0.09, 0.12
Molybdenum filament	727, 2590	0.096, 0.292
Ni-Cu alloy, oxidized at 600°C	200, 600	0.41, 0.46
Nickel:		
Electroplated on polished iron, then polished	23	0.045
Technically pure (98.9% Ni by weight, + Mn), polished	227, 377	0.07, 0.087
Electroplated on pickled iron, not polished	20	0.11
Wire	187, 1007	0.096, 0.186
Plate, oxidized by heating at 600°C	200, 600	0.37, 0.48
Nickel oxide	650, 1255	0.59, 0.86
Nickel alloys:		
Cr-Ni alloy	52, 1035	0.64, 0.76
(18–32% Ni, 55–68% Cu, 20% Zn by weight), gray oxidized	21	0.262
Alloy steel (8% Ni, 18% Cr); light silvery, rough; brown after heating	215, 490	0.44, 0.36
Same, after 24 hr heating at 527°C	215, 527	0.62, 0.73

Table B-9   (Continued)

SURFACES	°C	ε
A. METALS AND THEIR OXIDES		
Alloy (20% Ni, 25% Cr), brown, splotched, oxidized from service	215, 527	0.90, 0.97
Alloy (60% Ni, 12% Cr), smooth, black, firm adhesive oxide coat from service	271, 563	0.89, 0.82
Platinum:		
Pure, polished plate	227, 627	0.054, 0.104
Strip	927, 1627	0.12, 0.17
Filament	27, 1227	0.036, 0.192
Wire	227, 1377	0.073, 0.182
Silver:		
Polished, pure	227, 627	0.0198, 0.0324
Polished	38, 371	0.0221, 0.0312
Steel, see Iron		
Tantalum filament	1327, 2527	0.193, 0.31
Tin, bright, tinned iron sheet	24	0.043, 0.064
Tungsten:		
Filament, aged	24, 3315	0.032, 0.35
Filament	3315	0.39
Zinc:		
Commercial, 99.1% pure, polished	227, 327	0.045, 0.053
Oxidized by heating at 400°C	400	0.11
Galvanized sheet iron:		
Fairly bright	28	0.228
Gray, oxidized	24	0.276
B. REFRACTORIES, BUILDING MATERIALS, PAINTS, AND MISCELLANEOUS		
Asbestos board	23	0.96
Asbestos paper	38, 371	0.93, 0.945
Brick:		
Red, rough, but no gross irregularities	21	0.93
Silica, unglazed, rough	1000	0.80
Silica, glazed, rough	1100	0.85
Grog brick, glazed	1100	0.75
See Refractory materials, below		
Carbon:		
T-carbon, 0.9% ash	127, 627	0.81, 0.79
Carbon filament	1037, 1404	0.526
Candle soot	127, 271	0.952
Lampblack:		
Water-glass coating	98, 183	0.959, 0.947
Water-glass coating	127, 227	0.957, 0.952
Thin layer on iron plate	21	0.927
Thick coat	20	0.967
0.003 in. or thicker	38, 371	0.945
Enamel, white, fused on iron	19	0.897
Glass, smooth	22	0.937
Gypsum, 0.02 in. thick or smooth on blackened plate	21	0.903
Marble, light gray, polished	22	0.931

Table B-9 (Continued)

SURFACES	°C	ε
B. REFRACTORIES, BUILDING MATERIALS, PAINTS, AND MISCELLANEOUS		
Oak, planed	21	0.895
Oil layers on polished nickel (lubricating oil):		
Polished surface alone	20	0.045
+0.001 in. oil	20	0.27
+0.002 in. oil	20	0.46
+0.005 in. oil	20	0.72
+∞	20	0.82
Oil layers on aluminum foil (linseed oil):		
Aluminum foil	100	0.087
+1 coat oil	100	0.561
+2 coats oil	100	0.574
Paints, lacquers, varnishes:		
Snow-white enamel varnish on rough iron plate	23	0.906
Black shiny lacquer, sprayed on iron	24	0.875
Black shiny shellac on tinned iron sheet	21	0.821
Black-matte shellac	77, 147	0.91
Black lacquer	38, 93	0.80, 0.95
Flat black lacquer	38, 93	0.96, 0.98
White lacquer	38, 93	0.80, 0.95
Oil paints, 16 different, all colors	100	0.92, 0.96
Aluminum paints and lacquers:		
10% Al, 22% lacquer body, on rough or smooth surface	100	0.52
26% Al, 27% lacquer body, on rough or smooth surface	100	0.30
Other aluminum paints, varying age and Al content	100	0.27, 0.67
Aluminum lacquer, varnish binder, on rough plate	21	0.39
Aluminum paint, after heating to 327°C	150, 315	0.35
Paper, thin:		
Pasted on tinned iron plate	19	0.924
Pasted on rough iron plate	19	0.929
Pasted on black lacquered plate	19	0.944
Plaster, rough, lime	10, 88	0.91
Porcelain, glazed	22	0.924
Quartz, rough, fused	21	0.932
Refractory materials, 40 different		
Poor radiators	600, 1000	0.65, 0.75 0.70
Good radiators	600, 1000	0.80, 0.85 0.85, 0.90
Roofing paper	21	0.91
Rubber:		
Hard, glossy plate	23	0.945
Soft, gray, rough (reclaimed)	24	0.859
Serpentine, polished	23	0.900
Water	0, 100	0.95, 0.963

SOURCE: Adapted from data by H. C. Hottel in W. H. McAdams, *Heat Transmission*, 3rd ed., McGraw-Hill, New York, 1954.

Table B-10 OVERALL HEAT TRANSFER COEFFICIENTS FOR VARIOUS FORMS OF BUILDING CONSTRUCTION

	THICKNESS		OVERALL HEAT TRANSFER COEFFICIENT (U)		
	[(in.)]	[cm]	PLAIN	PLASTERED	FURRED, LATHED, AND PLASTERED
WALLS					
Solid					
Adobe brick	(12)	30.5	1.5	1.4	1.1
Brick	(8)	20.3	2.8	2.6	1.7
	(12)	30.5	2.0	1.9	1.4
Cement					
Hollow block (cinder)	(8)	20.3	2.3	2.2	1.5
Poured (140 lb/ft^3)	(8)	20.3	3.9	3.7	2.1
Glass Block (6 in. ×6 in. ×4 in. thick)	—	—	3.5	—	—
(8 in. ×8 in. ×4 in. thick)	—	—	3.3	—	—
Stone	(12)	30.5	3.2	3.0	2.0
Tile, Hollow	(8)	20.3	2.1	2.0	1.4
FRAME					
Clapboard, 1-in. wood sheathing, lathed and plastered		1.4			
Clapboard, 25/32-in. rigid insulation, lathed and plastered		1.1			
Clapboard, 1-in. wood sheathing, 3$\frac{5}{8}$-in. rock wool fill between studing, lathed and plastered		0.4			
WINDOWS					
Ordinary glass windows (single pane)		6.4			
Double windows		3.7			

Table B-10 (Continued)

		OVERALL HEAT TRANSFER COEFFICIENT (U)		
THICKNESS [(in.)]	[cm]	PLAIN	PLASTERED	FURRED, LATHED, AND PLASTERED

DOORS

	THICKNESS [cm]
Wood, 4.0 cm ($1\frac{5}{8}$-in.)	3.0
Wood, glass upper part	4.3
Wood frame, all glass	5.4
Steel	6.8
Steel, glass upper part	6.5
Steel frame, all glass	6.0

FLOORS AND CEILINGS

Wood Joists

	DIRECTION OF HEAT FLOW	
	UP	DOWN
Lathed and plastered ceiling, no floor above	3.5	2.5
Lathed and plastered ceiling, no floor above		
$3\frac{5}{8}$-in. rock wool fill between joists	0.5	0.4
Lathed and plastered ceiling, 1-in. pine floor above	1.6	1.4
No ceiling, 1-in. pine floor above	2.0	2.5
Concrete floor on ground 15.2-cm (6-in.) thick	1.4	—

ROOFS

PLAIN	WITH 2.5-cm (1-in.) SOLID INSULATION	WITH 2.5-cm (1-in.) SOLID INSULATION AND SUSPENDED PLASTER CEILING

Flat

Concrete, reinforced			
15-cm (6-in.) thick	2.9	1.2	0.9
Metal	3.8	1.3	1.0
Wood			
5-cm (2-in.) thick	1.6	0.9	0.7

	DIRECTION OF HEAT FLOW	
	UP	DOWN
Pitched		
Rafters		
Asphalt shingles		
2-cm (¾-in.) wood sheathing and roofing felt		
Open (no ceiling)	1.9	1.7
With lath and plaster	1.5	1.3
Corrugated iron on battens	10	—
Slate or tiles		
On battens	8	—
Battens and roofing felt	4	—
2-cm (¾-in.) wood sheathing and roofing felt	1.4	1.2
Wood shingles		
Battens and roofing felt		
Open (no ceiling)	3.4	3.1
With lath and plaster ceiling	1.8	1.7
2-cm (¾-in.) wood sheathing and roofing felt		
Open (no ceiling)	2.2	2.0
With lath and plaster ceiling	1.5	1.3
Single glass sky light	6.8	—

Table B-11 THERMAL RESISTANCES OF AIR SPACES AND AIR SURFACE FILMS

	AIR SPACE			
AXIS NORMAL TO WIDTH	DIRECTION OF HEAT FLOW	AIRSPACE [(in.)]	WIDTH [cm]	THERMAL RESISTANCE ($\mathcal{R}$) [$m^2 \cdot \Delta\,°C/W$]
Horizontal	Up	$(\frac{3}{4}-4)$	2–10	0.15
Horizontal	Down	$(\frac{3}{4})$	2	0.19
Horizontal	Down	(4)	10	0.22
Vertical	Horizontal	$(\frac{3}{4}-4)$	2–10	0.17
Sloping at 45°	Up	$(\frac{3}{4}-4)$	2–10	0.16
Sloping at 45°	Down	$(\frac{3}{4}-4)$	2–10	0.18

AIR SURFACE FILMS FOR STILL AIR		
Surface position		
Horizontal	Up	0.11
Horizontal	Down	0.16
Vertical	Horizontal	0.12
Sloping at 45°	Up	0.11
Sloping at 45°	Down	0.13

AIR SURFACE FILMS

As a function of wind velocity and surface texture

SOURCE: F. B. Rowley, A. B. Algren, and J. L. Blackshaw, "Surface Conductance as Affected by Air Velocity, Temperature, and Character of Surface," *Trans. ASHVE* 36 (1930).

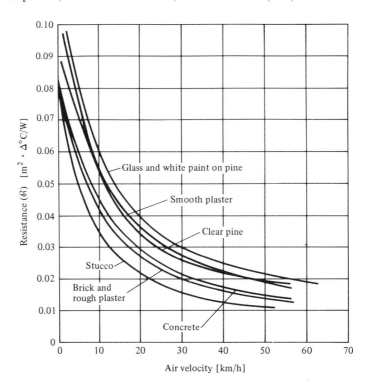

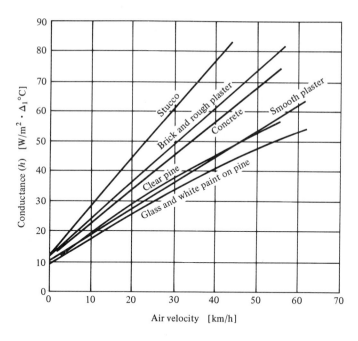

Appendix C
Conversion Factors and
Equations of Equality

C-1 EQUATIONS OF EQUALITY

C-1.1. Dimensions

Length (l)

1 in. = 0.0254 cm

1 in. = 2.540 cm

1 ft = 0.3048 m

1 mi = 1.609 km

Area (l^2)

1 in.^{2} = 0.64516 × 10^{-3} m^2

1 in.^{2} = 6.4516 cm^2

1 ft^2 = 0.09290 m^2

1 m^2 = 10,000 cm^2

Volume (l^3)

1 in.^{3} = 16.387 cm^3

1 ft^3 = 0.0283 m^3

1 gal $= 3.7854$ liters
1 gal $= 0.0037854$ m^3
1 m$^3 = 1,000,000$ cm^3
1 m$^3 = 1000$ liters
1 liter $= 1000$ cm^3

C-1.2 Fluid Flow, Mass, Density, and Pressure

Flow Velocity (V)
1 ft/sec $= 0.3048$ m/s
1 ft/sec $= 1097$ m/h
1 mi/hr $= 1.609$ km/h

Flow Rates – Volume
1 ft^3/sec $= 0.0283$ m^3/s
1 ft^3/sec $= 101.88$ m^3/h
1 gal/min $= 0.2271$ m^3/h

Flow Rates – Mass
1 lb$_m$/sec $= 0.4536$ kg/s
1 lb$_m$/sec $= 1632.96$ kg/h
1 lb$_m$/hr $= 0.000126$ kg/s

Mass
1 lb$_m = 0.4536$ kg
1 metric ton $= 1000$ kg

Force
1 lb$_f = 4.4482$ N

Density
1 lb$_m$/ft$^3 = 16.0185$ kg/m^3
1 lb$_m$/ft$^3 = 0.016019$ g/cm^3

Pressure
1 lb$_f$/in.$^2 = 6894.76$ Pa
1 lb$_f$/ft$^2 = 47.88$ Pa
1 atm $= 101352$ Pa $= 0.10135$ MPa
1 in. Hg $= 3.38$ kPa

Viscosity — Absolute (μ)
1 lb$_m$/ft·hr $= 0.0004133$ Pa·s
1 lb$_m$/ft·sec $= 1.488$ Pa·s
1 lb$_f$·sec/ft$^2 = 47.88$ Pa·s
1 centipoise $= 1 \times 10^{-3}$ Pa·s

Viscosity — Kinematic (ν)
$$1 \text{ ft}^2/\text{sec} = 0.0929 \text{ m}^2/\text{s}$$
$$1 \text{ centistoke} = 1 \times 10^{-6} \text{ m}^2/\text{s}$$

C-1.3 Latent and Specific Heats and Temperature

Latent Heat
$$1 \text{ Btu/lb} = 2.3254 \text{ J/g} = 2.324 \text{ kJ/kg}$$
$$1 \text{ g} \cdot \text{cal/g} = 4.1865 \text{ J/gm}$$

Specific Heat
$$1 \text{ Btu/lb} \cdot \Delta_1 {}^\circ\text{F} = 4186 \text{ J/kg} \cdot \Delta_1 {}^\circ\text{C}$$
$$1 \text{ g} \cdot \text{cal/g} \cdot \Delta_1 {}^\circ\text{C} = 4186 \text{ J/kg} \cdot \Delta_1 {}^\circ\text{C}$$

Temperature Level
$$^\circ\text{C} = 0.555 \, (^\circ\text{F} - 32)$$
$$\text{K} = {}^\circ\text{C} + 273.2$$
$$\text{K} = 5/9 {}^\circ\text{R}$$
$$^\circ\text{R} = {}^\circ\text{F} + 459.7$$

Temperature Change
$$1 \, \Delta {}^\circ\text{C} = 1 \, \Delta\text{K}$$
$$5 \, \Delta {}^\circ\text{C} = 9 \, \Delta {}^\circ\text{F}$$
$$5 \, \Delta\text{K} = 9 \, \Delta {}^\circ\text{R}$$

C-1.4 Energy, Heat, and Work

$$1 \text{ Btu} = 1054.8 \text{ J}$$
$$1 \text{ Btu} = 778.3 \text{ ft} \cdot \text{lb}_f$$
$$1 \text{ ft} \cdot \text{lb}_f = 1.3553 \text{ J} = 1.3553 \text{ N} \cdot \text{m}$$
$$1 \text{ g} \cdot \text{cal} = 4.186 \text{ J}$$
$$1 \text{ J} = 1 \text{ W} \cdot \text{s}$$
$$1 \text{ J} = \text{N} \cdot \text{m}$$
$$1 \text{ kWh} = 3.6 \times 10^6 \text{ N} \cdot \text{m}$$
$$1 \text{ kWh} = 3.6 \times 10^6 \text{ J}$$
$$1 \text{ horsepower hr} = 0.7457 \text{ kWh}$$
$$10^7 \text{ erg} = 1 \text{ J}$$

C-1.5 Enthalpy, Entropy

$$(h) \; 1 \text{ Btu/lb}_m = 2.3266 \text{ kJ/kg}$$
$$(s) \; 1 \text{ Btu/lb}_m \cdot \Delta_1 {}^\circ\text{R} = 4.1875 \text{ kJ/kg} \cdot \Delta_1 \text{K}$$

C-1.6 Power

$$1 \text{ Btu/sec} = 1.05448 \text{ kW}$$
$$1 \text{ Btu/hr} = 0.293 \text{ W}$$
$$1 \text{ Btu/min} = 17.58 \text{ W}$$
$$1 \text{ ft} \cdot \text{lb}_f / \text{sec} = 1.356 \text{ W}$$
$$1 \text{ horsepower} = 0.7457 \text{ kW}$$
$$1 \text{ ton refrigeration} = 3.516 \text{ kW}$$
$$10^7 \text{ erg/sec} = 1 \text{ W}$$
$$1 \text{ J/sec} = 1 \text{ W}$$
$$1 \text{ g} \cdot \text{cal/sec} = 4.184 \text{ W}$$
$$1 \text{ K} \cdot \text{cal/hr} = 1.1622 \text{ W}$$

C-1.7 Thermal Conductivities, Conductances, Film Coefficients, and Heat Flux

Thermal Conductivity (k)
$$1 \text{ Btu/hr} \cdot \text{ft} \cdot \Delta_1 {}^\circ\text{F} = 1.73 \text{ W/m} \cdot \Delta_1 {}^\circ\text{C}$$
$$1 \text{ Btu/hr} \cdot \text{ft}^2 \cdot \Delta_1 {}^\circ\text{F/in.} = 0.144 \text{ W/m} \cdot \Delta_1 {}^\circ\text{C}$$
$$1 \text{ W/in.}^2 \cdot \Delta_1 {}^\circ\text{C/in.} = 39.37 \text{ W/m} \cdot \Delta_1 {}^\circ\text{C}$$
$$1 \text{ W/m} \cdot \Delta_1 {}^\circ\text{C} = 100 \text{ W/m}^2 \cdot \Delta_1 {}^\circ\text{C/cm}$$
$$1 \text{ kg} \cdot \text{cal/s} \cdot \text{m} \cdot \Delta_1 {}^\circ\text{C} = 4.186 \text{ kW/m} \cdot \Delta_1 {}^\circ\text{C}$$
$$1 \text{ g} \cdot \text{cal/s} \cdot \text{cm}^2 \cdot \Delta_1 {}^\circ\text{C/cm} = 418.6 \text{ W/m} \cdot \Delta_1 {}^\circ\text{C}$$

Thermal Conductance $(\mathcal{K})$
$$1 \text{ Btu/hr} \cdot \text{ft}^2 \cdot \Delta {}^\circ\text{F} = 5.68 \text{ W/m}^2 \cdot \Delta_1 {}^\circ\text{C}$$

Thermal Conductance $(\overline{\mathcal{K}})$
$$1 \text{ Btu/hr} \cdot \Delta_1 {}^\circ\text{F} = 0.5274 \text{ W/} \Delta_1 {}^\circ\text{C}$$

Film Coefficient (h)
$$1 \text{ Btu/hr} \cdot \text{ft}^2 \cdot \Delta_1 {}^\circ\text{F} = 5.68 \text{ W/m}^2 \cdot \Delta_1 {}^\circ\text{C}$$
$$1 \text{ K} \cdot \text{cal/hr} \cdot \text{m}^2 \cdot \Delta_1 {}^\circ\text{C} = 1.1622 \text{ W/m}^2 \cdot \Delta_1 {}^\circ\text{C}$$

Heat Flux (Q/A)
$$1 \text{ Btu/hr} \cdot \text{ft}^2 = 3.154 \text{ W/m}^2$$
$$1 \text{ kg} \cdot \text{cal/h} \cdot \text{m}^2 = 1.1622 \text{ W/m}^2$$

Thermal Resistance $(\mathcal{R})$
$$1 \text{ ft}^2 \cdot \Delta_1 {}^\circ\text{F/Btu/hr} = 0.176 \text{ m}^2 \cdot \Delta {}^\circ\text{C/W}$$

Thermal Resistance $(\overline{\mathcal{R}})$
$$1 \Delta {}^\circ\text{F/Btu/hr} = 1.896 \Delta {}^\circ\text{C/W}$$

Table C-2   TEMPERATURE CONVERSION TABLE

°F	°C	°C	°F
−50	−45.6	−50	−58.0
−40	−40.0	−40	−40.0
−30	−34.4	−30	−22.0
−20	−28.9	−20	−4.0
−10	−23.3	−10	+14.0
0	−17.9	0	32.0
+10	−12.2	+10	50.0
20	−6.7	20	68.0
30	−1.1	30	86.0
40	4.4	40	104.0
50	10.0	50	122.0
60	15.6	60	140.0
70	21.1	70	158.0
80	26.7	80	176.0
90	32.2	90	194.0
100	37.8	100	212.0
120	48.9	120	248.0
140	60.0	140	284.0
160	71.1	160	320.0
180	82.2	180	356.0
200	93.3	200	392.0
220	104.4	220	428.0
240	115.6	240	464.0
260	126.7	260	500.0
280	137.8	280	536.0
300	148.9	300	572.0
350	176.7	350	662.0
400	204.4	400	752.0
450	232.2	450	842.0
500	260.0	500	932.0
600	315.6	600	1112.0
650	343.4	650	1202.0
700	371.1	700	1292.0
750	398.9	750	1382.0
800	426.7	800	1472.0
850	454.4	850	1562.0
900	482.2	900	1652.0
950	510.0	950	1742.0
1000	537.8	1000	1832.0
1050	565.6	1050	1922.0
1100	593.3	1100	2012.0
1150	621.1	1150	2102.0
1200	648.9	1200	2192.0
1300	704.4	1300	2372.0
1400	760.0	1400	2552.0
1500	815.5	1500	2732.0
2000	1093.6	2000	3632.0
2500	1371.1	2500	4532.0
3000	1648.8	3000	5432.0
3500	1926.7	3500	6332.0
4000	2204.4	4000	7232.0

Appendix D
An Approach to Problem Solving

The solution of a heat transfer problem can be divided into two parts, namely

- planning
- numerical solution

The planning part can be considered analogous to "programming" a computer. The planning delineates how the numerical solution is to be obtained. The numerical solution is analogous to computer computations which follow the program directions. Consequently, the authors of this text have placed an emphasis upon the planning aspects of the solution of a problem. Proper planning, in essence, solves the problem, leaving just the mechanics of computation to obtain the numerical solution.

A generalized approach to planning and numerical solution of heat transfer problems is given below.

Planning

1. *Visualize the problem* — Determine as appropriate:
- the geometry of the heat transfer configuration; make a sketch to indicate what is involved,

519

- the type(s) of heat transfer involved,
- the nature of the answer to the problem.

2. *Establish a mathematical equation for the problem answer, with units.* The identification in step 1 of the type(s) of heat transfer involved directs the student to the appropriate section of the text for the selection of the mathematical equation or the means for solving the problem. For example, is the heat transfer conduction (Chapter 2), convection (Chapters 3 and 4), radiation (Chapter 7), or heat exchangers (Chapter 8)? If it is conduction, is the path configuration one-, two-, or three-dimensional simple linear (Section 2-2.1, page 44), series path (Section 2-2.1, page 54), or parallel path (Section 2-2.1, page 48)? If it is convection, is it free (Chapter 3) or forced (Chapter 4) convection? If it is radiation, is the problem involved with emissions (Section 7-2), incident energy (Section 7-3), or net transfer between surfaces (Section 7-4)? If it is a heat exchanger (Chapter 8) is latent heat transfer (Section 8-2.3) or sensible heat transfer (Section 8-2.3, page 419) involved? If sensible heat is being transferred, is the physical configuration parallel flow, counter flow, cross flow, or tube-in-shell? After determining the applicable chapter and section in the text:
- select or develop an equation for the problem unknown when a numerical answer is required,
- indicate the units of your intended numerical answer,
- review the problem to verify that the established equation does give the answer to the problem,
- add the equation parameter symbols to your sketch to help clarify the solution.

3. *Plan the method for the evaluation of the equation parameters:*
- delineate the means for the numerical evaluation of the equation parameters. Some values may be given, some may be obtained from tables, and some may require determination by the use of another equation, which in turn requires a delineation of the means for numerical evaluation.
- give each equation solution a reference calculation number such as (C-1), (C-2), and so on, which can serve to indicate the planned sequence of your solution as well as providing a convenient reference to indicate the source of the numerical value when it is used in the numerical solution.

Step 3 completes the planning (programming) of your solution. The problem has actually been solved at this point if the planning is complete. Then obtaining the numerical answer becomes simply a matter of computational mechanics following the planning.

• *Numerical Solution* The authors have stressed a format in the numerical solution which results in computations that can readily be checked, and from the experience of the authors this format is good engineering office practice.

For Each Computation

- Write the equation (as appropriate) with units.
- List the equation parameters directly under the equation in the sequence the parameters appear in the equation, giving the
  - numerical value
  - units
  - source of the numerical value
- If a secondary equation is required to evaluate a parameter, list the equation for the parameter and solve the equation for the parameter value.
- After numerical values have been determined for all of the equation parameters, make two separate substitutions in the same sequence as listed in the equation:
  - one for the numerical values
  - one for the units
- Verify that the unit substitution reduces to the stated equation answer units. If it does not, modify the parameter numerical values and units accordingly, or add the appropriate conversion factor.
- Solve the numerical substitution, and list the numerical answer appropriately marked with its units.

Final Problem Solution

- When the parameters of the established equation for the problem answer have been obtained, the substitution of the values in the equation along with the unit verification and solving the numerical substitution give the numerical answer.

Appendix E
Mathematical Tables and Dimensional Data

Table E-1   NATURAL LOGARITHMS

The natural logarithm of a number is the index of the power to which the base e (= 2.7182818) must be raised in order to equal the number.

EXAMPLE: $\log_e 4.12 = \ln 4.12 = 1.4159$.

The table gives the natural logarithms of numbers from 1.00 to 9.99 directly, and permits the finding of the logarithms of numbers outside of that range by the addition or subtraction of the natural logarithms of powers of 10.

EXAMPLES: $\log_e 679. = \log_e 6.79 + \log_e 10^2 = 1.9155 + 4.6052 = 6.5207$.
$\log_e .0679 = \log_e 6.79 - \log_e 10^2 = 1.9155 - 4.6052 = -2.6897$.

Natural Logarithms of Powers of 10

$\log_e 10 = 2.302\ 585$	$\log_e 10^4 = 9.210\ 340$	$\log_e 10^7 = 16.118\ 096$
$\log_e 10^2 = 4.605\ 170$	$\log_e 10^5 = 11.512\ 925$	$\log_e 10^8 = 18.420\ 681$
$\log_e 10^3 = 6.907\ 755$	$\log_e 10^6 = 13.815\ 511$	$\log_e 10^9 = 20.723\ 266$

To obtain the common logarithm, the natural logarithm is multiplied by $\log_{10} e$, which is 0.434 294, or $\log_{10} N = 0.434\ 294 \log_e N$.

A negative number or number less than zero has no real logarithm.

N	0	1	2	3	4	5	6	7	8	9
1.0	**0.0000**	**0.0100**	**0.0198**	**0.0296**	**0.0392**	**0.0488**	**0.0583**	**0.0677**	**0.0770**	**0.0862**
1.1	0.0953	0.1044	0.1133	0.1222	0.1310	0.1398	0.1484	0.1570	0.1655	0.1740
1.2	0.1823	0.1906	0.1989	0.2070	0.2151	0.2231	0.2311	0.2390	0.2469	0.2546
1.3	0.2624	0.2700	0.2776	0.2852	0.2927	0.3001	0.3075	0.3148	0.3221	0.3293
1.4	0.3365	0.3436	0.3507	0.3577	0.3646	0.3716	0.3784	0.3853	0.3920	0.3988
1.5	0.4055	0.4121	0.4187	0.4253	0.4318	0.4383	0.4447	0.4511	0.4574	0.4637
1.6	0.4700	0.4762	0.4824	0.4886	0.4947	0.5008	0.5068	0.5128	0.5188	0.5247
1.7	0.5306	0.5365	0.5423	0.5481	0.5539	0.5596	0.5653	0.5710	0.5766	0.5822
1.8	0.5878	0.5933	0.5988	0.6043	0.6098	0.6152	0.6206	0.6259	0.6313	0.6366
1.9	0.6419	0.6471	0.6523	0.6575	0.6627	0.6678	0.6729	0.6780	0.6831	0.6881
2.0	**0.6931**	**0.6981**	**0.7031**	**0.7080**	**0.7129**	**0.7178**	**0.7227**	**0.7275**	**0.7324**	**0.7372**
2.1	0.7419	0.7467	0.7514	0.7561	0.7608	0.7655	0.7701	0.7747	0.7793	0.7839
2.2	0.7885	0.7930	0.7975	0.8020	0.8065	0.8109	0.8154	0.8198	0.8242	0.8286
2.3	0.8329	0.8372	0.8416	0.8459	0.8502	0.8544	0.8587	0.8629	0.8671	0.8713
2.4	0.8755	0.8796	0.8838	0.8879	0.8920	0.8961	0.9002	0.9042	0.9083	0.9123
2.5	0.9163	0.9203	0.9243	0.9282	0.9322	0.9361	0.9400	0.9439	0.9478	0.9517
2.6	0.9555	0.9594	0.9632	0.9670	0.9708	0.9746	0.9783	0.9821	0.9858	0.9895
2.7	0.9933	0.9969	1.0006	1.0043	1.0080	1.0116	1.0152	1.0188	1.0225	1.0260
2.8	1.0296	1.0332	1.0367	1.0403	1.0438	1.0473	1.0508	1.0543	1.0578	1.0613
2.9	1.0647	1.0682	1.0716	1.0750	1.0784	1.0818	1.0852	1.0886	1.0919	1.0953
3.0	**1.0986**	**1.1019**	**1.1053**	**1.1086**	**1.1119**	**1.1151**	**1.1184**	**1.1217**	**1.1249**	**1.1282**
3.1	1.1314	1.1346	1.1378	1.1410	1.1442	1.1474	1.1506	1.1537	1.1569	1.1600
3.2	1.1632	1.1663	1.1694	1.1725	1.1756	1.1787	1.1817	1.1848	1.1878	1.1909
3.3	1.1939	1.1969	1.2000	1.2030	1.2060	1.2090	1.2119	1.2149	1.2179	1.2208
3.4	1.2238	1.2267	1.2296	1.2326	1.2355	1.2384	1.2413	1.2442	1.2470	1.2499
3.5	1.2528	1.2556	1.2585	1.2613	1.2641	1.2669	1.2698	1.2726	1.2754	1.2782
3.6	1.2809	1.2837	1.2865	1.2892	1.2920	1.2947	1.2975	1.3002	1.3029	1.3056
3.7	1.3083	1.3110	1.3137	1.3164	1.3191	1.3218	1.3244	1.3271	1.3297	1.3324
3.8	1.3350	1.3376	1.3403	1.3429	1.3455	1.3481	1.3507	1.3533	1.3558	1.3584
3.9	1.3610	1.3635	1.3661	1.3686	1.3712	1.3737	1.3762	1.3788	1.3813	1.3838
4.0	**1.3863**	**1.3888**	**1.3913**	**1.3938**	**1.3962**	**1.3987**	**1.4012**	**1.4036**	**1.4061**	**1.4085**
4.1	1.4110	1.4134	1.4159	1.4183	1.4207	1.4231	1.4255	1.4279	1.4303	1.4327
4.2	1.4351	1.4375	1.4398	1.4422	1.4446	1.4469	1.4493	1.4516	1.4540	1.4563
4.3	1.4586	1.4609	1.4633	1.4656	1.4679	1.4702	1.4725	1.4748	1.4770	1.4793
4.4	1.4816	1.4839	1.4861	1.4884	1.4907	1.4929	1.4951	1.4974	1.4996	1.5019
4.5	1.5041	1.5063	1.5085	1.5107	1.5129	1.5151	1.5173	1.5195	1.5217	1.5239
4.6	1.5261	1.5282	1.5304	1.5326	1.5347	1.5369	1.5390	1.5412	1.5433	1.5454
4.7	1.5476	1.5497	1.5518	1.5539	1.5560	1.5581	1.5602	1.5623	1.5644	1.5665
4.8	1.5686	1.5707	1.5728	1.5748	1.5769	1.5790	1.5810	1.5831	1.5851	1.5872
4.9	1.5892	1.5913	1.5933	1.5953	1.5974	1.5994	1.6014	1.6034	1.6054	1.6074

SOURCE: O. W. Eshbach, *Handbook of Engineering Fundamentals*, Wiley, New York, 1936.

Table E-1 (Continued)

N	0	1	2	3	4	5	6	7	8	9
5.0	**1.6094**	**1.6114**	**1.6134**	**1.6154**	**1.6174**	**1.6194**	**1.6214**	**1.6233**	**1.6253**	**1.6273**
5.1	1.6292	1.6312	1.6332	1.6351	1.6371	1.6390	1.6409	1.6429	1.6448	1.6467
5.2	1.6487	1.6506	1.6525	1.6544	1.6563	1.6582	1.6601	1.6620	1.6639	1.6658
5.3	1.6677	1.6696	1.6715	1.6734	1.6752	1.6771	1.6790	1.6808	1.6827	1.6845
5.4	1.6864	1.6882	1.6901	1.6919	1.6938	1.6956	1.6974	1.6993	1.7011	1.7029
5.5	1.7047	1.7066	1.7084	1.7102	1.7120	1.7138	1.7156	1.7174	1.7192	1.7210
5.6	1.7228	1.7246	1.7263	1.7281	1.7299	1.7317	1.7334	1.7352	1.7370	1.7387
5.7	1.7405	1.7422	1.7440	1.7457	1.7475	1.7492	1.7509	1.7527	1.7544	1.7561
5.8	1.7579	1.7596	1.7613	1.7630	1.7647	1.7664	1.7681	1.7699	1.7716	1.7733
5.9	1.7750	1.7766	1.7783	1.7800	1.7817	1.7834	1.7851	1.7867	1.7884	1.7901
6.0	**1.7918**	**1.7934**	**1.7951**	**1.7967**	**1.7984**	**1.8001**	**1.8017**	**1.8034**	**1.8050**	**1.8066**
6.1	1.8083	1.8099	1.8116	1.8132	1.8148	1.8165	1.8181	1.8197	1.8213	1.8229
6.2	1.8245	1.8262	1.8278	1.8294	1.8310	1.8326	1.8342	1.8358	1.8374	1.8390
6.3	1.8405	1.8421	1.8437	1.8453	1.8469	1.8485	1.8500	1.8516	1.8532	1.8547
6.4	1.8563	1.8579	1.8594	1.8610	1.8625	1.8641	1.8656	1.8672	1.8687	1.8703
6.5	1.8718	1.8733	1.8749	1.8764	1.8779	1.8795	1.8810	1.8825	1.8840	1.8856
6.6	1.8871	1.8886	1.8901	1.8916	1.8931	1.8946	1.8961	1.8976	1.8991	1.9006
6.7	1.9021	1.9036	1.9051	1.9066	1.9081	1.9095	1.9110	1.9125	1.9140	1.9155
6.8	1.9169	1.9184	1.9199	1.9213	1.9228	1.9242	1.9257	1.9272	1.9286	1.9301
6.9	1.9315	1.9330	1.9344	1.9359	1.9373	1.9387	1.9402	1.9416	1.9430	1.9445
7.0	**1.9459**	**1.9473**	**1.9488**	**1.9502**	**1.9516**	**1.9530**	**1.9544**	**1.9559**	**1.9573**	**1.9587**
7.1	1.9601	1.9615	1.9629	1.9643	1.9657	1.9671	1.9685	1.9699	1.9713	1.9727
7.2	1.9741	1.9755	1.9769	1.9782	1.9796	1.9810	1.9824	1.9838	1.9851	1.9865
7.3	1.9879	1.9892	1.9906	1.9920	1.9933	1.9947	1.9961	1.9974	1.9988	2.0001
7.4	2.0015	2.0028	2.0042	2.0055	2.0069	2.0082	2.0096	2.0109	2.0122	2.0136
7.5	2.0149	2.0162	2.0176	2.0189	2.0202	2.0215	2.0229	2.0242	2.0255	2.0268
7.6	2.0281	2.0295	2.0308	2.0321	2.0334	2.0347	2.0360	2.0373	2.0386	2.0399
7.7	2.0412	2.0425	2.0438	2.0451	2.0464	2.0477	2.0490	2.0503	2.0516	2.0528
7.8	2.0541	2.0554	2.0567	2.0580	2.0592	2.0605	2.0618	2.0631	2.0643	2.0656
7.9	2.0669	2.0681	2.0694	2.0707	2.0719	2.0732	2.0744	2.0757	2.0769	2.0782
8.0	**2.0794**	**2.0807**	**2.0819**	**2.0832**	**2.0844**	**2.0857**	**2.0869**	**2.0882**	**2.0894**	**2.0906**
8.1	2.0919	2.0931	2.0943	2.0956	2.0968	2.0980	2.0992	2.1005	2.1017	2.1029
8.2	2.1041	2.1054	2.1066	2.1078	2.1090	2.1102	2.1114	2.1126	2.1138	2.1150
8.3	2.1163	2.1175	2.1187	2.1199	2.1211	2.1223	2.1235	2.1247	2.1258	2.1270
8.4	2.1282	2.1294	2.1306	2.1318	2.1330	2.1342	2.1353	2.1365	2.1377	2.1389
8.5	2.1401	2.1412	2.1424	2.1436	2.1448	2.1459	2.1471	2.1483	2.1494	2.1506
8.6	2.1518	2.1529	2.1541	2.1552	2.1564	2.1576	2.1587	2.1599	2.1610	2.1622
8.7	2.1633	2.1645	2.1656	2.1668	2.1679	2.1691	2.1702	2.1713	2.1725	2.1736
8.8	2.1748	2.1759	2.1770	2.1782	2.1793	2.1804	2.1815	2.1827	2.1838	2.1849
8.9	2.1861	2.1872	2.1883	2.1894	2.1905	2.1917	2.1928	2.1939	2.1950	2.1961
9.0	**2.1972**	**2.1983**	**2.1994**	**2.2006**	**2.2017**	**2.2028**	**2.2039**	**2.2050**	**2.2061**	**2.2072**
9.1	2.2083	2.2094	2.2105	2.2116	2.2127	2.2138	2.2148	2.2159	2.2170	2.2181
9.2	2.2192	2.2203	2.2214	2.2225	2.2235	2.2246	2.2257	2.2268	2.2279	2.2289
9.3	2.2300	2.2311	2.2322	2.2332	2.2343	2.2354	2.2364	2.2375	2.2386	2.2396
9.4	2.2407	2.2418	2.2428	2.2439	2.2450	2.2460	2.2471	2.2481	2.2492	2.2502
9.5	2.2513	2.2523	2.2534	2.2544	2.2555	2.2565	2.2576	2.2586	2.2597	2.2607
9.6	2.2618	2.2628	2.2638	2.2649	2.2659	2.2670	2.2680	2.2690	2.2701	2.2711
9.7	2.2721	2.2732	2.2742	2.2752	2.2762	2.2773	2.2783	2.2793	2.2803	2.2814
9.8	2.2824	2.2834	2.2844	2.2854	2.2865	2.2875	2.2885	2.2895	2.2905	2.2915
9.9	2.2925	2.2935	2.2946	2.2956	2.2966	2.2976	2.2986	2.2996	2.3006	2.3016

SOURCE: O. W. Eshbach, *Handbook of Engineering Fundamentals*, Wiley, New York, 1936.

Table E-2 HYPERBOLIC FUNCTIONS

x	Natural Values				
	e^x	e^{-x}	Sinh x	Cosh x	Tanh x
0.00	**1.0000**	**1.0000**	**0.0000**	**1.0000**	**.00000**
0.01	1.0101	.99005	0.0100	1.0001	.01000
0.02	1.0202	.98020	0.0200	1.0002	.02000
0.03	1.0305	.97045	0.0300	1.0005	.02999
0.04	1.0408	.96079	0.0400	1.0008	.03998
0.05	1.0513	.95123	0.0500	1.0013	.04996
0.06	1.0618	.94176	0.0600	1.0018	.05993
0.07	1.0725	.93239	0.0701	1.0025	.06989
0.08	1.0833	.92312	0.0801	1.0032	.07983
0.09	1.0942	.91393	0.0901	1.0041	.08976
0.10	**1.1052**	**.90484**	**0.1002**	**1.0050**	**.09967**
0.11	1.1163	.89583	0.1102	1.0061	.10956
0.12	1.1275	.88692	0.1203	1.0072	.11943
0.13	1.1388	.87810	0.1304	1.0085	.12927
0.14	1.1503	.86936	0.1405	1.0098	.13909
0.15	1.1618	.86071	0.1506	1.0113	.14889
0.16	1.1735	.85214	0.1607	1.0128	.15865
0.17	1.1853	.84366	0.1708	1.0145	.16838
0.18	1.1972	.83527	0.1810	1.0162	.17808
0.19	1.2092	.82696	0.1911	1.0181	.18775
0.20	**1.2214**	**.81873**	**0.2013**	**1.0201**	**.19738**
0.21	1.2337	.81058	0.2115	1.0221	.20697
0.22	1.2461	.80252	0.2218	1.0243	.21652
0.23	1.2586	.79453	0.2320	1.0266	.22603
0.24	1.2712	.78663	0.2423	1.0289	.23550
0.25	1.2840	.77880	0.2526	1.0314	.24492
0.26	1.2969	.77105	0.2629	1.0340	.25430
0.27	1.3100	.76338	0.2733	1.0367	.26362
0.28	1.3231	.75578	0.2837	1.0395	.27291
0.29	1.3364	.74826	0.2941	1.0423	.28213
0.30	**1.3499**	**.74082**	**0.3045**	**1.0453**	**.29131**
0.31	1.3634	.73345	0.3150	1.0484	.30044
0.32	1.3771	.72615	0.3255	1.0516	.30951
0.33	1.3910	.71892	0.3360	1.0549	.31852
0.34	1.4049	.71177	0.3466	1.0584	.32748
0.35	1.4191	.70469	0.3572	1.0619	.33638
0.36	1.4333	.69768	0.3678	1.0655	.34521
0.37	1.4477	.69073	0.3785	1.0692	.35399
0.38	1.4623	.68386	0.3892	1.0731	.36271
0.39	1.4770	.67706	0.4000	1.0770	.37136
0.40	**1.4918**	**.67032**	**0.4108**	**1.0811**	**.37995**
0.41	1.5068	.66365	0.4216	1.0852	.38847
0.42	1.5220	.65705	0.4325	1.0895	.39693
0.43	1.5373	.65051	0.4434	1.0939	.40532
0.44	1.5527	.64404	0.4543	1.0984	.41364
0.45	1.5683	.63763	0.4653	1.1030	.42190
0.46	1.5841	.63128	0.4764	1.1077	.43008
0.47	1.6000	.62500	0.4875	1.1125	.43820
0.48	1.6161	.61878	0.4986	1.1174	.44624
0.49	1.6323	.61263	0.5098	1.1225	.45422
0.50	**1.6487**	**.60653**	**0.5211**	**1.1276**	**.46212**
0.51	1.6653	.60050	0.5324	1.1329	.46995
0.52	1.6820	.59452	0.5438	1.1383	.47770
0.53	1.6989	.58860	0.5552	1.1438	.48538
0.54	1.7160	.58275	0.5666	1.1494	.49299
0.55	1.7333	.57695	0.5782	1.1551	.50052
0.56	1.7507	.57121	0.5897	1.1609	.50798
0.57	1.7683	.56553	0.6014	1.1669	.51536
0.58	1.7860	.55990	0.6131	1.1730	.52267
0.59	1.8040	.55433	0.6248	1.1792	.52990
0.60	**1.8221**	**.54881**	**0.6367**	**1.1855**	**.53705**

SOURCE: O. W. Eshbach, *Handbook of Engineering Fundamentals*, Wiley, New York, 1936.

Table E-2 (Continued)

x	Natural Values				
	e^x	e^{-x}	Sinh x	Cosh x	Tanh x
0.60	1.8221	.54881	0.6367	1.1855	.53705
0.61	1.8404	.54335	0.6485	1.1919	.54413
0.62	1.8589	.53794	0.6605	1.1984	.55113
0.63	1.8776	.53259	0.6725	1.2051	.55805
0.64	1.8965	.52729	0.6846	1.2119	.56490
0.65	1.9155	.52205	0.6967	1.2188	.57167
0.66	1.9348	.51685	0.7090	1.2258	.57836
0.67	1.9542	.51171	0.7213	1.2330	.58498
0.68	1.9739	.50662	0.7336	1.2402	.59152
0.69	1.9937	.50158	0.7461	1.2476	.59798
0.70	2.0138	.49659	0.7586	1.2552	.60437
0.71	2.0340	.49164	0.7712	1.2628	.61068
0.72	2.0544	.48675	0.7838	1.2706	.61691
0.73	2.0751	.48191	0.7966	1.2785	.62307
0.74	2.0959	.47711	0.8094	1.2865	.62915
0.75	2.1170	.47237	0.8223	1.2947	.63515
0.76	2.1383	.46767	0.8353	1.3030	.64108
0.77	2.1598	.46301	0.8484	1.3114	.64693
0.78	2.1815	.45841	0.8615	1.3199	.65271
0.79	2.2034	.45384	0.8748	1.3286	.65841
0.80	2.2255	.44933	0.8881	1.3374	.66404
0.81	2.2479	.44486	0.9015	1.3464	.66959
0.82	2.2705	.44043	0.9150	1.3555	.67507
0.83	2.2933	.43605	0.9286	1.3647	.68048
0.84	2.3164	.43171	0.9423	1.3740	.68581
0.85	2.3396	.42741	0.9561	1.3835	.69107
0.86	2.3632	.42316	0.9700	1.3932	.69626
0.87	2.3869	.41895	0.9840	1.4029	.70137
0.88	2.4109	.41478	0.9981	1.4128	.70642
0.89	2.4351	.41066	1.0122	1.4229	.71139
0.90	2.4596	.40657	1.0265	1.4331	.71630
0.91	2.4843	.40252	1.0409	1.4434	.72113
0.92	2.5093	.39852	1.0554	1.4539	.72590
0.93	2.5345	.39455	1.0700	1.4645	.73059
0.94	2.5600	.39063	1.0847	1.4753	.73522
0.95	2.5857	.38674	1.0995	1.4862	.73978
0.96	2.6117	.38289	1.1144	1.4973	.74428
0.97	2.6379	.37908	1.1294	1.5085	.74870
0.98	2.6645	.37531	1.1446	1.5199	.75307
0.99	2.6912	.37158	1.1598	1.5314	.75736
1.00	2.7183	.36788	1.1752	1.5431	.76159
1.01	2.7456	.36422	1.1907	1.5549	.76576
1.02	2.7732	.36059	1.2063	1.5669	.76987
1.03	2.8011	.35701	1.2220	1.5790	.77391
1.04	2.8292	.35345	1.2379	1.5913	.77789
1.05	2.8577	.34994	1.2539	1.6038	.78181
1.06	2.8864	.34646	1.2700	1.6164	.78566
1.07	2.9154	.34301	1.2862	1.6292	.78946
1.08	2.9447	.33960	1.3025	1.6421	.79320
1.09	2.9743	.33622	1.3190	1.6552	.79688
1.10	3.0042	.33287	1.3356	1.6685	.80050
1.11	3.0344	.32956	1.3524	1.6820	.80406
1.12	3.0649	.32628	1.3693	1.6956	.80757
1.13	3.0957	.32303	1.3863	1.7093	.81102
1.14	3.1268	.31982	1.4035	1.7233	.81441
1.15	3.1582	.31664	1.4208	1.7374	.81775
1.16	3.1899	.31349	1.4382	1.7517	.82104
1.17	3.2220	.31037	1.4558	1.7662	.82427
1.18	3.2544	.30728	1.4735	1.7808	.82745
1.19	3.2871	.30422	1.4914	1.7957	.83058
1.20	3.3201	.30119	1.5095	1.8107	.83365

SOURCE: O. W. Eshbach, *Handbook of Engineering Fundamentals*, Wiley, New York, 1936.

Table E-2 (Continued)

x			Natural Values		
	e^x	e^{-x}	Sinh x	Cosh x	Tanh x
1.20	**3.3201**	**.30119**	**1.5095**	**1.8107**	**.83365**
1.21	3.3535	.29820	1.5276	1.8258	.83668
1.22	3.3872	.29523	1.5460	1.8412	.83965
1.23	3.4212	.29229	1.5645	1.8568	.84258
1.24	3.4556	.28938	1.5831	1.8725	.84546
1.25	3.4903	.28650	1.6019	1.8884	.84828
1.26	3.5254	.28365	1.6209	1.9045	.85106
1.27	3.5609	.28083	1.6400	1.9208	.85380
1.28	3.5966	.27804	1.6593	1.9373	.85648
1.29	3.6328	.27527	1.6788	1.9540	.85913
1.30	**3.6693**	**.27253**	**1.6984**	**1.9709**	**.86172**
1.31	3.7062	.26982	1.7182	1.9880	.86428
1.32	3.7434	.26714	1.7381	2.0053	.86678
1.33	3.7810	.26448	1.7583	2.0228	.86925
1.34	3.8190	.26185	1.7786	2.0404	.87167
1.35	3.8574	.25924	1.7991	2.0583	.87405
1.36	3.8962	.25666	1.8198	2.0764	.87639
1.37	3.9354	.25411	1.8406	2.0947	.87869
1.38	3.9749	.25158	1.8617	2.1132	.88095
1.39	4.0149	.24908	1.8829	2.1320	.88317
1.40	**4.0552**	**.24660**	**1.9043**	**2.1509**	**.88535**
1.41	4.0960	.24414	1.9259	2.1700	.88749
1.42	4.1371	.24171	1.9477	2.1894	.88960
1.43	4.1787	.23931	1.9697	2.2090	.89167
1.44	4.2207	.23693	1.9919	2.2288	.89370
1.45	4.2631	.23457	2.0143	2.2488	.89569
1.46	4.3060	.23224	2.0369	2.2691	.89765
1.47	4.3492	.22993	2.0597	2.2896	.89958
1.48	4.3929	.22764	2.0827	2.3103	.90147
1.49	4.4371	.22537	2.1059	2.3312	.90332
1.50	**4.4817**	**.22313**	**2.1293**	**2.3524**	**.90515**
1.51	4.5267	.22091	2.1529	2.3738	.90694
1.52	4.5722	.21871	2.1768	2.3955	.90870
1.53	4.6182	.21654	2.2008	2.4174	.91042
1.54	4.6646	.21438	2.2251	2.4395	.91212
1.55	4.7115	.21225	2.2496	2.4619	.91379
1.56	4.7588	.21014	2.2743	2.4845	.91542
1.57	4.8066	.20805	2.2993	2.5073	.91703
1.58	4.8550	.20598	2.3245	2.5305	.91860
1.59	4.9037	.20393	2.3499	2.5538	.92015
1.60	**4.9530**	**.20190**	**2.3756**	**2.5775**	**.92167**
1.61	5.0028	.19989	2.4015	2.6013	.92316
1.62	5.0531	.19790	2.4276	2.6255	.92462
1.63	5.1039	.19593	2.4540	2.6499	.92606
1.64	5.1552	.19398	2.4806	2.6746	.92747
1.65	5.2070	.19205	2.5075	2.6995	.92886
1.66	5.2593	.19014	2.5346	2.7247	.93022
1.67	5.3122	.18825	2.5620	2.7502	.93155
1.68	5.3656	.18637	2.5896	2.7760	.93286
1.69	5.4195	.18452	2.6175	2.8020	.93415
1.70	**5.4739**	**.18268**	**2.6456**	**2.8283**	**.93541**
1.71	5.5290	.18087	2.6740	2.8549	.93665
1.72	5.5845	.17907	2.7027	2.8818	.93786
1.73	5.6407	.17728	2.7317	2.9090	.93906
1.74	5.6973	.17552	2.7609	2.9364	.94023
1.75	5.7546	.17377	2.7904	2.9642	.94138
1.76	5.8124	.17204	2.8202	2.9922	.94250
1.77	5.8709	.17033	2.8503	3.0206	.94361
1.78	5.9299	.16864	2.8806	3.0492	.94470
1.79	5.9895	.16696	2.9112	3.0782	.94576
1.80	**6.0496**	**.16530**	**2.9422**	**3.1075**	**.94681**

SOURCE: O. W. Eshbach, *Handbook of Engineering Fundamentals*, Wiley, New York, 1936.

Table E-2 (Continued)

x	Natural Values				
	e^x	e^{-x}	Sinh x	Cosh x	Tanh x
1.80	**6.0496**	**.16530**	**2.9422**	**3.1075**	**.94681**
1.81	6.1104	.16365	2.9734	3.1371	.94783
1.82	6.1719	.16203	3.0049	3.1669	.94884
1.83	6.2339	.16041	3.0367	3.1972	.94983
1.84	6.2965	.15882	3.0689	3.2277	.95080
1.85	6.3598	.15724	3.1013	3.2585	.95175
1.86	6.4237	.15567	3.1340	3.2897	.95268
1.87	6.4883	.15412	3.1671	3.3212	.95359
1.88	6.5535	.15259	3.2005	3.3530	.95449
1.89	6.6194	.15107	3.2341	3.3852	.95537
1.90	**6.6859**	**.14957**	**3.2682**	**3.4177**	**.95624**
1.91	6.7531	.14808	3.3025	3.4506	.95709
1.92	6.8210	.14661	3.3372	3.4838	.95792
1.93	6.8895	.14515	3.3722	3.5173	.95873
1.94	6.9588	.14370	3.4075	3.5512	.95953
1.95	7.0287	.14227	3.4432	3.5855	.96032
1.96	7.0993	.14086	3.4792	3.6201	.96109
1.97	7.1707	.13946	3.5156	3.6551	.96185
1.98	7.2427	.13807	3.5523	3.6904	.96259
1.99	7.3155	.13670	3.5894	3.7261	.96331
2.00	**7.3891**	**.13534**	**3.6269**	**3.7622**	**.96403**
2.01	7.4633	.13399	3.6647	3.7987	.96473
2.02	7.5383	.13266	3.7028	3.8355	.96541
2.03	7.6141	.13134	3.7414	3.8727	.96609
2.04	7.6906	.13003	3.7803	3.9103	.96675
2.05	7.7679	.12873	3.8196	3.9483	.96740
2.06	7.8460	.12745	3.8593	3.9867	.96803
2.07	7.9248	.12619	3.8993	4.0255	.96865
2.08	8.0045	.12493	3.9398	4.0647	.96926
2.09	8.0849	.12369	3.9806	4.1043	.96986
2.10	**8.1662**	**.12246**	**4.0219**	**4.1443**	**.97045**
2.11	8.2482	.12124	4.0635	4.1847	.97103
2.12	8.3311	.12003	4.1056	4.2256	.97159
2.13	8.4149	.11884	4.1480	4.2669	.97215
2.14	8.4994	.11765	4.1909	4.3085	.97269
2.15	8.5849	.11648	4.2342	4.3507	.97323
2.16	8.6711	.11533	4.2779	4.3932	.97375
2.17	8.7583	.11418	4.3221	4.4362	.97426
2.18	8.8463	.11304	4.3666	4.4797	.97477
2.19	8.9352	.11192	4.4116	4.5236	.97526
2.20	**9.0250**	**.11080**	**4.4571**	**4.5679**	**.97574**
2.21	9.1157	.10970	4.5030	4.6127	.97622
2.22	9.2073	.10861	4.5494	4.6580	.97668
2.23	9.2999	.10753	4.5962	4.7037	.97714
2.24	9.3933	.10646	4.6434	4.7499	.97759
2.25	9.4877	.10540	4.6912	4.7966	.97803
2.26	9.5831	.10435	4.7394	4.8437	.97846
2.27	9.6794	.10331	4.7880	4.8914	.97888
2.28	9.7767	.10228	4.8372	4.9395	.97929
2.29	9.8749	.10127	4.8868	4.9881	.97970
2.30	**9.9742**	**.10026**	**4.9370**	**5.0372**	**.98010**
2.31	10.074	.09926	4.9876	5.0868	.98049
2.32	10.176	.09827	5.0387	5.1370	.98087
2.33	10.278	.09730	5.0903	5.1876	.98124
2.34	10.381	.09633	5.1425	5.2388	.98161
2.35	10.486	.09537	5.1951	5.2905	.98197
2.36	10.591	.09442	5.2483	5.3427	.98233
2.37	10.697	.09348	5.3020	5.3954	.98267
2.38	10.805	.09255	5.3562	5.4487	.98301
2.39	10.913	.09163	5.4109	5.5026	.98335
2.40	**11.023**	**.09072**	**5.4662**	**5.5569**	**.98367**

SOURCE: O. W. Eshbach, *Handbook of Engineering Fundamentals*, Wiley, New York, 1936.

Table E-2 (Continued)

x	Natural Values				
	e^x	e^{-x}	Sinh x	Cosh x	Tanh x
2.40	**11.023**	**.09072**	**5.4662**	**5.5569**	**.98367**
2.41	11.134	.08982	5.5221	5.6119	.98400
2.42	11.246	.08892	5.5785	5.6674	.98431
2.43	11.359	.08804	5.6354	5.7235	.98462
2.44	11.473	.08716	5.6929	5.7801	.98492
2.45	11.588	.08629	5.7510	5.8373	.98522
2.46	11.705	.08543	5.8097	5.8951	.98551
2.47	11.822	.08458	5.8689	5.9535	.98579
2.48	11.941	.08374	5.9288	6.0125	.98607
2.49	12.061	.08291	5.9892	6.0721	.98635
2.50	**12.182**	**.08208**	**6.0502**	**6.1323**	**.98661**
2.51	12.305	.08127	6.1118	6.1931	.98688
2.52	12.429	.08046	6.1741	6.2545	.98714
2.53	12.554	.07966	6.2369	6.3166	.98739
2.54	12.680	.07887	6.3004	6.3793	.98764
2.55	12.807	.07808	6.3645	6.4426	.98788
2.56	12.936	.07730	6.4293	6.5066	.98812
2.57	13.066	.07654	6.4946	6.5712	.98835
2.58	13.197	.07577	6.5607	6.6365	.98858
2.59	13.330	.07502	6.6274	6.7024	.98881
2.60	**13.464**	**.07427**	**6.6947**	**6.7690**	**.98903**
2.61	13.599	.07353	6.7628	6.8363	.98924
2.62	13.736	.07280	6.8315	6.9043	.98946
2.63	13.874	.07208	6.9008	6.9729	.98966
2.64	14.013	.07136	6.9709	7.0423	.98987
2.65	14.154	.07065	7.0417	7.1123	.99007
2.66	14.296	.06995	7.1132	7.1831	.99026
2.67	14.440	.06925	7.1854	7.2546	.99045
2.68	14.585	.06856	7.2583	7.3268	.99064
2.69	14.732	.06788	7.3319	7.3998	.99083
2.70	**14.880**	**.06721**	**7.4063**	**7.4735**	**.99101**
2.71	15.029	.06654	7.4814	7.5479	.99118
2.72	15.180	.06587	7.5572	7.6231	.99136
2.73	15.333	.06522	7.6338	7.6991	.99153
2.74	15.487	.06457	7.7112	7.7758	.99170
2.75	15.643	.06393	7.7894	7.8533	.99186
2.76	15.800	.06329	7.8683	7.9316	.99202
2.77	15.959	.06266	7.9480	8.0106	.99218
2.78	16.119	.06204	8.0285	8.0905	.99233
2.79	16.281	.06142	8.1098	8.1712	.99248
2.80	**16.445**	**.06081**	**8.1919**	**8.2527**	**.99263**
2.81	16.610	.06020	8.2749	8.3351	.99278
2.82	16.777	.05961	8.3586	8.4182	.99292
2.83	16.945	.05901	8.4432	8.5022	.99306
2.84	17.116	.05843	8.5287	8.5871	.99320
2.85	17.288	.05784	8.6150	8.6728	.99333
2.86	17.462	.05727	8.7021	8.7594	.99346
2.87	17.637	.05670	8.7902	8.8469	.99359
2.88	17.814	.05613	8.8791	8.9352	.99372
2.89	17.993	.05558	8.9689	9.0244	.99384
2.90	**18.174**	**.05502**	**9.0596**	**9.1146**	**.99396**
2.91	18.357	.05448	9.1512	9.2056	.99408
2.92	18.541	.05393	9.2437	9.2976	.99420
2.93	18.728	.05340	9.3371	9.3905	.99431
2.94	18.916	.05287	9.4315	9.4844	.99443
2.95	19.106	.05234	9.5268	9.5791	.99454
2.96	19.298	.05182	9.6231	9.6749	.99464
2.97	19.492	.05130	9.7203	9.7716	.99475
2.98	19.688	.05079	9.8185	9.8693	.99485
2.99	19.886	.05029	9.9177	9.9680	.99496
3.00	**20.086**	**.04979**	**10.018**	**10.068**	**.99505**

SOURCE: O. W. Eshbach, *Handbook of Engineering Fundamentals*, Wiley, New York, 1936.

Table E-2 (Continued)

x	Natural Values				
	e^x	e^{-x}	Sinh x	Cosh x	Tanh x
3.00	**20.086**	**.04979**	**10.018**	**10.068**	**.99505**
3.01	20.287	.04929	10.119	10.168	.99515
3.02	20.491	.04880	10.221	10.270	.99525
3.03	20.697	.04832	10.325	10.373	.99534
3.04	20.905	.04783	10.429	10.477	.99543
3.05	21.115	.04736	10.534	10.581	.99552
3.06	21.328	.04689	10.640	10.687	.99561
3.07	21.542	.04642	10.748	10.794	.99570
3.08	21.758	.04596	10.856	10.902	.99578
3.09	21.977	.04550	10.966	11.011	.99587
3.10	**22.198**	**.04505**	**11.077**	**11.122**	**.99595**
3.11	22.421	.04460	11.188	11.233	.99603
3.12	22.646	.04416	11.301	11.345	.99611
3.13	22.874	.04372	11.415	11.459	.99618
3.14	23.104	.04328	11.530	11.574	.99626
3.15	23.336	.04285	11.647	11.689	.99633
3.16	23.571	.04243	11.764	11.807	.99641
3.17	23.807	.04200	11.883	11.925	.99648
3.18	24.047	.04159	12.003	12.044	.99655
3.19	24.288	.04117	12.124	12.165	.99662
3.20	**24.533**	**.04076**	**12.246**	**12.287**	**.99668**
3.21	24.779	.04036	12.369	12.410	.99675
3.22	25.028	.03996	12.494	12.534	.99681
3.23	25.280	.03956	12.620	12.660	.99688
3.24	25.534	.03916	12.747	12.786	.99694
3.25	25.790	.03877	12.876	12.915	.99700
3.26	26.050	.03839	13.006	13.044	.99706
3.27	26.311	.03801	13.137	13.175	.99712
3.28	26.576	.03763	13.269	13.307	.99717
3.29	26.843	.03725	13.403	13.440	.99723
3.30	**27.113**	**.03688**	**13.538**	**13.575**	**.99728**
3.31	27.385	.03652	13.674	13.711	.99734
3.32	27.660	.03615	13.812	13.848	.99739
3.33	27.938	.03579	13.951	13.987	.99744
3.34	28.219	.03544	14.092	14.127	.99749
3.35	28.503	.03508	14.234	14.269	.99754
3.36	28.789	.03474	14.377	14.412	.99759
3.37	29.079	.03439	14.522	14.556	.99764
3.38	29.371	.03405	14.668	14.702	.99768
3.39	29.666	.03371	14.816	14.850	.99773
3.40	**29.964**	**.03337**	**14.965**	**14.999**	**.99777**
3.41	30.265	.03304	15.116	15.149	.99782
3.42	30.569	.03271	15.268	15.301	.99786
3.43	30.877	.03239	15.422	15.455	.99790
3.44	31.187	.03206	15.577	15.610	.99795
3.45	31.500	.03175	15.734	15.766	.99799
3.46	31.817	.03143	15.893	15.924	.99803
3.47	32.137	.03112	16.053	16.084	.99807
3.48	32.460	.03081	16.215	16.245	.99810
3.49	32.786	.03050	16.378	16.408	.99814
3.50	**33.115**	**.03020**	**16.543**	**16.573**	**.99818**
3.51	33.448	.02990	16.709	16.739	.99821
3.52	33.784	.02960	16.877	16.907	.99825
3.53	34.124	.02930	17.047	17.077	.99828
3.54	34.467	.02901	17.219	17.248	.99832
3.55	34.813	.02872	17.392	17.421	.99835
3.56	35.163	.02844	17.567	17.596	.99838
3.57	35.517	.02816	17.744	17.772	.99842
3.58	35.874	.02788	17.923	17.951	.99845
3.59	36.234	.02760	18.103	18.131	.99848
3.60	**36.598**	**.02732**	**18.285**	**18.313**	**.99851**

SOURCE: O. W. Eshbach, *Handbook of Engineering Fundamentals*, Wiley, New York, 1936.

Table E-2 (Continued)

x	Natural Values				
	e^x	e^{-x}	Sinh x	Cosh x	Tanh x
3.60	**36.598**	**.02732**	**18.285**	**18.313**	**.99851**
3.61	36.966	.02705	18.470	18.497	.99854
3.62	37.338	.02678	18.655	18.682	.99857
3.63	37.713	.02652	18.843	18.870	.99859
3.64	38.092	.02625	19.033	19.059	.99862
3.65	38.475	.02599	19.224	19.250	.99865
3.66	38.861	.02573	19.418	19.444	.99868
3.67	39.252	.02548	19.613	19.639	.99870
3.68	39.646	.02522	19.811	19.836	.99873
3.69	40.045	.02497	20.010	20.035	.99875
3.70	**40.447**	**.02472**	**20.211**	**20.236**	**.99878**
3.71	40.854	.02448	20.415	20.439	.99880
3.72	41.264	.02423	20.620	20.644	.99883
3.73	41.679	.02399	20.828	20.852	.99885
3.74	42.098	.02375	21.037	21.061	.99887
3.75	42.521	.02352	21.249	21.272	.99889
3.76	42.948	.02328	21.463	21.486	.99892
3.77	43.380	.02305	21.679	21.702	.99894
3.78	43.816	.02282	21.897	21.919	.99896
3.79	44.256	.02260	22.117	22.140	.99898
3.80	**44.701**	**.02237**	**22.339**	**22.362**	**.99900**
3.81	45.150	.02215	22.564	22.586	.99902
3.82	45.604	.02193	22.791	22.813	.99904
3.83	46.063	.02171	23.020	23.042	.99906
3.84	46.525	.02149	23.252	23.274	.99908
3.85	46.993	.02128	23.486	23.507	.99909
3.86	47.465	.02107	23.722	23.743	.99911
3.87	47.942	.02086	23.961	23.982	.99913
3.88	48.424	.02065	24.202	24.222	.99915
3.89	48.911	.02045	24.445	24.466	.99916
3.90	**49.402**	**.02024**	**24.691**	**24.711**	**.99918**
3.91	49.899	.02004	24.939	24.960	.99920
3.92	50.400	.01984	25.190	25.210	.99921
3.93	50.907	.01964	25.444	25.463	.99923
3.94	51.419	.01945	25.700	25.719	.99924
3.95	51.935	.01925	25.958	25.977	.99926
3.96	52.457	.01906	26.219	26.238	.99927
3.97	52.985	.01887	26.483	26.502	.99929
3.98	53.517	.01869	26.749	26.768	.99930
3.99	54.055	.01850	27.018	27.037	.99932
4.00	**54.598**	**.01832**	**27.290**	**27.308**	**.99933**
4.01	55.147	.01813	27.564	27.583	.99934
4.02	55.701	.01795	27.842	27.860	.99936
4.03	56.261	.01777	28.122	28.139	.99937
4.04	56.826	.01760	28.404	28.422	.99938
4.05	57.397	.01742	28.690	28.707	.99939
4.06	57.974	.01725	28.979	28.996	.99941
4.07	58.557	.01708	29.270	29.287	.99942
4.08	59.145	.01691	29.564	29.581	.99943
4.09	59.740	.01674	29.862	29.878	.99944
4.10	**60.340**	**.01657**	**30.162**	**30.178**	**.99945**
4.11	60.947	.01641	30.465	30.482	.99946
4.12	61.559	.01624	30.772	30.788	.99947
4.13	62.178	.01608	31.081	31.097	.99948
4.14	62.803	.01592	31.393	31.409	.99949
4.15	63.434	.01576	31.709	31.725	.99950
4.16	64.072	.01561	32.028	32.044	.99951
4.17	64.715	.01545	32.350	32.365	.99952
4.18	65.366	.01530	32.675	32.691	.99953
4.19	66.023	.01515	33.004	33.019	.99954
4.20	**66.686**	**.01500**	**33.336**	**33.351**	**.99955**

SOURCE: O. W. Eshbach, *Handbook of Engineering Fundamentals*, Wiley, New York, 1936.

Table E-2 (Continued)

x	Natural Values				
	e^x	e^{-x}	Sinh x	Cosh x	Tanh x
4.20	**66.686**	**.01500**	**33.336**	**33.351**	**.99955**
4.21	67.357	.01485	33.671	33.686	.99956
4.22	68.033	.01470	34.009	34.024	.99957
4.23	68.717	.01455	34.351	34.366	.99958
4.24	69.408	.01441	34.697	34.711	.99958
4.25	70.105	.01426	35.046	35.060	.99959
4.26	70.810	.01412	35.398	35.412	.99960
4.27	71.522	.01398	35.754	35.768	.99961
4.28	72.240	.01384	36.113	36.127	.99962
4.29	72.966	.01370	36.476	36.490	.99962
4.30	**73.700**	**.01357**	**36.843**	**36.857**	**.99963**
4.31	74.440	.01343	37.214	37.227	.99964
4.32	75.189	.01330	37.588	37.601	.99965
4.33	75.944	.01317	37.966	37.979	.99965
4.34	76.708	.01304	38.347	38.360	.99966
4.35	77.478	.01291	38.733	38.746	.99967
4.36	78.257	.01278	39.122	39.135	.99967
4.37	79.044	.01265	39.515	39.528	.99968
4.38	79.838	.01253	39.913	39.925	.99969
4.39	80.640	.01240	40.314	40.326	.99969
4.40	**81.451**	**.01228**	**40.719**	**40.732**	**.99970**
4.41	82.269	.01216	41.129	41.141	.99970
4.42	83.096	.01203	41.542	41.554	.99971
4.43	83.931	.01191	41.960	41.972	.99972
4.44	84.775	.01180	42.382	42.393	.99972
4.45	85.627	.01168	42.808	42.819	.99973
4.46	86.488	.01156	43.238	43.250	.99973
4.47	87.357	.01145	43.673	43.684	.99974
4.48	88.235	.01133	44.112	44.123	.99974
4.49	89.121	.01122	44.555	44.566	.99975
4.50	**90.017**	**.01111**	**45.003**	**45.014**	**.99975**
4.51	90.922	.01100	45.455	45.466	.99976
4.52	91.836	.01089	45.912	45.923	.99976
4.53	92.759	.01078	46.374	46.385	.99977
4.54	93.691	.01067	46.840	46.851	.99977
4.55	94.632	.01057	47.311	47.321	.99978
4.56	95.583	.01046	47.787	47.797	.99978
4.57	96.544	.01036	48.267	48.277	.99979
4.58	97.514	.01025	48.752	48.762	.99979
4.59	98.494	.01015	49.242	49.252	.99979
4.60	**99.484**	**.01005**	**49.737**	**49.747**	**.99980**
4.61	100.48	.00995	50.237	50.247	.99980
4.62	101.49	.00985	50.742	50.752	.99981
4.63	102.51	.00975	51.252	51.262	.99981
4.64	103.54	.00966	51.767	51.777	.99981
4.65	104.58	.00956	52.288	52.297	.99982
4.66	105.64	.00947	52.813	52.823	.99982
4.67	106.70	.00937	53.344	53.354	.99982
4.68	107.77	.00928	53.880	53.890	.99983
4.69	108.85	.00919	54.422	54.431	.99983
4.70	**109.95**	**.00910**	**54.969**	**54.978**	**.99983**
4.71	111.05	.00900	55.522	55.531	.99984
4.72	112.17	.00892	56.080	56.089	.99984
4.73	113.30	.00883	56.643	56.652	.99984
4.74	114.43	.00874	57.213	57.222	.99985
4.75	115.58	.00865	57.788	57.796	.99985
4.76	116.75	.00857	58.369	58.377	.99985
4.77	117.92	.00848	58.955	58.964	.99986
4.78	119.10	.00840	59.548	59.556	.99986
4.79	120.30	.00831	60.147	60.155	.99986
4.80	**121.51**	**.00823**	**60.751**	**60.759**	**.99986**

SOURCE: O. W. Eshbach, *Handbook of Engineering Fundamentals*, Wiley, New York, 1936.

Table E-2 (Continued)

x	Natural Values				
	e^x	e^{-x}	Sinh x	Cosh x	Tanh x
4.80	**121.51**	**.00823**	**60.751**	**60.760**	**.99986**
4.81	122.73	.00815	61.362	61.370	.99987
4.82	123.97	.00807	61.979	61.987	.99987
4.83	125.21	.00799	62.601	62.609	.99987
4.84	126.47	.00791	63.231	63.239	.99987
4.85	127.74	.00783	63.866	63.874	.99988
4.86	129.02	.00775	64.508	64.516	.99988
4.87	130.32	.00767	65.157	65.164	.99988
4.88	131.63	.00760	65.812	65.819	.99988
4.89	132.95	.00752	66.473	66.481	.99989
4.90	**134.29**	**.00745**	**67.141**	**67.149**	**.99989**
4.91	135.64	.00737	67.816	67.823	.99989
4.92	137.00	.00730	68.498	68.505	.99989
4.93	138.38	.00723	69.186	69.193	.99990
4.94	139.77	.00715	69.882	69.889	.99990
4.95	141.17	.00708	70.584	70.591	.99990
4.96	142.59	.00701	71.293	71.300	.99990
4.97	144.03	.00694	72.010	72.017	.99990
4.98	145.47	.00687	72.734	72.741	.99991
4.99	146.94	.00681	73.465	73.472	.99991
5.00	**148.41**	**.00674**	**74.203**	**74.210**	**.99991**
5.01	149.90	.00667	74.949	74.956	.99991
5.02	151.41	.00660	75.702	75.710	.99991
5.03	152.93	.00654	76.463	76.470	.99991
5.04	154.47	.00647	77.232	77.238	.99992
5.05	156.02	.00641	78.008	78.014	.99992
5.06	157.59	.00635	78.792	78.798	.99992
5.07	159.17	.00628	79.584	79.590	.99992
5.08	160.77	.00622	80.384	80.390	.99992
5.09	162.39	.00616	81.192	81.198	.99992
5.10	**164.02**	**.00610**	**82.008**	**82.014**	**.99993**
5.11	165.67	.00604	82.832	82.838	.99993
5.12	167.34	.00598	83.665	83.671	.99993
5.13	169.02	.00592	84.506	84.512	.99993
5.14	170.72	.00586	85.355	85.361	.99993
5.15	172.43	.00580	86.213	86.219	.99993
5.16	174.16	.00574	87.079	87.085	.99993
5.17	175.91	.00568	87.955	87.960	.99994
5.18	177.68	.00563	88.839	88.844	.99994
5.19	179.47	.00557	89.732	89.737	.99994
5.20	**181.27**	**.00552**	**90.633**	**90.639**	**.99994**
5.21	183.09	.00546	91.544	91.550	.99994
5.22	184.93	.00541	92.464	92.470	.99994
5.23	186.79	.00535	93.394	93.399	.99994
5.24	188.67	.00530	94.332	94.338	.99994
5.25	190.57	.00525	95.281	95.286	.99994
5.26	192.48	.00520	96.238	96.243	.99995
5.27	194.42	.00514	97.205	97.211	.99995
5.28	196.37	.00509	98.182	98.188	.99995
5.29	198.34	.00504	99.169	99.174	.99995
5.30	**200.34**	**.00499**	**100.17**	**100.17**	**.99995**
5.31	202.35	.00494	101.17	101.18	.99995
5.32	204.38	.00489	102.19	102.19	.99995
5.33	206.44	.00484	103.22	103.22	.99995
5.34	208.51	.00480	104.25	104.26	.99995
5.35	210.61	.00475	105.30	105.31	.99995
5.36	212.72	.00470	106.36	106.36	.99996
5.37	214.86	.00465	107.43	107.43	.99996
5.38	217.02	.00461	108.51	108.51	.99996
5.39	219.20	.00456	109.60	109.60	.99996
5.40	**221.41**	**.00452**	**110.70**	**110.71**	**.99996**

SOURCE: O. W. Eshbach, *Handbook of Engineering Fundamentals*, Wiley, New York, 1936.

Table E-2 (Continued)

x	Natural Values				
	e^x	e^{-x}	Sinh x	Cosh x	Tanh x
5.40	**221.41**	**.00452**	**110.70**	**110.71**	**.99996**
5.41	223.63	.00447	111.81	111.82	.99996
5.42	225.88	.00443	112.94	112.94	.99996
5.43	228.15	.00438	114.07	114.08	.99996
5.44	230.44	.00434	115.22	115.22	.99996
5.45	232.76	.00430	116.38	116.38	.99996
5.46	235.10	.00425	117.55	117.55	.99996
5.47	237.46	.00421	118.73	118.73	.99996
5.48	239.85	.00417	119.92	119.93	.99997
5.49	242.26	.00413	121.13	121.13	.99997
5.50	**244.69**	**.00409**	**122.34**	**122.35**	**.99997**
5.51	247.15	.00405	123.57	123.58	.99997
5.52	249.64	.00401	124.82	124.82	.99997
5.53	252.14	.00397	126.07	126.07	.99997
5.54	254.68	.00393	127.34	127.34	.99997
5.55	257.24	.00389	128.62	128.62	.99997
5.56	259.82	.00385	129.91	129.91	.99997
5.57	262.43	.00381	131.22	131.22	.99997
5.58	265.07	.00377	132.53	132.54	.99997
5.59	267.74	.00374	133.87	133.87	.99997
5.60	**270.43**	**.00370**	**135.21**	**135.22**	**.99997**
5.61	273.14	.00366	136.57	136.57	.99997
5.62	275.89	.00362	137.94	137.95	.99997
5.63	278.66	.00359	139.33	139.33	.99997
5.64	281.46	.00355	140.73	140.73	.99997
5.65	284.29	.00352	142.14	142.15	.99998
5.66	287.15	.00348	143.57	143.58	.99998
5.67	290.03	.00345	145.02	145.02	.99998
5.68	292.95	.00341	146.47	146.48	.99998
5.69	295.89	.00338	147.95	147.95	.99998
5.70	**298.87**	**.00335**	**149.43**	**149.44**	**.99998**
5.71	301.87	.00331	150.93	150.94	.99998
5.72	304.90	.00328	152.45	152.45	.99998
5.73	307.97	.00325	153.98	153.99	.99998
5.74	311.06	.00321	155.53	155.53	.99998
5.75	314.19	.00318	157.09	157.10	.99998
5.76	317.35	.00315	158.67	158.68	.99998
5.77	320.54	.00312	160.27	160.27	.99998
5.78	323.76	.00309	161.88	161.88	.99998
5.79	327.01	.00306	163.51	163.51	.99998
5.80	**330.30**	**.00303**	**165.15**	**165.15**	**.99998**
5.81	333.62	.00300	166.81	166.81	.99998
5.82	336.97	.00297	168.48	168.49	.99998
5.83	340.36	.00294	170.18	170.18	.99998
5.84	343.78	.00291	171.89	171.89	.99998
5.85	347.23	.00288	173.62	173.62	.99998
5.86	350.72	.00285	175.36	175.36	.99998
5.87	354.25	.00282	177.12	177.13	.99998
5.88	357.81	.00279	178.90	178.91	.99998
5.89	361.41	.00277	180.70	180.70	.99998
5.90	**365.04**	**.00274**	**182.52**	**182.52**	**.99998**
5.91	368.71	.00271	184.35	184.35	.99999
5.92	372.41	.00269	186.20	186.21	.99999
5.93	376.15	.00266	188.08	188.08	.99999
5.94	379.93	.00263	189.97	189.97	.99999
5.95	383.75	.00261	191.88	191.88	.99999
5.96	387.61	.00258	193.80	193.81	.99999
5.97	391.51	.00255	195.75	195.75	.99999
5.98	395.44	.00253	197.72	197.72	.99999
5.99	399.41	.00250	199.71	199.71	.99999
6.00	**403.43**	**.00248**	**201.71**	**201.72**	**.99999**

SOURCE: O. W. Eshbach, *Handbook of Engineering Fundamentals*, Wiley, New York, 1936.

Table E-3   MENSURATION (Solids Having Curved Surfaces)

Notation. Lines, a, b, c, . . . ; altitude (perpendicular height), h, h_1, . . . ; slant height, s; radius, r; perimeter of base, p_b; perimeter of a right section, p_r; angle in radians, ϕ; arc, s; chord of segment, l; rise, h; area of base, A_b or A_B; area of a right section, A_r; total area of convex surface, A_l; total area of all surfaces, A_t; volume, V.

31. Right Circular Cylinder (and Truncated Right Circular Cylinder)	*For Right Circular Cylinder:* $A_l = 2\pi rh; \quad A_t = 2\pi r\,(r + h);$ $V = \pi r^2 h.$ *For Truncated Right Circular Cylinder:* $A_l = \pi r\,(h_1 + h_2); \quad A_t = \pi r\left[h_1 + h_2 + r + \sqrt{r^2 + \left(\dfrac{h_1 - h_2}{2}\right)^2}\,\right];$ $V = \dfrac{\pi r^2}{2}\,(h_1 + h_2).$
32. Ungula (Wedge) of Right Circular Cylinder	$A_l = \dfrac{2\,rh}{b}\,[a + (b - r)\phi];$ $V = \dfrac{h}{3b}\,[a\,(3r^2 - a^2) + 3\,r^2\,(b - r)\phi]$ $\quad = \dfrac{hr^3}{b}\left[\sin\phi - \dfrac{\sin^3\phi}{3} - \phi\cos\phi\right].$ *For Semicircular Base (letting $a = b = r$):* $A_l = 2\,rh; \quad V = \dfrac{2\,r^2h}{3}.$
33. General Cylinder	$A_l = p_b h = p_r s;$ $V = A_b h = A_r s.$
34. Right Circular Cone (and Frustum of Right Circular Cone) 	*For Right Circular Cone:* $A_l = \pi r_B s = \pi r_B \sqrt{r_B{}^2 + h^2}; \quad A_t = \pi r_B\,(r_B + s);$ $V = \dfrac{\pi r_B{}^2 h}{3}.$ *For Frustum of Right Circular Cone:* $s = \sqrt{h_1{}^2 + (r_B - r_b)^2}; \quad A_l = \pi s\,(r_B + r_b);$ $V = \dfrac{\pi h_1}{3}\,(r_B{}^2 + r_b{}^2 + r_B\,r_b).$
35. General Cone (and Frustum of General Cone)	*For General Cone:* $V = \dfrac{A_B h}{3}.$ *For Frustum of General Cone:* $V = \dfrac{h_1}{3}\left(A_B + A_b + \sqrt{A_B A_b}\right).$
36. Sphere	Let diameter $= d$. $A_t = 4\pi r^2 = \pi d^2;$ $V = \dfrac{4\pi r^3}{3} = \dfrac{\pi d^3}{6}.$
37. Spherical Sector (and Hemisphere)	*For Spherical Sector:* $A_t = \dfrac{\pi r}{2}\,(4h + l); \quad V = \dfrac{2\pi r^2 h}{3}.$ *For Hemisphere (letting $h = \dfrac{l}{2} = r$):* $A_t = 3\pi r^2; \quad V = \dfrac{2\pi r^3}{3}.$

SOURCE: O. W. Eshbach, *Handbook of Engineering Fundamentals*, Wiley, New York, 1936.

Table E-4 NOMINAL DIMENSIONS OF STEEL PIPE

NOMINAL PIPE SIZE [(in.)]	OUTSIDE DIAMETER [(in.)]	[cm]	SCHEDULE NO.*	WALL THICKNESS [cm]	INSIDE DIAMETER [cm]	INSIDE AREA (CROSS-SECTIONAL) [cm²]
$(\frac{1}{8})$	(0.405)	1.029	40	0.173	0.683	0.367
			80	0.241	0.546	0.231
$(\frac{1}{4})$	(0.540)	1.372	40	0.224	0.925	0.671
			80	0.302	0.767	0.602
$(\frac{3}{8})$	(0.675)	1.715	40	0.231	1.252	1.232
			80	0.320	1.074	0.907
$(\frac{1}{2})$	(0.840)	2.134	40	0.277	1.580	1.960
			80	0.373	1.387	1.511
			160	0.475	1.184	1.100
$(\frac{3}{4})$	(1.050)	2.667	40	0.287	2.093	3.440
			80	0.391	1.885	2.790
			160	0.554	1.560	1.910
(1)	(1.315)	3.340	40	0.338	2.664	5.576
			80	0.455	2.431	4.641
			160	0.635	2.070	3.366
$(1\frac{1}{4})$	(1.660)	4.216	40	0.356	3.505	9.650
			80	0.485	3.246	8.276
			160	0.635	2.946	6.818
$(1\frac{1}{2})$	(1.900)	4.826	40	0.368	4.089	13.13
			80	0.508	3.81	11.40
			160	0.714	3.399	9.071
(2)	(2.375)	6.033	40	0.391	5.250	21.65
			80	0.554	4.925	19.051
			160	0.871	4.290	14.46
$(2\frac{1}{2})$	(2.875)	7.303	40	0.516	6.271	30.89
			80	0.701	5.900	27.34
			160	0.953	5.398	22.88
(3)	(3.500)	8.890	40	0.549	7.793	47.69
			80	0.762	7.366	42.61
			160	1.110	6.670	34.94
$(3\frac{1}{2})$	(4.000)	10.16	40	0.574	9.012	63.79
			80	0.808	8.545	57.34
(4)	(4.500)	11.43	40	0.602	10.23	82.13
			80	0.856	9.718	74.17
			120	1.110	9.210	66.62
			160	1.349	8.733	59.89

SOURCE: Based on A.S.A. Standards B 36.16.
*Schedule 40 equivalent to former "standard" size and schedule 80 equivalent to former "extra strong" size.

Table E-5   NOMINAL DIMENSIONS OF STEEL TUBING

OUTSIDE DIAMETER [cm (in.)]	WALL THICKNESS			INSIDE DIAMETER [cm]	INSIDE CROSS-SECTIONAL AREA [cm²]	SURFACE PER METER LENGTH	
	GAUGE	[(in.)]	[cm]			OUTSIDE [m²/m]	INSIDE [m²/m]
(1) 2.540	13	(0.095)	0.241	2.057	3.323	0.0798	0.0646
(1¼) 3.175	13	(0.095)	0.241	2.692	5.693	0.0997	0.0846
(1½) 3.810	13	(0.095)	0.241	3.327	8.696	0.1197	0.1045
(1¾) 4.445	13	(0.095)	0.241	3.962	12.331	0.1396	0.1243
(2) 5.090	10	(0.134)	0.340	4.398	15.200	0.1596	0.1382
	12	(0.109)	0.277	4.526	16.091	0.1596	0.1422
	13	(0.095)	0.241	4.597	16.600	0.1596	0.1444
(2½) 6.350	9	(0.148)	0.376	5.598	24.614	0.1995	0.1759
	12	(0.109)	0.277	5.796	26.387	0.1994	0.1821
(3) 7.620	9	(0.148)	0.376	6.868	37.049	0.2399	0.2158
	10	(0.134)	0.346	6.939	37.820	0.2399	0.2180
	11	(0.120)	0.305	7.010	38.599	0.2399	0.2202
	12	(0.109)	0.277	7.066	39.217	0.2399	0.2220
(3¼) 8.255	11	(0.120)	0.305	7.645	45.908	0.2593	0.2402
(3½) 8.890	9	(0.148)	0.376	8.138	52.016	0.2793	0.2557
	11	(0.120)	0.305	8.280	53.85	0.2793	0.2601

SOURCE: Ducommun Metals Co., Los Angeles, California.

Table E-6 NOMINAL DIMENSIONS OF BRASS TUBING

OUTSIDE DIAMETER	WALL THICKNESS GAUGE BWG	[(in.)]	[cm]	INSIDE DIAMETER [cm]	AREA [cm²]	HEAT TRANSFER SURFACE PER METER LENGTH OUTSIDE [m²/m]	INSIDE [m²/m]
$(\frac{1}{4})$	16	(0.065)	0.165	0.305	0.0731	0.01995	0.00958
0.365	18	(0.049)	0.124	0.387	0.1176		0.01216
	20	(0.035)	0.089	0.457	0.1640		0.01436
	22	(0.028)	0.071	0.493	0.1909		0.01549
	24	(0.022)	0.056	0.523	0.2148	0.01995	0.01643
$(\frac{5}{16})$	16	(0.065)	0.165	0.464	0.1691	0.02017	0.01458
0.794	18	(0.049)	0.124	0.546	0.2341		0.01717
	20	(0.035)	0.089	0.616	0.2980		0.01935
	22	(0.028)	0.071	0.652	0.3339		0.02048
	24	(0.022)	0.056	0.682	0.3653	0.02017	0.02143
$(\frac{3}{8})$	16	(0.065)	0.165	0.623	0.3048	0.02994	0.01957
0.953	18	(0.049)	0.124	0.705	0.3904		0.02215
	20	(0.035)	0.089	0.775	0.4717		0.02435
	22	(0.028)	0.071	0.811	0.5166		0.02548
	24	(0.022)	0.056	0.841	0.5555	0.02994	0.02642
$(\frac{7}{16})$	16	(0.065)	0.165	0.781	0.4791	0.03490	0.02454
1.111	18	(0.049)	0.124	0.863	0.5849		0.02711
	20	(0.035)	0.089	0.933	0.6837		0.02931
	22	(0.028)	0.071	0.969	0.7375		0.03044
	24	(0.022)	0.056	0.999	0.7838	0.03490	0.03138
$(\frac{1}{2})$	16	(0.065)	0.165	0.940	0.6940	0.03990	0.02953
1.270	18	(0.049)	0.124	1.022	0.8203		0.03211
	20	(0.035)	0.089	1.092	0.9366		0.03431
	22	(0.028)	0.071	1.128	0.9993		0.03544
	24	(0.022)	0.056	1.158	1.0532	0.03990	0.03638
$(\frac{5}{8})$	14	(0.083)	0.211	1.166	1.0678	0.04989	0.03663
1.588	16	(0.065)	0.165	1.258	1.2429		0.03952
	18	(0.049)	0.124	1.340	1.4103		0.04210
	19	(0.042)	0.107	1.374	1.4827		0.04317
	21	(0.032)	0.081	1.426	1.5971	0.04989	0.04480
$(\frac{3}{4})$	14	(0.083)	0.211	1.483	1.7273	0.05985	0.04659
1.905	16	(0.065)	0.165	1.575	1.9483		0.04948
	18	(0.049)	0.124	1.657	2.1564		0.05206
	19	(0.042)	0.107	1.691	2.2458		0.05312
	21	(0.032)	0.081	1.743	2.3861	0.05985	0.05476
$(\frac{7}{8})$	14	(0.083)	0.211	1.801	2.5475	0.06984	0.05658
2.223	16	(0.065)	0.165	1.893	2.8144		0.05947
	18	(0.049)	0.124	1.975	3.0638		0.06265
	19	(0.042)	0.107	2.009	3.1700		0.06311
	21	(0.032)	0.081	2.061	3.3362	0.06984	0.06473
(1)	14	(0.083)	0.211	2.112	3.5033	0.07980	0.06635
2.540	16	(0.065)	0.165	2.210	3.8360		0.06943
	18	(0.049)	0.124	2.292	4.1259		0.07201
	19	(0.042)	0.107	2.326	4.2492		0.07309
	21	(0.032)	0.081	2.378	4.4413	0.07980	0.07471

Table E-6   (Continued)

OUTSIDE DIAMETER	WALL THICKNESS GAUGE BWG	[(in.)]	[cm]	INSIDE DIAMETER [cm]	AREA [cm²]	HEAT TRANSFER SURFACE PER METER LENGTH OUTSIDE [m²/m]	INSIDE [m²/m]
$(1\frac{1}{4})$	14	(0.083)	0.211	2.753	5.9525	0.09975	0.08488
3.175	16	(0.065)	0.165	2.845	6.3570		0.08938
	18	(0.049)	0.124	2.927	6.7288		0.09195
	19	(0.042)	0.107	2.961	6.8860		0.09302
	21	(0.032)	0.081	3.013	7.1300	0.09975	0.09466
$(1\frac{1}{2})$	14	(0.083)	0.211	3.380	8.9727	0.1197	0.1062
3.810	16	(0.065)	0.165	3.480	9.5115		0.1093
	18	(0.049)	0.124	3.562	9.9650		0.1119
	19	(0.042)	0.107	3.596	10.1562		0.1130
	21	(0.032)	0.081	3.648	10.4520	0.1197	0.1146
$(1\frac{3}{4})$	14	(0.083)	0.211	4.023	12.7113	0.1397	0.1264
4.445	16	(0.065)	0.165	4.115	13.2993		0.1293
	18	(0.049)	0.124	4.197	13.8346		0.1319
	19	(0.042)	0.107	4.231	14.0597		0.1329
	21	(0.032)	0.081	4.283	14.4074	0.1397	0.1346
(2)	14	(0.083)	0.211	4.678	17.1874	0.1596	0.1470
5.080	16	(0.065)	0.165	4.750	17.7205		0.1492
	18	(0.049)	0.124	4.832	18.3377		0.1518
	19	(0.042)	0.107	4.866	18.6012	0.1596	0.1529
$(2\frac{1}{2})$	14	(0.083)	0.211	5.928	27.600	0.1995	0.1862
6.350	16	(0.065)	0.165	6.020	28.463		0.1891
	18	(0.049)	0.124	6.102	29.244		0.1917
	19	(0.042)	0.107	6.136	29.571	0.1995	0.1928
(3)	14	(0.083)	0.211	7.198	40.692	0.2394	0.2261
7.620	16	(0.065)	0.165	7.290	41.739		0.2290
	18	(0.049)	0.124	7.372	42.694		0.2316
	19	(0.042)	0.107	7.406	43.078	0.2394	0.2327

SOURCE: Ducommun Metals Co., Los Angeles, California

Index

82 83 84 85 9 8 7 6 5 4 3 2 1